A. Neville Jones

EUTROPHICATION:

RESEARCH AND APPLICATION TO WATER SUPPLY

Edited by

DAVID W. SUTCLIFFE AND J. GWYNFRYN JONES

Published by the Freshwater Biological Association

Invited papers from a specialised conference held in London
on 10-11 December 1991 by the
Freshwater Biological Association,
The Ferry House, Far Sawrey, Ambleside, Cumbria LA22 0LP
and
International Water Supply Association,
1 Queen Anne's Gate, London SW1H 9BT

ISBN 0-900386-52-5

Many problems are caused in the water-supply industry as a result of eutrophication. The Freshwater Biological Association, UK, and the International Water Supply Association, brought together some of the world's leading experts on the subject, in order to exchange the latest knowledge between interested parties at a specialised conference held in London on 10-11 December 1991. The conference programme combined a balance between practice and theory so that wide coverage of all the major aspects of the subject received attention. This volume contains all but one of the invited papers from speakers at the conference, dealing with the causes and control of large algal populations (blooms) resulting from the inputs of nutrients to rivers, lakes and reservoirs that are used to supply drinking-water. The articles provide an informed and critical insight into the state-of-the-art research and its application in Belgium, Britain, Denmark, France, Germany, Switzerland and The Netherlands.

This volume will be of great importance to all those engaged in research, freshwater ecology, water treatment, reservoir management, and anyone interested in current solutions to a suite of problems associated with the chemistry and control of nutrient inputs, the growth of algal populations and their control of zooplankton and fish (including deliberate biomanipulation of lakes and reservoirs), and mathematical modelling as descriptive and management aids for controlling eutrophication.

CONTENTS

Introduction and opening remarks

H. Bernhardt
Wahnbachtalsperrenverband, 5200 Siegburg, FRG

For many decades, both theoretical and applied limnology have been dedicated to investigating the phenomenon of a man-made eutrophication and the resulting adverse effects on the quality of flowing and standing waters. In the past, eutrophication had generally been regarded from a phenomenological point of view. The aim of the studies was to determine which factors were responsible for occasional mass developments of algae in lakes or a production of predominantly green or blue-green algae rather than diatoms. The primary thrust of limnological research was directed towards investigations on the diverse behaviour of lakes.

In recent years, however, research into eutrophication has gained new dimensions. Limitations in the use of eutrophic waters – especially for the production of drinking-water – called for measures to control the eutrophication process. Which factors are responsible for a temporary prolific growth of algae in lakes? Is it possible to restrict the development of blue-green algae, if necessary in favour of a different type of algae? Which technical measure would prove most successful in the restriction of eutrophication? Questions soon became increasingly practice-oriented and both basic and applied research were viewed as equally important. It was clear that the phenomenon of eutrophication had to be fully understood so as to be able to control this process in what may be termed a "lake therapy".

There is a close connection between the history of eutrophication research and that of limnology. From the discussion by Thienemann and Nauman on the causes of eutrophication we learn that as early as the 1920s, the nutrient supply and the morphology of standing waters were considered to be two essential factors for eutrophication. In addition, it was also discovered that the surrounding area of a lake, i.e. the interrelation between the catchment area and the lake itself, exerts a considerable influence. Ohle pointed out the significance of the structure of a catchment area for the biomass production and thus for the trophic state of the adjacent lake.

As time went on, it became increasingly possible to quantify such quality-related connections by expressing limnological processes in the form of mathematical models. Particularly useful in this connection were statistical models such as those developed by Vollenweider and numerous other scientists.

From what we know about the nutrient supply, morphology, hydrodynamics and geographic characteristics, (climate, temperature of water and air, light intensity etc.) of a standing water-body, we are today able to predict its trophic condition within limits. This likewise allows us to adopt specific water protection measures which are required to attain a certain trophic state as a restoration target. Accordingly, it can be justly stated that we are able to implement a "lake therapy" with reasonable success.

In practice, an attempt was made in recent years to control the development of phytoplankton in a lake through a biomanipulation using a specifically designed food chain. In addition to measures aimed at restricting eutrophication through limiting the input of nutrients, this approach also allowed us to control eutrophication by biological processes.

For the purpose of drinking-water production, a restriction of algal growth in lakes and reservoirs has become indispensible. As we know that naturally occurring dissolved and suspended organic matter emanating from algae and their metabolites acts as a precursor for the formation of trihalomethanes during disinfection with chlorine, raw water should contain only low concentrations of plankton and algogenic matter. Bacterial regrowth in distribution systems and problems arising from substances affecting the odour and taste of drinking-water are associated with the presence of certain types of algae. It should also be noted at this point that some algae release toxins which are critical to public health even at very low concentrations. The risk of microbial growth in drinking-water supply systems is markedly dependent on the extent of the plankton production and the amount of algal metabolites in the lake. For these reasons and also with respect to existing stringent quality requirements for drinking-water, we need, more than ever, lakes and reservoirs which show no major algal developments and are not likely to produce large amounts of blue-green algae. Bearing this in mind, we therefore depend on methods to counteract to the best of our technical abilities a eutrophication of standing waters, i.e. an excessively strong production of algae and a development of particular species.

Still in their infancy are purely biological processes for the control of phytoplankton biomass by zooplankton as well as the formation and operation of an effective long-term food chain. This kind of biomanipulation of lakes still has to be advanced through extensive research work in the years to come so that it can be successfully applied in practice.

There was a time, at least in Germany, when research was thought to have solved the problem of eutrophication. At the end of the 1970s, Schindler demonstrated on Canadian lakes, on a 1:1 scale, that the extent of plankton production is dependent on the supply of phosphorus compounds and that the ratio of phosphorus to nitrogen decides whether or not blue-green algae prevail in a lake. I believe that we still need to push on with eutrophication research. Any other opinion on this subject fails to realize the actual necessities.

Over the past years, several large-scale restoration projects have been implemented with the aim of turning eutrophic lakes or reservoirs into oligotrophic water-bodies. Their effects on the respective waters, i.e. the relationship between input and response, were compiled in a volume coordinated by H. Sas (1989): *Lake Restoration by Reduction of Nutrient Loading,* Academic Verlag Richarz, St Augustin. This discussion of the individual results shows that in several instances the respective lakes reacted to the restoration measures just as we had expected. Only in some cases were the results contrary to our expectations. Further differences were noticed with regard to the time which had to elapse before an effect became clearly discernible. Nevertheless, a lot of research still has to be done. Despite the long and intensive investigations in this field it will certainly take a while before eutrophication can be fully controlled.

With respect to a protection strategy for flowing waters, the question remains open as to whether it is possible to develop input/response models which are just as efficient as those now applied, with certain limitations, for the adjustment of the trophic state of lakes or reservoirs. Against the background that rivers and streams have to be used for discharging waste-water and various other materials, we have to know which phosphorus and nitrogen inputs can be tolerated without endangering the use of these waters for the production of drinking-water or recreation purposes (the latter being governed by the EC Bathing Water Directive). The limnology for flowing waters still has to provide the required knowledge.

How much do we actually know about eutrophication after decades of research? In order to find an answer, the six sessions of this conference should put forward the latest results in the field of theoretical and applied limnology with special reference to eutrophication. We should discuss the different possible methods currently available to predict and manipulate the growth of phytoplankton and zooplankton.

The problems touched upon above will be put to the discussion at this conference by internationally-recognized scientists and engineers. I hope that the presented results will be discussed and evaluated with much interest. I can only warmly recommend that scientists from different areas of work – of which there are quite a few in limnology as an interdisciplinary science – should meet from time to time to exchange their views and experience. This will give everyone new incentives for further investigations, different perspectives on the overall problem of eutrophication, and vital information about recently disclosed connections which may find practical application in the area of lake protection.

Eutrophication and the management of planktonic algae: what Vollenweider couldn't tell us

C. S. REYNOLDS

Freshwater Biological Association and NERC Institute of Freshwater Ecology, Windermere Laboratory, Far Sawrey, Ambleside, Cumbria LA22 0LP, England

The "Vollenweider model" is a sophisticated mathematical statement about the long-range behaviour of (mainly temperate) lakes and their ability to support phytoplankton chlorophyll. Misapplication of the model, against which Vollenweider himself warned, has led to many misconceptions about the dynamics of plankton in lakes and reservoirs and about how best to manage systems subject to eutrophication. This contribution intends to introduce the topics addressed at this symposium on Eutrophication by framing them in context of the phosphorus-loading and phosphorus-limitation concepts. Emphasis is placed on the need to distinguish rate-limitation from capacity-limitation, to understand which is more manageable and why, to discern the mechanisms of internal recycling and their importance, and to appreciate the respective roles of physical and biotic components in local control of algal dynamics. Some general approaches to the management of water quality in lakes and reservoirs to eutrophication are outlined.

Introduction

Eutrophication is generally understood to refer to enrichment of biological systems by nutrient elements, notably nitrogen and phosphorus, and to the enhanced production of algal and higher plant biomass that the added loads stimulate. The concept has a firm factual foundation and has been understood well enough from the early days of comparative limnology (see, for instance, Thienemann 1918; Naumann 1919; Pearsall 1921). Public concern about the deterioration in recreational and drinking water quality, as a consequence of culturally accelerated eutrophication, prompted the development of a number of quantitative relationships, between selected indices of the respondent plant biomass and selected representations of the availability of nutrients to primary producers, including those of Sakamoto (1966), Lund (1970), Dillon & Rigler (1974), Oglesby & Schaffner (1975, 1978), Rast & Lee (1978). By far the best known and still the most widely applied, however, is the relationship presented by Richard Vollenweider (1968), as later extended and refined during an international collaborative project, convened by the Organisation for Economic Cooperation and Development (OECD), and ultimately published in definitive form by Vollenweider & Kerekes (1982).

These developments resolved many of the intellectual aspects of the issue and, perhaps more pertinently (Reynolds 1989a), staunched the financial support to maintain the philosophical investment in the problem. Clearly enough, the models demonstrated that the greater is the external load of phosphorus then the greater is the quantity of algal chlorophyll that will be supported. End of problem: science moves on.

Yet, no less clearly, the problem endures. Algae are still produced, pending some concerted scheme to reduce the external loading of nutrients. However, attempts to reverse the effects of eutrophication have had conspicuously variable results; not all have been so evidently successful as the Lake Washington scheme (reviewed in Edmondson 1977). In the United

Kingdom, toxic algal blooms have at last been recognised for what they are (Pearson *et al.* 1990) and, perhaps correctly, are popularly viewed as a symptom of eutrophication. Thus, although it is both ironic and humbling that managers and academics should return to this topic, the reasons for doing so are abundantly clear. While it is easy enough to diagnose the root causes for lake eutrophication and its consequences, it is quite another to be able to cure it or, indeed, to predict the extent and effectiveness of treatment measures. In this age of accountability, the cost-effectiveness and cost-benefit appraisals of further research have been difficult of substantiate. Who needs more models when you are already awash with blue-green algal scums?

Experimentally-verified philosophical guidance is required, however, if eutrophied systems are to be adequately managed for supply and amenity purposes and if the correct approaches are to be adopted in securing effective measures of amelioration of water-quality problems. This would presuppose there to be a consensus view of aquatic ecosystem functioning that is implicitly correct. Even cursory examination of the international literature shows this is far from being the case (see especially Carpenter 1988). It is not wholly academic to consider how given ecosystems are regulated, whether mainly internally or externally, whether from the "top down" or the "bottom up" or whether they are determined ("limited"?) by nutrients which arise externally or are recycled within. It is also essential to devise a strategic approach to the application of knowledge in resolving problems faced in the water-supply industry.

The objective of this paper is to identify the key issues and outstanding questions posed in the management of planktonic algal populations in eutrophied lakes and reservoirs and to contribute to a rationale for selecting the most appropriate approach to the solution of specific problems at specific sites.

What Vollenweider *did* tell us

I want to make it perfectly clear that the title of this article in no way implies any criticism of the Vollenweider Model or its authors. The model remains a most elegant and powerful statement about the high-level ordering of limnetic systems. After all the physical variability, chemical drives and myriad of potential biotic responses are averaged out, then the mean level of planktonic biomass (as measured by its chlorophyll content, chl^{y}; see note [1]) supported is broadly a function of the mean, in-lake availability of total phosphorus (P_i^y) through the year. In this sense, the model provided empirical weight to eloquent contemporaneous argument (e.g. Schindler 1977) that a good deal of the inevitable variability among lake behaviours was attributable to their phosphorus-dependent carrying capacities. Indeed, it is perhaps remarkable that the mean annual standing-crop biomass should remain so consistently proportional to phosphorus availability, even across the spectrum of lakes where the element has been shown to be continuously or frequently limiting the production of phytoplankton. However, it is important to emphasise that the regression equation which can be fitted to the accumulated data points (Fig. 1) is an *average of behaviours*; it is not a slope up or down which a given lake will progress during a period of artificial enrichment or deliberate restoration.

This last point has been amply demonstrated in the IMSA analysis (Sas 1989). In several of the restoration performances considered (e.g. Wahnbach Reservoir, Fig. 2a), reduced phosphorus loadings produced an immediate and predictable response, and the year-to-year slope was closer to 1.0 than to Vollenweider's 0.91. In other cases and to varying extents (e.g. Veluwemeer, Fig. 2b; Schlactensee, Fig. 2c) there was a noticeable lag during which mean phosphorus was sharply reduced but chlorophyll failed to respond. Later, as phosphorus was

[1] Usage of symbols follows, as far as possible, that of the Vollenweider/OECD and IMSA (Sas 1989) studies. In this scheme, P stands for total phosphorus.

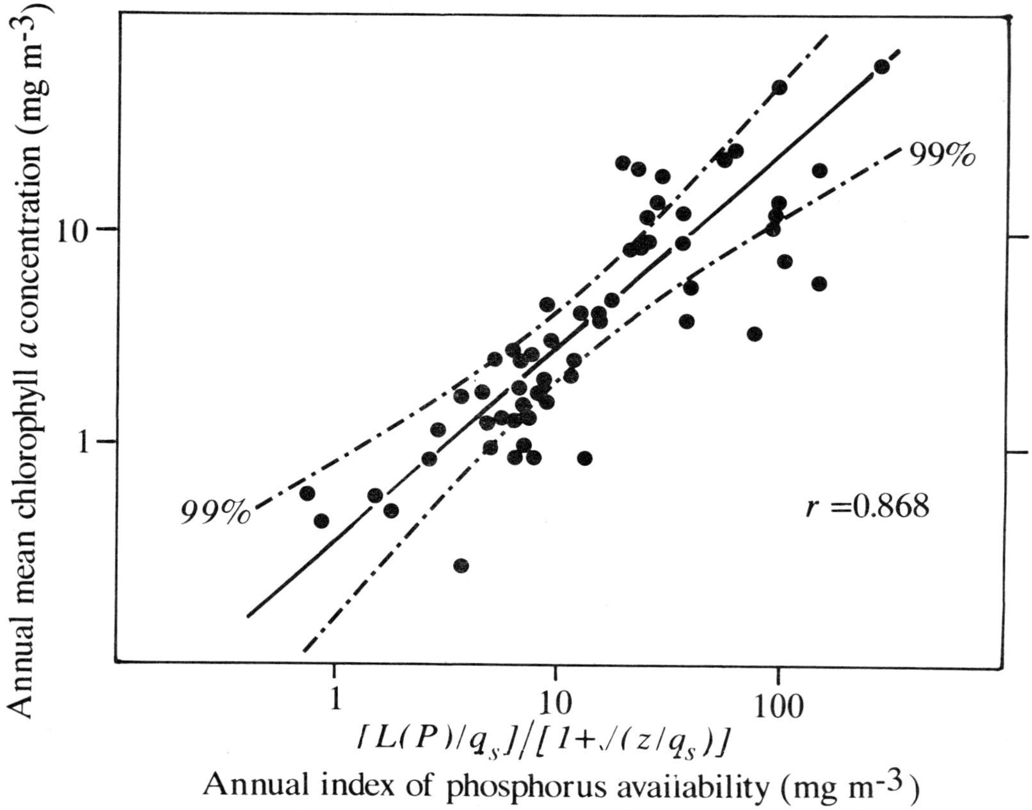

Figure 1. The Vollenweider–OECD relationship between average chlorophyll concentrations and P-loading characteristics, with 99% confidence limits inserted. The equation of the regression is log $(chl^y) = 0.91(\log P_1^y) - 0.435$, where $P_1^y = [L(P)/q_s]/[1+(z/q_s)^{1/2}]$ and $L(P)$ is the annual P load (in mg m^{-2}), q_s is the relative volume of water exchanged per year (m^3) and z is the mean depth of the water-body (m).

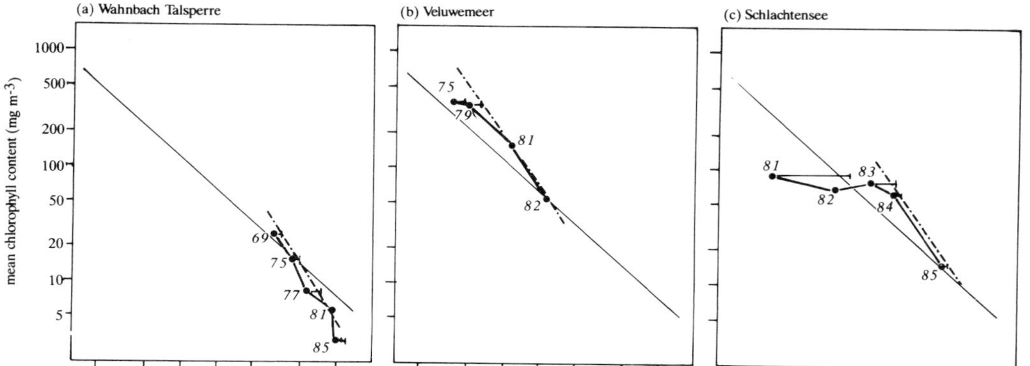

Figure 2. Stages in the restoration of three European lakes following reduction in external P loading. The heavy lines link the changing annual average chlorophyll concentration against the phosphorus available and are superimposed upon the Vollenweider–OECD regression. The bars to the right of each point represent the difference between *TP* (the total phosphorus) and *PP*, the particulate fraction. Note that this difference (equivalent to the dissolved P) is virtually exhausted before mean biomass chlorophyll declines.

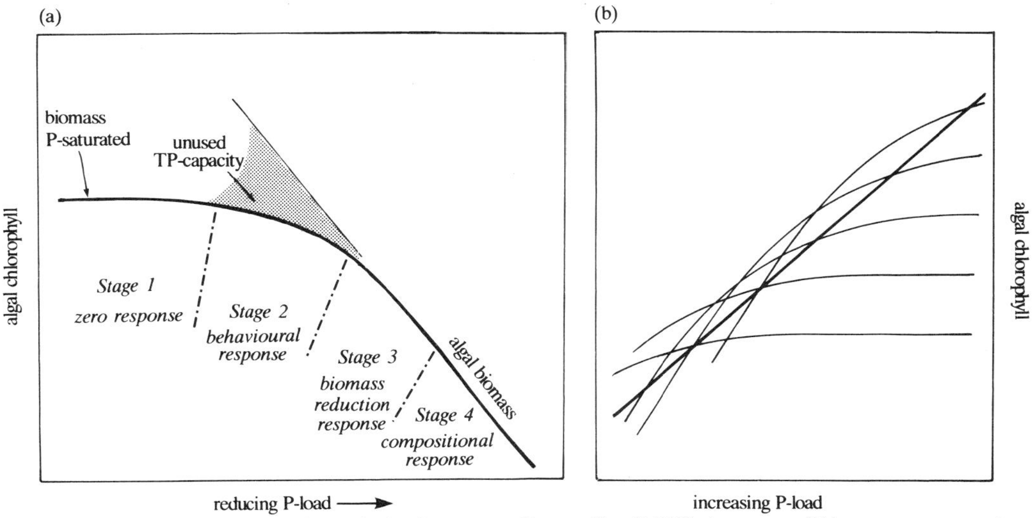

Figure 3. (a) The explanation of lake behaviour according to Sas (1990), in terms of biomass response to phosphorus reduction, moving from P-saturation and zero response to biomass reduction and compositional change. The effects of eutrophication would be approximated by reading from right to left. (b) Hypothetical P–Chl relationships from several separate lakes are represented: the Vollenweider–OECD regression is believed to fit this family of individual response curves.

still further reduced, then chlorophyll did fall, with a slope close to 1.0, against phosphorus. The shapes of the curves are generic, depending upon analogous responses, as interpreted in Sas (1989) (see Fig. 3a), but their co-ordinates are unique to individual lakes. It is not difficult to recognise that Vollenweider's regression may well describe a family of overlapping curves (Fig. 3b). Moreover, if more of the lakes in the original OECD data-set had been unresponsive to changes in P loading, it is likely that the regression coefficient would have been less than 0.91. Using a different data-set of (mainly) North American lakes, which is apparently less weighted in favour of phosphorus-deficient oligotrophic lakes, Rast & Lee (1978) were able to fit a slightly flatter regression slope (0.76).

The problems perceived with the use of the model are the same as those with any other – its uncritical application. Vollenweider (1976) himself explained that the models were intended to be illustrative, doubted that they could be usefully developed further and cautioned against too literal an interpretation without much finer resolution and understanding (p. 78). The principal application he envisaged was the better definition of "critical loads" of phosphorus as an aid to catchment-management decisions.

Those who would extend the use of the model need to be fully aware of its conditions and limitations and also to realise that if further questions arise about managing the phosphorus-chlorophyll relationship in a particular system (as they surely must), the next steps in deductive logic must be understood to be less general and increasingly subject to stochastic error.

Capacity limitation by phosphorus

The relatively tight coupling between the quantities of algal mass eventually produced in a given system, and the amounts of phosphorus available to support them, is often assumed to attest to "limitation" of algal growth by phosphorus, as opposed to any other factor, and is even described as such. It is also becoming common to read in the literature that "lake X is phosphorus-limited", frequently on the grounds of the elemental N:P ratio of the amounts of nutrient available (see later). Again, another implicit expectation which sometimes arises is that a given phosphorus concentration in water is necessarily accompanied by a *pro rata* concentration of chlorophyll, without allowance for the time taken by algal populations to increase to a level commensurate with the phosphorus supply. Understanding the mechanisms of the phosphorus–chlorophyll relationship, and their application to the management of eutrophication, is not assisted by sloppy conceptualisation.

For instance, it is important to be precise about what *is* limited and when. Reynolds (1992a) has distinguished capacity-limiting factors as being those setting the eventual population maximum: n_{max} is often a function of the available nutrient pool, (K), and the "minimum cell quota" (q_o), as conceived by Droop (1973, 1983), being the lowest amount of intracellular nutrient upon which the cell can still function. Then,

$$n_{max} = K/q_o$$

There is no assumption that this condition is unique to phosphorus. The fact remains that, if K

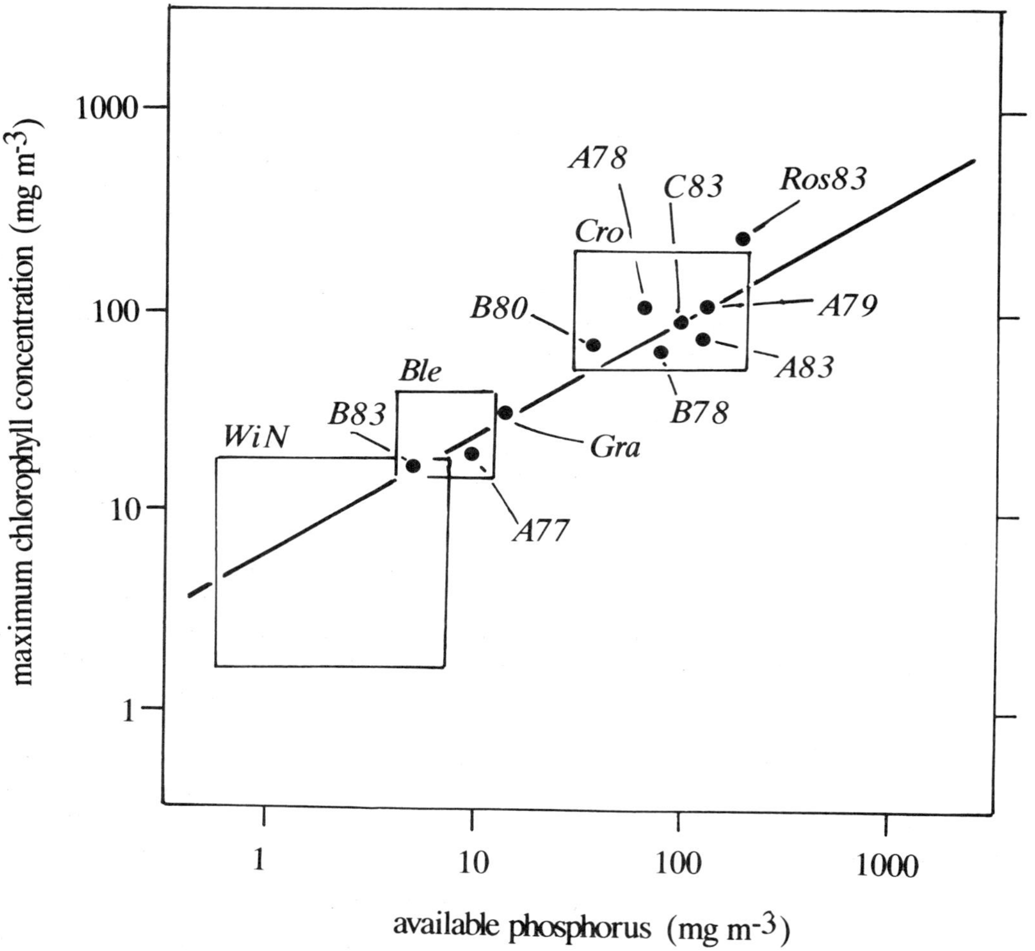

Figure 4. Reynolds' (1978) regression to determine maximum chlorophyll carrying capacity was solved to show behavioural differences among Windermere North (*WiN*), Blelham Tarn (*Ble*) and Crose Mere (*Cro*): the original points lie within the respective rectangles. The regression has been quite successful in predicting the maximum biomass observed in Blelham Enclosures (*A77*, *A78*, *A79*, *A83*, *B78*, *B80*, *B83*, *C83* are shown), Grasmere (*Gra*) and Rostherne Mere (*Ros*). The equation of the original regression is log (Chl_{max}) = 0.585 (log P_{max}) + 0.801.

is (say) silicon, n_{max} could be exceeded by non-diatoms, or if *K* is nitrogen, perhaps nitrogen-fixing cyanobacteria could expand to the limit of the phosphorus available. As no other proximal source of phosphorus is available to any phytoplankton, except the total phosphorus pool in the water, it follows that the ultimate chemical capacity is, indeed, often determined by the size of that pool (Schindler 1977).

In sub-capacity populations, the actual quota of phosphorous is likely to be rather greater than the minimum, generally falling within the range 7 – 30 mg P per g ash-free dry-cell mass or, assuming 50% of the dry organic mass to be elemental carbon, 5 – 23 × 10^{-3} mol cell P (mol cell C)$^{-1}$. As the typical chlorophyll-*a* content of cells has a similar range, 9 – 39 mg chl*a* (g dry cell mass)$^{-1}$, the ratio 1:1 is a useful guide to the phosphorus-determined carrying capacity. In reviewing the behaviour of populations raised in experimental enclosures, or "limnocorrals", Lund & Reynolds (1982) found that, though the ratio averaged nearer 1:2, the larger and more persistent maxima approached an efficiency of 1:1 at higher loads but this increased to >3:1 at low average phosphorus concentrations. These instances fit well (Fig. 4) to Reynolds' (1978) earlier regression of chlorophyll yield upon initial phosphorus availability, which remains the only chl*a*–P model that predicts the scale of the maximum (as opposed to average) phosphorus-determined yield of algal chlorophyll. The slope of the regression reflects the efficiency of utilization at chronically low phosphorus concentrations and the increasing "profligacy" of consumption at higher ones. Nevertheless, even this yield may well be unattainable if other limitations on capacity are imposed at some intermediate level of attainable yield. A hypothetical instance of this plateauing effect, with respect to the phosphorus availability, is

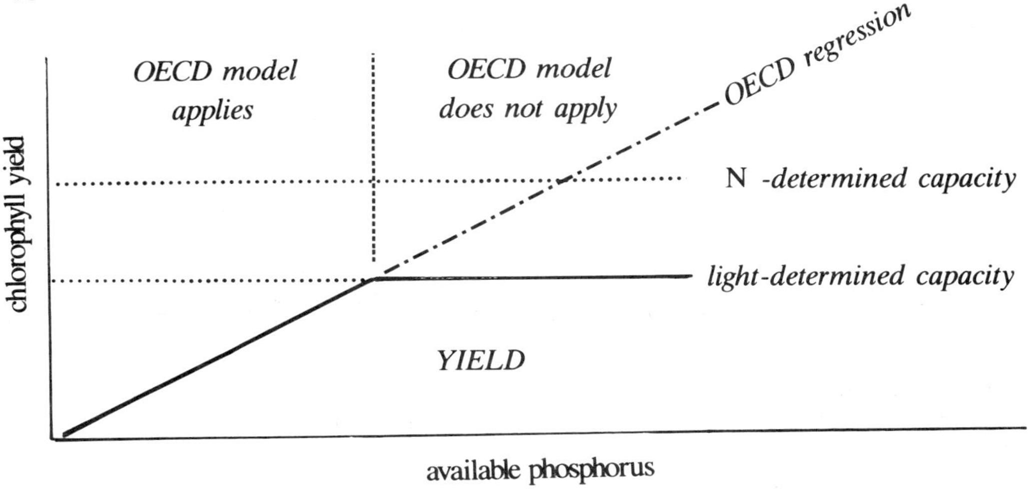

Figure 5. Conceptual figure showing the relationship of chlorophyll to phosphorus and the principle of a second capacity factor (here, light) intercepting the OECD response at higher concentrations of phosphorus. Changes in chlorophyll are not predictable to the right of this point.

inserted in Figure 5 to represent the intervention of the capacity of another nutrient or, as is commonly the case (see e.g. Stauffer 1991), of the exceedance of the carrying capacity of the underwater light availability and of Talling's (1957) column-compensation. The effect is enhanced in deep, mixed water layers where a substantial lower portion is dark or only dimly insolated by day, or in shallower waters where large quantities of non-algal particles (e.g. of suspended clay) contribute to elevated light extinction coefficients and achieve an analogous effect. The light-limited carrying capacity is reached when the diurnal integral of photosynthetic production, ΣnP, is balanced by the daily respirational consumption, within the same mixed layer, $\sum nR$, (= 24 nRh_m, where n is the phytoplankton population expressed in terms of its chlorophyll, R is the hourly respiration rate and h_m is the depth of the layer), i.e., when $(\sum nP)/(24\ nRh_m) = 1$.

Evaluation of the volume-integrated photosynthesis, $\sum nP$, is scarcely simple but may be approximately derived from a model relationship, the development of which also owes, in part, to Vollenweider (e.g. 1965):

$$\sum nP \simeq [0.75\ nP_{max} . \Gamma . \ln(0.7\ I'_{o\ max} / 0.5\ I_k)]/\varepsilon$$

where P_{max} is the light-saturated rate of photosynthesis, Γ is the length of day (sunrise to sunset), $I'_{o\ max}$ is the maximum flux density of photons passing into the water and I_k is the intensity required to saturate photosynthesis. ε is the vertical coefficient of light extinction and is the sum of components attributable to water and solutes (ε_w), the non-living particulates (ε_p), and the absorption and scatter due to chlorophyll ($\varepsilon_q = nk$, where k is the area subtended per unit of chlorophyll). Whence, n is solved:

$$n = (1/k)\ [0.75\ (P_{max}/R) . \Gamma/24 . \ln(0.7\ I'_{o\ max} / 0.5\ I_k)(1/h_m) - (\varepsilon_w + \varepsilon_p)]$$

The carrying capacity can vary with season and among the factors but the capacity is not necessarily continuously filled nor even at all. The case illustrated in Figure 6 concerns Rostherne Mere, a small (0.48 km^2) but relatively deep (H_{max}, 30 m; H, 13.4 m) and turbid

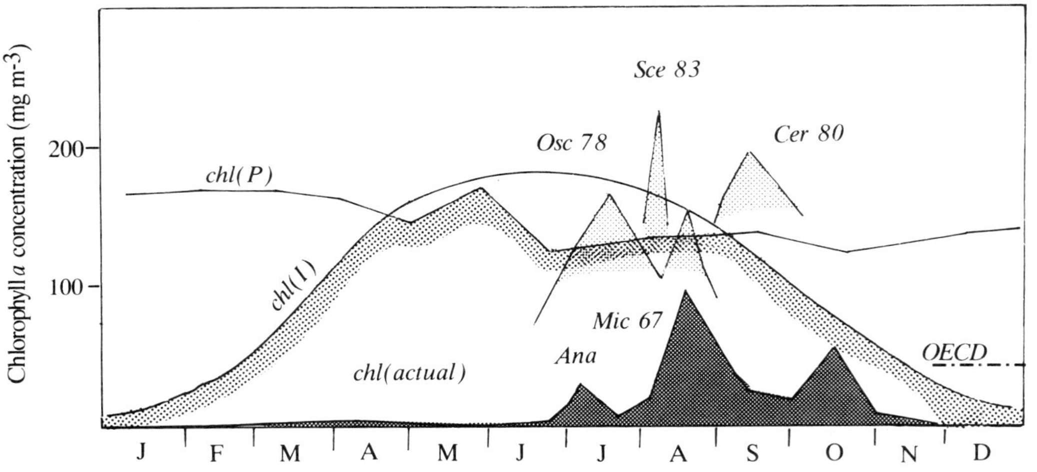

Figure 6. The chlorophyll content (dark silhouette) in Rostherne Mere through 1967, in relation to the capacities allowed by the available phosphorus, *Chl(P)*, in the trophogenic layer and the light income, *Chl(I)*. Chlorophyll did not come close to filling the *combined* capacity (heavy stipple), although there have been occasions when other populations did realise these capacities (*Oscillatoria* in 1978, *Ceratium* in 1980 and *Scenedesmus* in 1983). The Vollenweider-OECD prediction of *average* biomass is inserted for interest (right hand ordinate). Simplified from Fig. 2 of Reynolds (1992a).

($\varepsilon_w + \varepsilon_p > 0.45$ m^{-1}) natural kataglacial lake near Manchester, England. It is naturally rich in phosphorus but the supply is augmented by the output of a small sewage treatment plant and an agricultural contribution. The plot of instantaneous evaluations of the phosphorus-determined chlorophyll-carrying capacity, as calculated according to the regression equation in Figure 4 from measurements made during 1967, is superimposed upon the calculated light-compensated chlorophyll-carrying capacity through the year. The actual measurements of chlorophyll in the lake through the same year imply that the carrying capacity was never close to realisation, although maxima recorded in 1978, 1980 and 1983 suggest that the phytoplankton populations are not prevented from achieving the theoretical capacity on occasions.

Each of these capacity crops is known to have become established over lengthy periods of relatively constant weather conditions (cool, windy weather with frequent mixing in 1978 provided conditions favourable to *Oscillatoria* growth; uninterrupted near-surface stratification in 1983 allowed the epilimnion to resemble the environment of a shallow, rich pool). Reynolds & Bellinger (1992) deduced (i) that the period over which growth of an eventual dominant species is sustained can be as relevant to the scale of the realised population as is the initial carrying capacity, and (ii) that the greater is the unfilled capacity then the more that statement (i) will apply.

Rate limitation by phosphorus

It is also clear that for the attainment of a capacity-filling population, an inoculum of cells requires either to be able to maintain a rapid rate of increase under optimal conditions or, if slower, to be able to sustain net increase for longer.

In situ species-specific increase depends upon the *in situ* rates of cell replication net of any losses through mortality. Too few reliable data on *in situ* cell division are yet available, although improving techniques for assessing mitotic separation and their wider application (e.g. Heller 1977]; Braunwarth & Sommer 1985) are yielding results that generally accord with division rates reconstructed from intensive field measurements of the dynamics of individual populations (Sommer 1981; Reynolds 1982, 1986) as well as model predictions based on the behaviour of isolates in laboratory experiments (e.g. Reynolds 1989b, 1990). It is of particular interest that neither the light-saturated photosynthetic carbon fixation rates nor the maximal nutrient-uptake rates (except for certain more conservative components, such as silicate and sulphate), establish more than a resource-acquisition capability. When the supply of resources is adequate, the rates of cell replication are ultimately constrained by intracellular synthesis, translocation and assembly into new biomass. Not surprisingly, species-specific replication rates and their sensitivity to temperature and photoperiods are apparently correlated with cell morphology (Reynolds 1989c). Thus, supposing water temperatures and day-length characteristics to be those anticipated in temperate lakes (say, 20°C, 16h d^{-1}), reported *in situ* cell replication rates (Reynolds 1982) in the order of magnitude from 1.2 d^{-1} (for nano-planktonic non-motile unicells of *Chlorella*) to 0.12 d^{-1} (for large flagellate cells of *Ceratium*), seem to accord with respective maximal rates observed in culture, i.e. *c*. 21.3 x 10^{-6}, 2.4 x 10^{-6} mol C fixed (mol cell C)$^{-1}$ s^{-1}. In comparison to the 9 h or so required to complete one *Chlorella* division, the same cell could be expected to be able to fix sufficient carbon in some 5.4 h but to take up from a rate-saturating concentration sufficient phosphorus to provide for the next cell division in only 7 minutes.

Although the dependence of nutrient uptake upon external concentration conforms to Monod-type kinetics (Fig. 7), algae differ perceptibly in their rates of absorption and assimilation (see, for instance, Rhee 1982; Tilman, Kilham & Kilham 1982; Turpin 1988). It is possible to discriminate species, on the basis of the ratio between maximal rates of uptake and growth, as being (velocity-) adapted to capitalise on favourable supplies, (storage-) adapted to fluctuating resource levels, or (affinity-) adapted to operate at chronically low nutrient resources (Sommer 1984; Reynolds 1987a). Nevertheless, the ultimate requirement for the cell is to obtain sufficient nutrients to supply the cell material for the next generation. In this, there are only modest deviations from the ideal (Redfield) molecular ratio of key nutrients, viz. 106C:16N:1P. Thus, whether the demand is phased over the full replication period or whether it is garnered in a shorter period of opportunity, the amount of phosphorus required to sustain the next generation is some 9.43 x 10^{-3} mol P (mol cell C)$^{-1}$. To maintain the hypothesised growth rate of *Chlorella* (21.3 x 10^{-6} mol C (mol cell C)$^{-1}$ s^{-1}), for instance, without depleting the intracellular store, would require the assimilation of the equivalent of 0.20 x 10^{-6} mol P

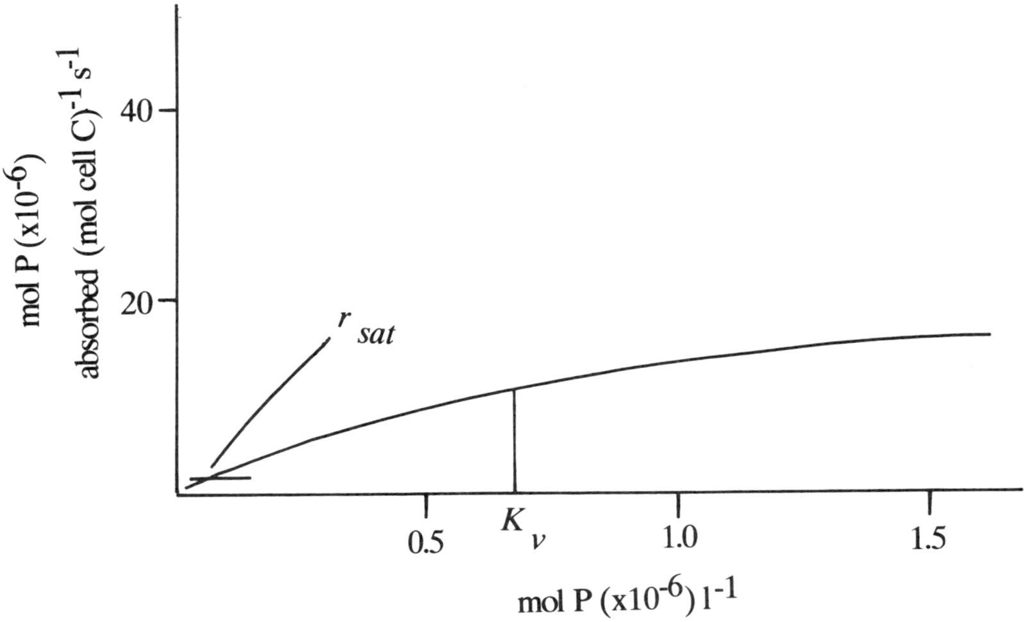

Figure 7. Cell mass-specific phosphorus uptake by P-starved *Chlorella* cells as a function of external concentration. K_v is the concentration required to half-saturate uptake. The intersection with r_{sat} denotes the phosphorus concentration (~7 x 10^{-8} mol P l^{-1}) required to balance growth rate at 20°C.

(mol cell C)$^{-1}$ s^{-1}. The concentration of phosphorus required to maintain this growth rate of *Chlorella* under the supposed natural conditions is marked on the curve in Figure 7; only 0.07 x 10^{-6} mol P l^{-1} i.e. about 2µg l^{-1}. When calculated in this way, moreover, the corresponding requirements of other common algae are also found to be substantially met by soluble reactive fractions of phosphorus (SRP) at concentrations in the order of *c.* 10^{-7} mol P l^{-1} or less (Reynolds 1992a).

The corollary of this deduction is that so long as the concentration of SRP exceeds *c.* 0.1 x 10^{-6} mol P l^{-1} (i.e. *c.* 3µg P l^{-1}), it is most improbable that algal growth is phosphorus limited. Conversely, the SRP concentration must be reduced to <3µg P l^{-1} before cell growth is likely to be limited by phosphorus availability. Even then, there may be a lag before the internal store (q) falls faster than it can be replenished towards the minimum quota (q_o). Other than at steady state (Burmaster 1979), the Droop formulation is likely to yield a more realistic model of the phosphorus-deficient growth kinetics than is the Monod model (see also Sommer 1991).

Three cases of vernal, generally diatom-dominated growth in northern hemisphere temperate lakes are illustrated (Fig. 8) in support of this principle. In winter, increase in chlorophyll is normally regulated by light availability – the duration, as well as the intensity, of irradiance and the interaction with vertical mixing, especially while the water remains unstratified. As the days lengthen, the light-limited capacity increases so, provided other resources are available, the spring increase gets under way accordingly. This is shown to apply in the case of an experimental enclosure (known elsewhere as "limnocorrals" and "mesocosms") in Blelham Tarn, English Lake District, where dissolved phosphate, silicate, nitrate and EDTA-complexed iron were artificially replenished weekly: the chlorophyll concentration and total phosphorus increased rapidly towards a maximum in April, when the column stratified and the dominant *Asterionella* was replaced by the colonial green flagellate

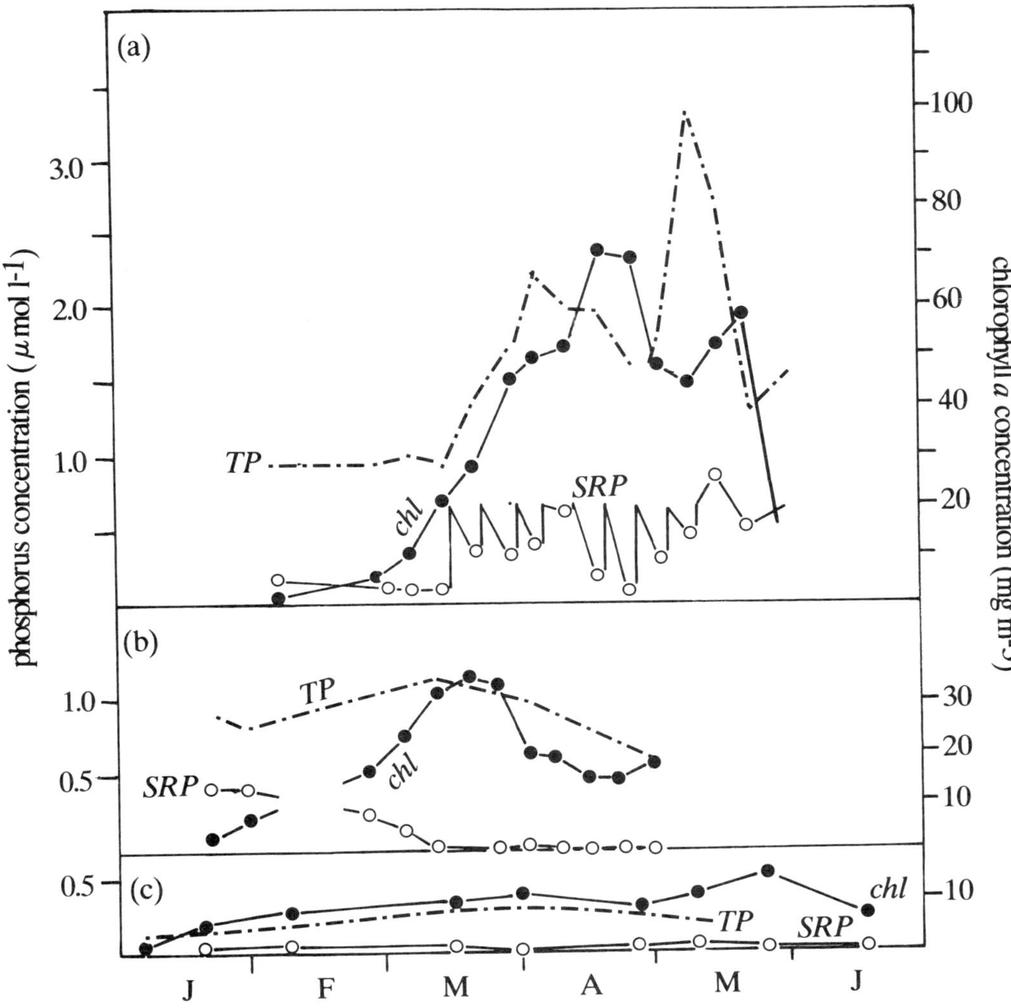

Figure 8. Changes in vernal (mainly diatom) chlorophyll in (a) Blelham Enclosure B, in the 1978 experiment, (b) Esthwaite Water and (c) Lake Michigan. While soluble reactive phosphorus (SRP) is replenished (a), algal chlorophyll increases together with the total phosphorus (TP) in suspension; in (b) the removal of SRP to algal mass sustains an increase in chlorophyll (but not in TP) until the supply is reduced to a truly limiting concentration (<0.1 μmol P l^{-1}); in (c) phosphorus limits almost continuously.

Eudorina (for full details see Reynolds & Wiseman 1982). In the case of the natural growth in nearby Esthwaite Water, the vernal increase in chlorophyll concentration proceeds at an initially similar rate but at the clear expense of the SRP pool; its reduction to *c.* 1 μg P l^{-1}, i.e. some 3 x 10^{-8} mol l^{-1}, indicates the onset of phosphorus limitation, with a plateauing and then a decline of the chlorophyll following in close sequence. In the case of Lake Michigan, where the SRP fraction scarcely exceeds 1 μg P l^{-1}, both the rate of lake spring chlorophyll increase and the carrying capacity of the lake are arguably phosphorus-limited throughout.

Nutrient ratios

The deduction that phosphorus does not limit phytoplankton growth until it has almost all but disappeared from solution calls into question certain other widely-held tenets concerning the supposed control of phytoplankton by nutrients, including those in the area of competitive exclusion and the selective importance of nutrient ratios. Although the requirements for (say) nitrogen and carbon are also universal among phytoplankton algae, the forms available and the optimal quantities are sufficiently dissimilar for it to be supposed that the growth of one species might become limited significantly before that of another. The deduction that, other things being equal, this would provide a competitive advantage to the latter over the former would then seem quite logical. Moreover, were the second species to be simultaneously limited by the availability of a second element, while the first was not, then they would cease to be direct competitors and thus be able to co-exist. This, indeed, is the basis of what has become known as Tilman's resource-based competition theory (Tilman 1977). Ultimately, it predicts that the greater the number of potentially limiting resources, then the greater is the number of potentially co-existing species (see also Petersen 1975).

Compelling experimental evidence has been advanced in support of this theory in the case of nitrogen-fixing cyanobacteria over, say, chlorophytes when external sources of combined nitrogen are low (Rhee 1978; Rhee & Gotham 1980), or of cyanobacteria over certain chrysophycophytes, and any chlorophytes that may be present, when free carbon dioxide has been exhausted and the ability to utilize bicarbonate is at a premium (Shapiro 1990a) or, again in the case of the non-diatoms, exemptions from requiring relatively large amounts of skeletal silica which allows them to supersede diatoms when soluble sources of reactive silicon (SRSi) have been exhausted (Holm & Armstrong 1981). Nor can exception be taken to the fact, ably demonstrated in the carefully controlled experiments of (e.g.) Tilman & Kilham (1976), that if the ratios of limiting resources can be maintained in some kind of steady state, the bias will inevitably bring about either the dominance of A over B, or of B over A, or the co-existence of A and B.

The real problem for the ecologist comes in detecting the relevance of this mechanism in nature. Certainly, one might predict that in a nutrient-poor lake having (say) chronically depleted levels of SRP and SRSi, some selection upon resource gradients is likely to occur. The argument is well supported among some larger lakes where the growing season is sufficient to permit long-term selections over long periods (Kilham & Kilham 1990). Indeed, this principle ought to be invoked to explain the assembly of the species associations and their typical dominant species which characterize lakes at particular points on the scale of trophy. It is perhaps more difficult to accept the argument that gradients in SRSi/SRP ratio that have been detected near fluvial inflows in Lake Michigan are either sufficiently robust or sufficiently permanent to maintain significant differences in the algal community; the challenge (Sell, Carney & Fahnenstiel 1984) that the species differences are due to the non-competitive effects of different compositional mixes introduced by the river inflows, probably has equal merit. The answer would appear to depend upon the interpretation of the dynamics of fluvial flow, limnetic dissipation and advection in relation to the doubling-times of nutrient-limited phytoplankton cells (Tilman, Kilham & Kilham 1984; see also Auer, Kieser & Canale 1986).

The explanation becomes even more implausible if the critical condition "that the cells are nutrient-limited" is ignored. For the growing cell, the only factor of relevance is whether there is enough of each of its various nutrient requirements available to sustain its next cell division. It is immaterial whether the N:P ratio is 50:1 or 3:1 if both nitrogen and phosphorus supersaturate the growth requirements for the current generation. It cannot be denied that when one or the other has been seriously depleted, the ratio may surge quite dramatically. This must be seen for what is it: a fortuitous consequence of one nutrient falling to a low availability, by which time it is the nutrient itself, not the ratio, that is critical. Moreover, if the ratio between two elements changes abruptly, as one or other is exhausted, the question must arise whether the gradient principle has an opportunity to take effect: we cannot expect a subdominant population B to supplant a large nutrient-limited population of A when insufficient resource persists to sustain (perhaps) four or more divisions of B while A declines. In other words, the inoculum effect is the dominant criterion at such stages and sufficiently so to overwhelm a subtly altered selective bias (see Reynolds 1986 for examples). The usefulness of citing nutrient ratios relative to likely requirements of algae probably lies in determining which, if at all, is likely to become limiting during the increase phase. This may be used, for instance, in a lake with high phosphorus loading and only moderate nitrogen loading to predict a summer dominance of nitrogen-fixing blue-green algae. Even here, the ratio itself is consequential upon and not the determinant of events; thus its general application, as envisaged by Smith (1983), has proved less reliable than other criteria (Trimbee & Prepas 1987; Shapiro 1990a).

Seasonality of phosphorus availability and utilisation

If, as argued here, soluble reactive phosphorus must fall to 10^{-7} molar or less before phytoplankton growth is likely to be phosphorus-limited, and the quantities in which phosphorus is required are (weight for weight) generally similar amongst the algae, then it may justly be asked why the presence of bloom-forming populations of blue-green algae bear such a consistent affinity for the more eutrophic lakes and reservoirs, at least in temperate regions (Rodhe 1948; Rawson 1956; Gorham *et al.*, 1974) and, indeed, why they respond positively to anthropogenic phosphorus enrichment (Edmondson 1977; Trimbee & Prepas 1987; Sas 1989). Even if distinction is drawn between the progressive build-up of *Oscillatoria* spp. (especially of the *O. agardhii* – *O. rubescens* – *O. proflica* sub-group) in larger, deeper lakes or the increased abundance of summer growth of *Anabaena*, *Aphanizomenon* or *Microcystis* spp. in smaller, physically variable basins, the organisms are generally rather slow-growing and are particularly sensitive to temperature (see Reynolds 1989c). Their growth periods thus tend to be confined to the warmer months of the year. Evidence gleaned from comparisons between years (A. E. Irish, in Harris 1986; Talling & Heaney 1988) and between lakes (Kalff & Knoechel 1978; Reynolds 1984a) strongly indicates that lake eutrophication often leads first to an "infilling" of the period between the spring and autumnal phases of algal abundance by intermediate population developments (Fig. 9). Several factors may contribute to this effect but the increased supportive capacity may often be attributable to a surplus of dissolved phosphorus in excess of the ability of the spring bloom to consume it. The opportunities to blue-green algae might then depend upon the simultaneous fulfilment of their environmental requirements for resources and for their exploitation. Other algae, of course, share these opportunities, notably the motile chrysophytes, the mucilaginous colonial species of chlorophyta and the large-celled dinoflagellates. The outcome in terms of species composition, dominance and successional changes, may well depend upon the interactions with other factors, including the subsequent spatial "segregation" of resources (Reynolds 1976, 1992b), the depletion of carbon (Shapiro 1973, 1990a) and nitrogen (Riddolls 1985), selective grazing by *Cladocera*-dominated zooplankton (Hrbaček 1964; Lynch 1980; Sommer *et al.* 1986; Sterner 1989) and the intervention of physical factors (floods, wind-mixing events; Reynolds 1987a). Particularly instructive in this context, however, is the reproducible behaviour, described in Sas (1989), of erstwhile dominant populations of blue-green algae self-regulating to greater depths in the light gradients of lakes as the summer surplus of phosphorus is reduced through restriction of external phosphorus loads.

The control of P-capacity: external sources

It is clear that in many instances, phosphorus availability sets an upper limit on algal-carrying capacity in a water-body, even though this may be achieved infrequently if at all. Where it does

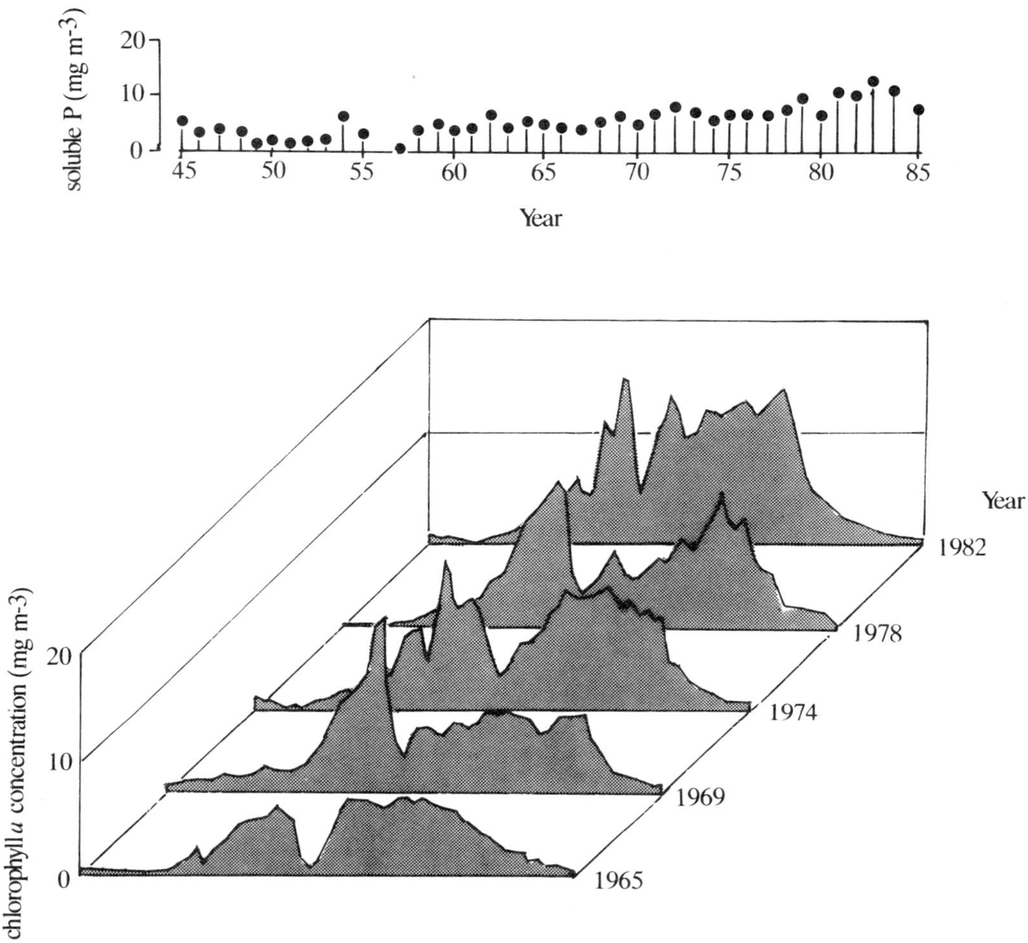

Figure 9. Stages in the enrichment of the north basin of Windermere, shown by prevernal soluble phosphorus concentrations (above) and the progressive increase in algal chlorophyll during the summer period. Data from Talling & Heaney (1988) and Talling (1992).

so continuously or often, there must be a strong incentive to protect that *status quo* and to control the external loads of phosphorus on the system. Elsewhere, the implicit objective would seem to be to reduce the external loadings in order to lower the carrying capacity and hence to impose the rate limitation earlier in the ascendancy of phytoplankton biomass. This form of words avoids making an assumption that less phosphorus always means fewer algae. This is not, however, a matter of semantics but a recognition of the functional relationships ("subsystems"; Sas 1989) operating within the "blackbox" of the Vollenweider–OECD equation. Moreover, an adequate reduction in loading is often very difficult to achieve: some prior understanding of the relative scales of the various sources, as well as the feasibility of controlling them, is essential to determining the best options for management.

Much of the phosphorus presently perceived to contribute to the cultural eutrophication of lakes and reservoirs is not only anthropogenically derived but it is of comparatively recent origin (the last 50 years or so). The principal reason for this is the enhanced delivery of dissolved phosphates to natural waters, following the widespread introduction of secondary treatment of domestic sewage and, more recently, detergents built on hydrolysable polyphosphates. The relative loadings due to these materials clearly differ with the scales of the sources and their dilution by the receiving waters, although estimates attributing the generation of between 0.5 and 1.0 g P per person per day seem to have a sound basis in human physiology (e.g. Best & Taylor 1958). Estimates of the fraction of limnetic loads attributable to domestic detergent range between 9 and 25% of the total (Department of Environment 1991). Industrial and non-physiological sources of phosphorus can further contribute to a *per capita* soluble-P generation perhaps averaging about 2.5 g d^{-1}. Secondary-treated sewage effluent may contain some 25 (± 15) mg SRP l^{-1} and *c.* 40 (± 25) mg l^{-1} inorganic N.

Sewage inputs are largely identifiable "point sources", in the sense of being (at least, in theory) readily controllable through tertiary (P-stripping) treatment of the effluent before discharge from the sewage treatment works. An established, simple technology, relying upon coaggulation and precipitation with (usually) ferric salts, has been applied at a number of sensitive sites in the UK, as indeed it has in other IWSA-member nations.

"Agricultural run-off" is often branded, though rarely with more than prejudice, as a contribution to eutrophication. There can be little doubt that certain forms of intensive animal husbandry and market-gardening are capable of generating relatively high concentrations of phosphate in drainage waters. On the other hand, the natural phosphates incorporated into bulk inorganic fertilisers are, generally, insufficiently soluble for it to be supposed that their widespread use provides a major resource of dissolved reactive phosphorus. However, the particulate flux from ploughed, arable soils, which in any case presents greatly expanded leachable surface area to run-off water and drainage, can be very greatly enhanced. Depending upon grain size, much of this material will find its way into streams and rivers and, thence, eventually enter lakes and reservoirs as part of their total phosphorus (TP) load. As such, it may add substantially to the diffuse, or non-point, natural load of particulate phosphate, derived from the weathering of phosphatic minerals (e.g. apatite) present in the catchment area. A fraction of finer particles may remain in aqueous suspension long enough to permit solution but, even though the bulk of this material may settle out of suspension, more or less quickly, it is not necessarily lost irretrievably to the pelagic system; rather, it remains potentially recyclable within it (see following section).

The impact of altering this balance between the external loads of dissolved and particulate loads of phosphorus may not have been wholly appreciated. For instance, the present external TP-load to Windermere, English Lake District (estimated to be *c.* 16 x 10^3 kg $year^{-1}$; FBA unpublished data) is roughly equally divided between particulate (mainly catchment-derived) and dissolved sources (largely attributable to sewage-treatment plants, commissioned since 1960; Talling & Heaney 1988). Thus, it is deduced that while the TP-load has probably nearly doubled over the same period, the dissolved phosphorus pool in the lake, where it is readily available to phytoplankton, is now some 7- to 10-fold greater than it was four decades ago. The concomitant changes in the yield of phytoplankton biomass (Talling 1992) must reflect this ongoing elevation in "soft sources" of phosphorus rather more than the increased external load *per se*.

Approaches to P-capacity reduction through control of external loading

The example of Windermere is a good illustration of a case suitable for treatment through external-load reduction: the recent eutrophication has been brief and is still relatively mild, the natural SRP load is minimal and the major contribution is mainly processed through two substantial secondary treatment plants and half-a-dozen or so small ones (some already with tertiary P-removal). A cost-effective return on investment, in terms of an improved algal quality, may be anticipated. Certainly, some of the more successful lake-restoration attempts include instances in which diversion or tertiary treatment has indeed brought about a greatly reduced external phosphorus load to Lake Washington (Edmondson 1977), and the several cases reviewed by Marsden (1989) and Sas (1989) furnish good examples.

The application of this philosophy to water-supply reservoirs affected by eutrophication is not always straightforward, especially where they are located in well-populated and/or intensively agricultural lowlands with dispersed or diffuse sources of phosphorus. In the case of pumped-storage schemes, where (enriched) water is abstracted from riverine source(s), it may be practicable to strip phosphorus from the proximal inflow in preference to a (probably) ineffective treatment of the individual point-sources in the catchment. Where the reservoir is an impoundment of the river, the solution may be to construct or adapt a pre-reservoir where the water may be subject to precipitation of phosphorus before entering the main store. One of the most impressively successful examples of this approach is the Wahnbach Reservoir, Germany. The enriched waters of the inflowing River Sieg are rehabilitated to the standard of a near-pristine oligotrophic source. The impact of this routine treatment of the source water upon the erstwhile character of the reservoir plankton, and the ongoing advantages in terms of the cost of treating the abstracted water for profitability have been well documented (for references and reviews, see Bernhardt 1987). By analogy, feeding nutrient and silt-laden riverine inflows through shallow reedswamp-dominated lagoons can have a similarly protective effect upon receiving lakes and reservoirs of lower trophic status (Szilágyi *et al* 1990).

While, in theory, it is possible to lower or to impose a constraint on phosphorus-determined phytoplankton carrying-capacity of *any* lake or reservoir system through control of the external loading, it is neither the only, nor the most practical, nor the most cost-effective approach in every instance, nor even is it always desirable, if others factors interfere. For instance, it is well understood in the management of fishponds that the "crop" production is enhanced by expanding the nutrient base (Korínek *et al.* 1987) and the deliberate addition of fertilisers (eutrophication) is, at least on site, beneficial so long as the biological yield is carefully managed to be "more of the same" (Reynolds 1987b). The same principle applies, albeit on a less intensive scale, to increasing the productivity of harvestable fish from different kinds of lake. Who is to say that the biology of some extreme base-poor, nutrient-deficient waters could not be enhanced by a modest increase in phosphorus loading? Phosphorus *addition* has also been shown to be positively beneficial to the restoration of acidified waters, by promoting biological production, altering the balance of carbon dioxide metabolism and invasion, generating hydroxyl ions and so raising the acid neutralizing capacity of the lake; the ambient pH may be moved positively and permanently towards neutrality (Davison *et al.* 1989). Davison (1987) has shown that phosphorus is 47 times more effective as a counter to acidity than the molecular equivalent of calcium carbonate. It is worth querying the policy of phosphate removal from inflows into lakes in, for example, the granitic, acid-rain sensitive areas of Adger and Telemark, southern Norway, as being inappreciative of the role played by phosphorus in maintaining the fish populations that they continue to lose through strong-acid precipitation (Henriksen, Lien & Traaen 1990).

Time-scales of P-capacity reduction: the "internal load" problem.

Supposing that restoration of a water body by the reduction of phosphorus-determined capacity is both desirable and technically feasible, some preconception of the extent and rate of recovery is desirable. While the explicit assumption of the Vollenweider–OECD regression is an output in terms of an average crop maintenance for a (continuous) supply of nutrients, there is an implicit assumption that the primary product-yield will be simultaneously eliminated, together with any unused resource, either down an outflow, or through consumption by higher trophic levels or, eventually, as a residual net particulate flux to the permanent (and usually chronostratigraphic) lake sediment. Several processes, both positive and negative, contribute to a net flux which, on balance over a period of time, is in the direction of sedimentary recruitment. Were this not so, it would have been more difficult than it has been to interpret past limnological environments from the contemporaneous stratigraphic variations in the chemistry of the sediment. The problem for the water manager is to anticipate the establishment of a new balance after one of the components in the current exchanges of phosphorus across the sediment–water interface is abruptly altered.

The topic is a large one with a substantial history of study (Einsele 1936, 1938; Mortimer 1941, 1942; Kamp-Nielsen 1974, 1975; Boström, Jansson & Forsberg 1982; Baccini 1985). Any attempt to emulate the recent review by Boström *et al.* (1988) is inappropriate here, although it is helpful to summarise the main fractions of sedimentary phosphorus, their solvency and the circumstances of their liberation into open water (see Fig. 10). Between-lake differences in behaviour reflect differences in loading, composition and retentiveness of the sedimentary phosphates and opportunities for their mobilisation. The nature of the binding to sedimentary particles is also relevant to post-restoration behaviour. The low total phosphorus content of sediments in oligotrophic lakes is dominated by refractory organic species, while the HCl-extractable (i.e calcium-bound) phosphorus of sediments in calcareous lakes provides a good buffer against subsequent internal loading. Elsewhere, in highly eutrophic lakes, the binding capacity may have been saturated, so that the pool of labile and loosely-bound (NH_4Cl-

INPUT TO SEDIMENT	RELATIVE BINDING CAPACITY *WEAK* ←——→ *STRONG*	RETURN TO WATER
External phosphorus minerals	*NH4-extractable* *labile and loose-bound*	desorption dissolution
Adsorbed/coprecipitated with Fe/Mn	*NaOH-extractable* *loose- or ligand-bound*	ligand exchange
Adsorbed/coprecipitated with clays,oxyhydroxide	*HCl-extractable*	
Adsorbed/coprecipitated with carbonate		
Biogenic particulate P	*Citrate-dithionate-extractable* *residue*	mineralisation hydrolysis
Direct uptake by bottom-dwelling algae/bacteria	*loose- or ligand bound*	release transport
Direct adsorption onto sediment	*labile*	dissolution desorption

Figure 10. Scheme to show sources of phosphorus input to lake sediments, their internal binding (according to analytical sensitivity) and the method by which phosphorus is given up to the water. Dominant sources, binding ability and release will obviously vary from lake to lake but the table has been arranged to provide a tentative guide to the interrelationships.

extractable) phosphorus is readily mobilised. The mobility of NaOH-extractable phosphorus, supposed to correspond to redox-sensitive, ligand-exchangeable iron-, manganese- and aluminium- bound phosphorus, is sensitive to redox potential and (e.g.) the $Fe^{2+} \leftrightharpoons Fe^{3+}$ transformation.

The circumstances of phosphorus "release" from the sediments can be attributable to both chemical and mechanical processes. Besides the water–sediment gradients (Kamp-Nielsen 1974), chemical mobilisation is especially influenced by microbial activity (directly and indirectly), so it is sensitive to temperature, pH and redox potential. With lowered redox (<200 mV), the use of alternative electron-acceptors in organic mineralisation leads to the presence of reduced products (such as H_2S, CH_4) affecting phosphorus cycling. The reduction of Fe^{3+} to Fe^{2+}, for example, will return associated phosphorus to limnetic solution. Other limnochemical transformations influence the outcome (see Boström *et al.* 1988, for references); thus, although much of the phosphorus will coprecipitate with the iron re-oxidised to Fe^{3+} in the oxic part of the water column, some of the residual phosphorus may well be available to support renewed primary or microbial production (Golterman, Bakels & Jakobs-Möglin 1969).

This classical view of phosphorus recycling as a series of chemical redox reactions is widely held to be the major internal-load pathway. However, certain types of lake may rely extensively on hydromechanical factors (organism transport and sediment resuspension) to effect any kind of recycle. For instance, Boström *et al.* (1988) concluded that the migration into the water-column of bottom-dwelling cyanobacteria (e.g. overwintering *Microcystis*) moves significant loads of intracellular nitrogen and phosphorus out of the sediment. Additional evidence that mechanical disturbance by burrowing animals and spawning fish can return significant quantities of phosphorus to solution (Petr 1977) must be balanced against claims that simultaneously raising the redox potential and resuspending new binding capacity will enhance precipitation from the water (Davis, Thurlow & Brewster 1975). Depending upon the fluidity of the sediments, the extent to which the P-bonding capacity is saturated (and, hence, how much phosphorus is dissolved in the interstitial water) and the strength and frequency with

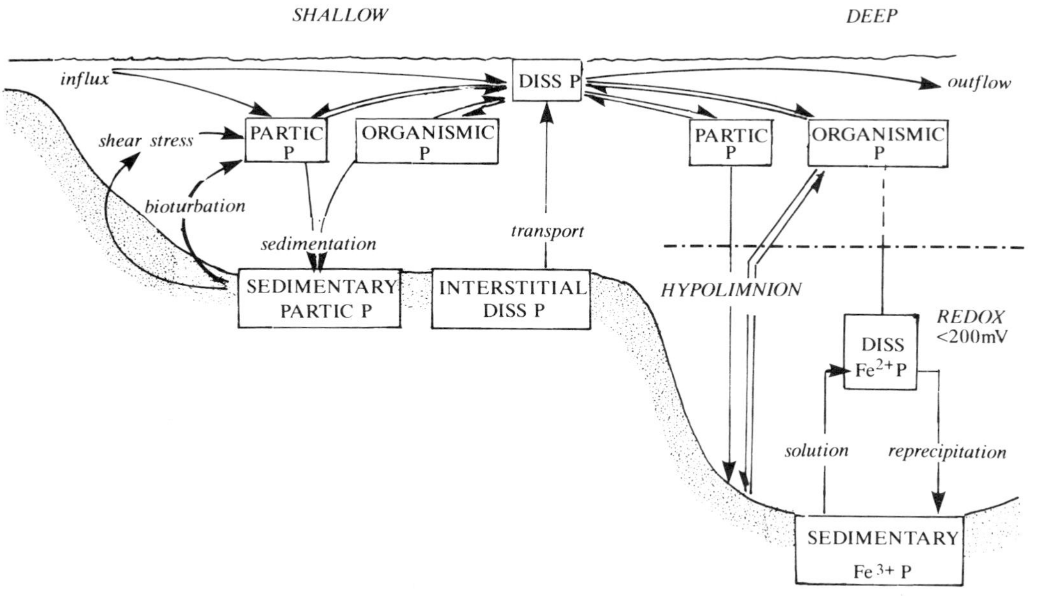

Figure 11. Diagram to show main pathways for phosphorus in lakes, contrasting events in shallow and deep (especially low-redox) regions.

which dissipative energy is transferred, physical resuspension may lead to significant phosphorus releases from shallow margins. Wind-induced turbulent shear, dissipating energy at rates in the range 10^{-6} to 10^{-5} m^2 s^{-3}, is capable of mixing perhaps the upper 30 mm (Nixon 1988), exceptionally 40 mm (data of Robbins & Edgington 1975), of the superficial sediments. In areal terms, shallow margins can be as important as deep, anoxic basins in phosphorus recycling; clearly, the relative areas may determine which is more important to the lake as a whole. In uniformly shallow waters, physical mixing of oxic sediments is the only such pathway. Certainly there is evidence that highly enriched, shallow waters (Kamp-Nielsen 1975; Osborne & Philips 1978) can develop (Fig. 11) substantial internal loads (up to 50 mg P m^{-2} d^{-1}; Boström *et al.* 1982) to an extent which is sufficient to have countered the effects of external load reductions for long periods (e.g. Norrviken: Ahlgren 1980) or until such time as additional restorative steps were taken (sediment removal at Lake Trummen, Sweden: Björk 1972, 1988; flushing of Veluwemeer, Netherlands: Hosper 1984). For full restoration, the particles and interstitial water associated with the phosphorus-rich stage require to be buried, or rather progressively "diluted out" (see Guinasso & Schink 1975), by the particulate flux and interstitial water of the later, phosphorus-deficient stage. Deeper sediments play little further role in the contemporary exchanges and lake ecologies.

The point is that sediment–water exchanges of phosphorus do not exclusively follow the Einsele–Mortimer model. There is a distinct challenge to resolve a sediment typology which should both account for and predict the relative significance of the various available pathways to an abrupt shift in the overall phosphorus metabolism of a lake or reservoir. This might clarify the criteria for management and restoration objectives. In the case of the frequently disturbed lake with high, unbound nutrient resources derived from proximally diffuse sources, the control of biological capacity might be difficult to achieve by the prescription of tertiary treatment of sewage effluents: alternative approaches are then required.

Controlling the rate of P-capacity attainment

The alternative approaches to managing planktonic algae in eutrophic systems depend upon the imposition of a further constraint at a lower threshold which effectively prevents, or delays indefinitely, the expansion of the biomass to its phosphorus-determined capacity. In one of these, the principle is to encourage the consumption of cells faster or no slower than they are produced; the other is to slow down the rate of cell replication by depriving it of sufficient light to saturate its growth requirement.

Biomanipulation

Biomanipulation is a term introduced by Shapiro, Lamarra & Lynch (1975). It is increasingly restricted to mean the control of organisms by deliberate adjustment of the abundance of the next trophic level of consuming organisms. A large literature and a great deal of debate have been directed towards this topic. It is inappropriate here to review the relative merits of the "top–down" trophic cascades theory (Carpenter, Kitchell & Hodgson 1985) as opposed to the "bottom–up" resource regulation, implicit in the Vollenweider–OECD model. It is not a case of which theory is correct – there is plenty of good evidence for either mechanism being effective at particular times. On the one hand, the initiation of active populations of secondary-produced herbivore populations cannot occur independently of primary production; on the other hand, the late vernal "clear-water" phase observed in many mesotrophic and eutrophic temperate lakes is often wholly, or in part, attributable to pelagic grazing which proceeds more rapidly than the rate at which food plants replace themselves (Carpenter 1988). Reynolds (1991) suggests that, at equilibrium "top–down" and "bottom–up" effects collapse into one another and become indistinguishable; thus, if one or other process is clearly dominant, it is symptomatic of a persistent non-steady state. Thus "biomanipulation" is a real and reasonable objective for controlling certain kinds of (edible) phytoplankton biomass. How to "make it stable" (Shapiro 1990b) may be illusory compared to "how often" the biology of the latter may be modified sufficiently to effect long-term practical benefits.

The most convincing demonstrations of lasting control of phytoplankton biomass have come from small, shallow waters (e.g. Shapiro *et al.* 1983; Moss 1990 and in this volume). Whereas in larger, deeper systems, where the zooplankton is more strictly pelagic and extremely responsive to the abundance and dynamics of the dominant food source (planktonic algae and bacteria: Ferguson, Thompson & Reynolds 1982; Sommer *et al.* 1986), the macrophytic vegetation and the benthic accumulation of organic detritus in the smaller or shallower water-bodies offer more refuges from predators and alternative food resources. In this way, a stock of herbivores adequate to control potential algal outbursts may be more plausibly maintained. Confirmation is required for the developing appreciation that the potential for management of planktonic algae by biomanipulative techniques may be greater among shallow lakes (McQueen 1990; Reynolds 1991), especially where nutrient limitation of the algae and tight regulation of the food chain are demonstrable (Elser & Goldman 1991). It is of interest that the fish ponds studied by Hrbaček *et al.* (1961), whose work inspired much of the subsequent interest in biomanipulation of pelagic systems as a means of managing phytoplankton, actually conform well to this generic description. At the other extreme, deep, weakly-flushed, enriched lakes are least likely to succumb to biomanipulative techniques, without (say) a simultaneous reduction in nutrient loading (Benndorf 1990). More zooplankton grazing only seems likely to drive the pelagic community towards dominance by self-regulating, grazing-resistant populations of algae, including cyanobacteria, assuming that their other environmental requirements are satisfied (Reynolds 1986, 1987c; Sterner 1989).

Physical manipulation

The remaining practical approaches to the control of algal biomass yields, species composition, and the rate of capacity attainment, have invoked the artificial enhancement of physical limitations. Methods for extending the period of full isothermal mixing in lakes deep enough to stratify, either by artificial destratification or by preventing stratification from occurring at all, have been available for over thirty years (see Irwin, Symons & Roebeck 1969; Dunst *et al.* 1974; Tolland 1977). The early experiences were perhaps founded upon observations that naturally-mediated variations in the extent of mixing generally have beneficial effects on the mass and species composition maintained in the phytoplankton, and a concomitant benefit in terms of stored water quality (Ridley 1970). Several authors, notably Steel (1972, 1975), have successfully modelled the quantitative responses and their effects upon maximum carrying capacity. The species-selection mechanisms have proved more difficult to explain, though broadly these seem to result in species associations similar to those in equivalent conditions promoted by natural mechanisms (see Reynolds *et al.* 1983; Reynolds & Reynolds 1985). Several of these responses have been imitated, on a pilot-scale, in the mesocosmic enclosures in Blelham Tarn (Reynolds, Wiseman & Clarke 1984; Reynolds 1986) and analysed in detail (Reynolds 1983a, b, 1984b, 1989c). Moreover, the responses appear to have plausible physiological explanations, in terms of strategic adaptations (Reynolds 1989c, 1992a; Dauta *et al.* 1990).

The arguments in favour of exploiting the dominance of physical factors in the dynamic behaviour of phytoplankton, as a practical management tool, have again been advocated by Reynolds (1992a): provided sufficiently abrupt alternatives in the physical environment can be effected and then maintained, a water column can be managed in a way which might favour the smaller, nanoplanktonic and edible species, of the desmid–diatom–*Oscillatoria* associations of well-mixed but increasingly light-deficient columns, and resisting for as long as possible the development of large nuisance populations of (e.g.) bloom-forming blue-green algae.

There is a danger of embracing any advocated method as a catch-all solution to the problem of eutrophication management. Artificial mixing is no more a panacea than is biomanipulation: it will not be successful in every circumstance. There is little to be gained, for instance, from destratifying a 5-m lake which habitually stratified at about 4m. If it were 10 m deep, destratification need not produce much change if the water column is generally clear. Moreover, constant mixing of even a deep reservoir may well successfully exclude (say) *Microcystis* but it may nevertheless promote large populations of diatoms, to the capacity predicted by Steel (1972). This need not be a problem if the diatom crops are preferred to the blue-green algae but it must be understood that artificial mixing does not obviate algae *per se*! Further, prolonged mixing would be expected to lead ultimately to *Oscillatoria* dominance (Reynolds 1989b): that it has not done so in the case of the routinely destratified Thames Valley reservoirs (Toms 1987) may be a function of their short retention times (Reynolds 1992c); a combination of destratification with 10 – 30 day retention times may be decisive.

In cases of excessive build-up of (say) diatoms and *Oscillatoria* it might be possible to consider one further technique, which is to destratify intermittently. Stopping artificial mixing might permit rapid stabilization of the water column, loss of diatoms from suspension and selection, initially, in favour of small, fast-growing nanoplankton. Conversely, mixing should be reimposed only after sufficient time had been allowed for a reduction of the earlier mixing-tolerant population to take place, but at the same time allowing the production of only a generation or two of the bloom-forming species. A cycle based on 2–3 *in situ* generation times (probably 7–14 days) would seem to be, in theory, an optimum period for a UK reservoir (see also data of Trimbee & Harris 1983).

Discussion

The ability to manage the quality of water in lakes and reservoirs depends upon an understanding of the underlying processes. By adopting a regression equation describing large-scale behaviour and using it as a basis for intervention with, essentially, large-scale counter-measures, the present attitudes to quality management tend to be correspondingly coarse and stochastic in their effects. This review is critical of three such attitudes, implicit in recent approaches to the eutrophication question.

(1). *That phosphorus is regarded as a pollutant* when it is a non-toxic life-supporting but scarce element which happens to be typically in chronically short supply. Cultural eutrophication is the oversupply of the element, the biological over-growth that may be promoted and which is prejudicial to the structure and health of the aquatic ecosystems so affected.

(2). *That supported biomass responds smoothly and proportionately to increase or decrease in phosphorus loading* when it manifestly acts in stepwise fashion; since phosphorus "limits" only when it is exhausted, a change may or may not alter the actual carrying capacity. While the biomass is within capacity, there is no ecological dependence upon the rate of phosphorus uptake, nor upon the ratio of phosphorus to another saturating element.

(3). *That the reduction of phosphorus will have an immediate result on abundance or species composition,* unless phosphorus is already continuously limiting. Phosphorus reduction, if feasible imposes a lowered capacity and, if appropriate, one which will act before the present crop sizes are achieved and do so for longer. On the other hand, if reduction is difficult to achieve, is inadequate or is offset by internal resilience, other management techniques may be needed.

Assiduous students of the literature will recognise that these points have been made before (Björk 1973; Harris 1980). There is no reason to believe that these strictures will be any more readily embraced this time. The suggestion may be made that a new scheme for lake management is required which recognises the individuality of lakes and their particular sensitivities. The diagram in Figure 12 is intended as a qualitative word model or decision tree, which might form the basis of a quantitative and deterministic approach to lake management. In a sense, this sort of scheme is already achievable, through the application of general behaviour models (e.g. of P-determined or light-determined carrying capacity) to operate the option switches. The outcome, however, would prescribe the "probable best approach" rather than entreat on "how much phosphorus must be removed". The further contributions in this symposium volume will assist in bringing the necessary quantitative resolution to Figure 12 and, in any case, give an enhanced understanding for the future management of freshwater ecosystems.

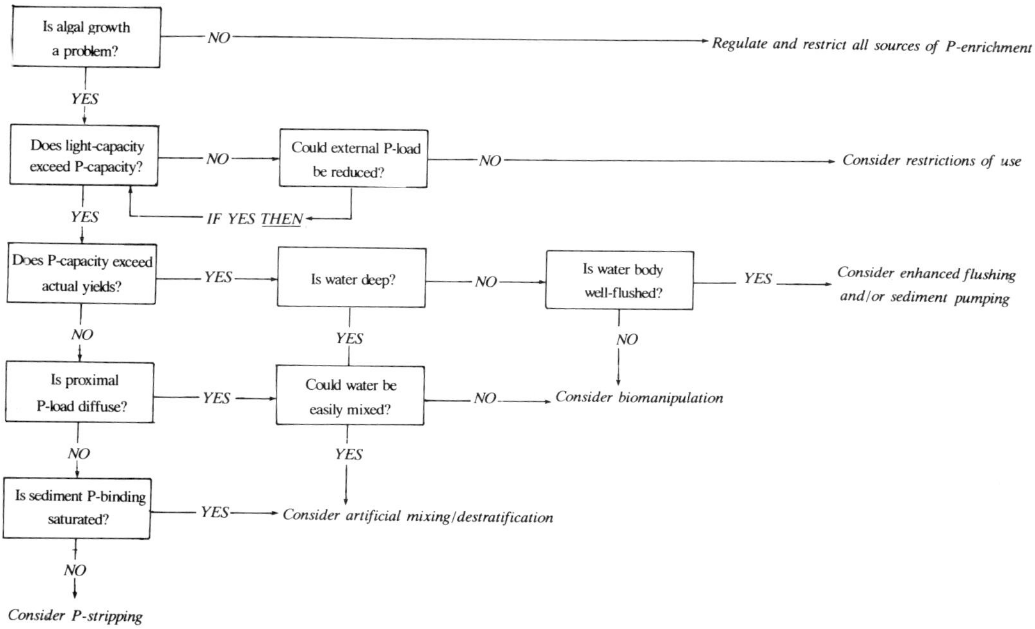

Figure 12. Provisional "decision tree" for lake restoration and management. Each of the boxes requires to be quantified and no option is exclusive. Future research might usefully concentrate upon the refinement of this approach and its application.

The present contribution draws upon experiences gained through research supported by the Natural Environment Research Council, the UK Department of the Environment, the National Rivers Authority and Essex Water Company. The opportunity provided by the Instituut voor Milieu-en SyteemAnalyse, Amsterdam, to confront with overseas colleagues many of the issues discussed here is gratefully acknowledged. The practical experience gained has accumulated over 25 years with the Freshwater Biological Association and, latterly, the NERC, during which I have benefited from the stimulus and companionship of a succession of enthusiastic and committed colleagues as well as from the trust and resources generously made available by the FBA's Council and Officers. I am grateful to Mr Jack Jeffery, Professor J. G. Jones and to Professor H. Bernhardt for the encouragement and opportunity to present the plenary review. Finally it is a pleasure to thank Mrs Jane Rhodes and Miss Kirsty Ross for their capable help in its preparation.

References

Ahlgren, I. (1980). A dilution model applied to a system of shallow eutrophic lakes after division of sewage effluents. *Archiv für Hydrobiologie*, **89**, 17-32.

Auer, M. T., Kieser, M. S. & Canale, R. P. (1986). Identification of critical nutrient levels through field verification of models of phosphorus and phytoplankton growth. *Canadian Journal of Fisheries and Aquatic Sciences*, **43**, 379-388.

Baccini, P. (1985). Phosphate interactions at the sediment-water interface. In *Chemical Processes in Lakes* (ed. W. Stumm), pp. 189-205. Wiley, New York.

Benndorf, J. (1990). Conditions for effective biomanipulation: conclusions derived from whole-lake experiments in Europe. *Hydrobiologia*, **200/201**, 187-203.

Bernhardt, H. (1987). Strategies of lake sanitation. *Schweizerische Zeitschrift für Hydrologie*, **49**, 202-217.

Best, C. H. & Taylor, N.B. (1958). *The Living Body*, 4th Edn. Chapman & Hall, London.

Björk, S. (1972) Swedish lake restoration program gets results. *Ambio*, **1**, 153-165.

Björk, S. (1973) Restoring lakes in Sweden. *Technology* Review, **76** (2), 1-10.

Björk, S. (1988) Redevelopment of lake ecosystems – a case study approach. *Ambio*, **17**, 90-98.

Boström, B., Andersen, J. M., Fleischer, S. & Jansson, M. (1988). Exchange of phosphorus across the sediment-water interface. *Hydrobiologia*, **170**, 229-244.

Boström, B., Jansson, M. & Forsberg C. (1982). Phosphorus release from lake sediments. *Ergebnisse der Limnologie*, **18**, 5-59.

Braunwarth, C. & Sommer, U. (1985). Analysis of the *in situ* growth rates of Cryptophyceae by use of the mitotic index technique. *Limnology and Oceanography*, **30**, 893-897.

Burmaster, D. (1979), The continuous culture of phytoplankton: mathematical equivalence among three steady-state models. *American Naturalist*, **113**, 123-134.

Carpenter, S. R. (1988). *Complex Interactions in Lake Communities*. Springer-Verlag, New York.

Carpenter, S. R., Kitchell, J. F & Hodgson, J. R. (1985). Cascading trophic interactions and lake productivity. *BioScience*, **35**, 634-639.

Dauta, A., Devaux, J., Piquemal, F. & Boumnich, L. (1990). Growth rate of four freshwater algae in relation to light and temperature. *Hydrobiologia*, **207**, 221-226.

Davis, R. B., Thurlow, D. L. & Brewster, F. E. (1975). Effects of burrowing tubificid worms on the exchange of phosphorus between lake sediment and overlying water. *Verhandlungen der Internationale Vereinigung für Theoretische und Angewandte Limnologie*, **19**, 382-394.

Davison, W. (1987). Internal elemental cycles affecting the long-term alkalinity status of lakes: implications for lake restoration. *Schweizerische Zeitschrift für Hydrologie*, **49**, 186-201.

Davison, W., Reynolds, C. S., Tipping, E. & Needham R. F. (1989). Reclamation of acid waters using sewage sludge. *Environmental Pollution*, **57**, 251-274.

Department of the Environment (1991). *Pollutants in cleaning agents: final report*. UK Department of Environment, London.

Dillon, P. J. & Rigler, F. H. (1974). The phosphorus-chlorophyll relationship in lakes. *Limnology and Oceanography*, **19,** 767-773.

Droop, M. R. (1973). Some thoughts on nutrient limitation in algae. *Journal of Phycology*, **9**, 264-272.

Droop, M. R. (1983). 25 years of algal growth kinetics. *Botanica Marina*, **26**, 99-112.

Dunst, R. C., Born, S. M., Uttormark, P. D., Smith, S. A., Nichols, S. A., Peterson, J. O., Knauer, D. R., Serns, S. L., Winter, D. R. & Wirth, T. L. (1974). Survey of lake rehabilitation techniques and experiences. *Technical Bulletins of the Department of Natural Resources, Wisconsin*, No. 75, 1-199. Wisconsin Department of Natural Resources, Madison.

Edmondson, W. T. (1977). Recovery of Lake Washington from eutrophication. In *Recovery and Restoration of Damaged Ecosystems* (eds J. Cairns, K. L. Dickson & E. E. Henricks) pp. 102-109. University Press of Virginia, Charlottesville.

Einsele, W. (1936). Uber die Beziehungen des Eisenkreislaufs zum Phosphatkreislauf im eutrophen See. *Archiv für Hydrobiologie*, **29**, 664-686.

Einsele, W. (1938). Uber chemische und kolloidchemische Vorgänge in Eisen-Phosphat-Systemen unter linmochemischen und limnogeologischen Gesichtspunkten. *Archiv für Hydrobiologie*, **33**, 361-387.

Elser, J. J. & Goldman, C. R. (1991). Zooplankton effects on phytoplankton in lakes of contrasting trophic status. *Limnology and Oceanography*, **36**, 64-90.

Ferguson, A. J. D., Thompson, J. M. & Reynolds, C. S. (1982). Structure and dynamics of zooplankton communities maintained in closed systems with special reference to the algal food supply. *Journal of Plankton Research*, **4**, 523-543.

Golterman, H. L., Bakels, C. C., & Jakobs-Möglin, J. (1969), Availability of mud phosphates for the growth of algae. *Verhandlungen der Internationale Vereinigung für Theoretische und Angewandte Limnologie*, **17**, 467-479.

Gorham, E., Lund, J. W. G., Sanger, J. E. & Dean, W. E. (1974). Some relationships between algal standing crop, water chemistry and sediment chemistry in the English Lakes. *Limnology and Oceanography*, **19**, 601-617.

Guinasso, N. L. & Schink, D. R. (1975). Quantitative estimates of biological mixing rates in abyssal sediments. *Journal of Geophysical Research*, **80**, 3022-3043.

Harris, G. P. (1980). Temporal and spatial scales in phytoplankton ecology. Mechanisms, methods, models and management. *Canadian Journal of Fisheries and Aquatic Sciences*, **37**, 877-900.

Harris, G. P. (1986). *Phytoplankton Ecology*. Chapman and Hall, London.

Heller, M. D. (1977). The phased division of the freshwater dinoflagellate *Ceratium hirundinella* and its use as a method of assessing growth in natural populations. *Freshwater Biology*, **7**, 527-533.

Henriksen, A., Lien, L. & Traaen, T. S. (1990). *Critical loads for surface waters – chemical criteria for inputs of strong acids (Acid Rain Research Report 22/1990)*. NIVA, Oslo.

Holm, N. P. & Armstrong, D. E. (1981). Role of nutrient limitation and competition in controlling the populations of *Asterionella formosa* and *Microcystis aeruginosa* in semicontinuous culture. *Limnology and Oceanography*, **26**, 622-634.

Hosper, S. H. (1984). Restoration of Lake Veluwe, The Netherlands, by reduction of phosphorus loading and flushing. *Water Science and Technology*, **17**, 757-768.

Hrbaček, J. (1964). Contribution to the ecology of water-bloom-forming blue-green algae, *Aphanizomenon flos-aquae* and *Microcystis aeruginosa*. *Verhandlungen der Internationale Vereinigung für Theoretische und Angewandte Limnologie*, **15**, 837-846.

Hrbaček, J., Dvoraková, M., Korínek, V. & Prochazková, L. (1961). Demonstration of the effect of the fish stock on the species composition of zooplankton and the intensity of metabolism of the whole plankton association. *Verhandlungen der Internationale Vereinigung für Theoretische und Angewandte Limnologie*, **14**, 192-195.

Irwin, W. H., Symons, J. M. & Roebeck, G. G. (1969). Water quality in impoundments and modifications from destratification. *Water Quality Behaviour in Reservoirs* (ed. J. M. Symons), pp. 363-388. U.S. Department of Health, Education and Welfare, Cincinnati.

Kalff, J. & Knoechel, R. (1978). Phytoplankton and their dynamics in oligotrophic and eutrophic lakes. *Annual Review of Ecology and Systematics*, **9**, 475-495.

Kamp-Nielsen, L. (1974). Mud-water exchange of phosphate and other ions in undisturbed sediment cores and factors affecting the exchange rates. *Archiv für Hydrobiologie*, **73**, 218-237.

Kamp-Nielsen, L. (1975). A kinetic approach to the aerobic sediment-water exchange of phosphorus in Lake Esrom. *Ecological Modelling*, **1**, 153-160.

Kilham, P. & Kilham, S. S. (1990). Endless summer: internal loading processes dominate nutrient cycling in tropical lakes. *Freshwater Biology*, **23**, 379-389.

Korínek, V., Fott, J., Fuksa, J., Lellák, J. & Pražková, M. (1987) Carp ponds of central Europe. In *Managed Aquatic Ecosystems* (ed. R. G. Michael), pp. 29-62. Elsevier, Amsterdam.

Lund, J. W.G. (1970). Primary production. *Water Treatment and Examination*, **19**, 332-358.

Lund, J. W. G. & Reynolds, C. S. (1982). The development and operation of large limnetic enclosures in Blelham Tarn, English Lake District, and their contribution to phytoplankton ecology. In *Progress in Phycological Research, Vol. 1* (eds, F. E. Round & D. J. Chapman), pp. 1-65, Elsevier, Amsterdam.

Lynch, M. (1980). *Aphanizomenon* blooms: alternative control and cultivation by *Daphnia pulex*. In *Evolution and Ecology of Zooplankton Populations* (ed. W. C. Kerfoot), pp. 299-304. University Press of New England, Hanover.

Marsden, M. W. (1989). Lake restoration by reducing external phosphorus loading: the influence of phosphorus release. *Freshwater Biology*, **21**, 139-162.

McQueen, D. J. (1990). Manipulating lake community structure: where do we go from here? *Freshwater Biology*, **23**, 613-620.

Mortimer, C. H. (1941). The exchange of dissolved substances between mud and water in lakes, *Journal of Ecology*, **29,** 280-329.

Mortimer, C. H. (1942). The exchange of dissolved substances between mud and water in lakes (continued). *Journal of Ecology*, **30**, 147-201.

Moss, B. (1990). Engineering and biological approaches to the restoration from eutrophication of shallow lakes in which aquatic plant communities are important components. *Hydrobiologia*, **200/201**, 367-377.

Naumann, E. (1919). Någřa Synpunkter Anagående limnoplanktons ökologi med särskild hänsyn till fytoplankton. *Svensk Botanisk Tidskrift*, **13**, 129-163.

Nixon, S. W. (1988). Physical energy inputs and the comparative ecology of lake and marine ecosystems. *Limnology and Oceanography*, **33**, 1005-1025.

Oglesby, R. T. & Schaffner, W. R. (1975). The response of lakes to phosphorus. In *Nitrogen and Phosphorus. Food*

Production, Waste and the Environment (ed. K. S. Porter), pp. 23-57, An Arbor Science, Ann Arbor.

Oglesby, R. T. & Schaffner, W. R. (1978). Phosphorus loadings to lakes and some of their responses. Part 2. Regresion models of summer phytoplankton standing crops, winter total P and transparency of New York lakes with known phosphorus loadings. *Limnology and Oceanography*, **23**, 135-145.

Osborne, P. L. & Phillips, G. L. (1978). Evidence for nutrient release from sediments of two shallow and productive lakes. *Verhandlungen der Internationale Vereinigung für Theoretische und Angewandte Limnologie*, **20**, 654-658.

Pearsall, W. H. (1921). The development of vegetation in English lakes, considered in relation to the general evolution of glacial lakes and rock basins. *Proceedings of the Royal Society of London*, B, **92**: 259-284.

Pearson, M. J., Ferguson, A. J. D., Codd, G. A., Reynolds, C. S., Fawell, J. K., Hamilton, R. M., Howard, S. R. & Attwood, M. R. (1990). *Toxic Blue-green Algae*. National Rivers Authority, London.

Petersen, R. (1975). The paradox of the plankton: an equilibrium hypothesis. *American Naturalist*, **109**, 35-49.

Petr, T. (1977). Bioturbation and exchange of chemicals in the mud water interface. In *Interactions between Sediments and Freshwater* (ed. H. L. Golterman), pp. 216-226. Junk, Den Haag.

Rast, W. & Lee, G. F. (1978). *Summary analysis of the North American (U.S. Portion) OECD eutrophication project* (EPA 600/3-78-008). U.S. Environmental Protection Agency, Corvallis.

Rawson, D. S. (1956). Algal indicators of trophic lake types. *Limnology and Oceanography*, **1**, 18-25.

Reynolds, C. S. (1976). Succession and vertical distribution of phytoplankton in response to thermal stratification in a lowland lake, with special reference to nutrient availability. *Journal of Ecology*, **64**, 529-551.

Reynolds, C. S. (1978). Phosphorus and the eutrophication of lakes – a personal view. In *Phosphorus in the Environment: its Chemistry and Biochemistry. (CIBA Foundation Symposium No. 57, New Series)* (eds. R. Porter & D. Fitzsimmons), pp. 201-228. Excerpta Medica, Amsterdam.

Reynolds, C. S. (1982). Phytoplankton periodicity: its motivation, mechanisms and manipulation. *Freshwater Biological Association Annual Report*, **50**, 60-75.

Reynolds, C. S. (1983a). A physiological interpretation of the dynamic responses of populations of a planktonic diatom to physical variability of the environment. *New Phytologist*, **95**, 41-53.

Reynolds, C. S. (1983b). Growth-rate responses of *Volvox aureus* Ehrenb. (Chlorophyta, Volvocales) to variability in the physical environment. *British Phycological Journal*, **18**, 433-442.

Reynolds, C. S. (1984a). *The Ecology of Freshwater Phytoplankton*. Cambridge University Press, Cambridge.

Reynolds, C. S. (1984b). Artificial induction of surface blooms of cyanobacteria. *Verhandlungen der Internationale Vereinigung für Theoretische und Angewandte Limnologie*, **22**, 638-643.

Reynolds, C. S. (1986). Experimental manipulations of the phytoplankton periodicity in large limnetic enclosures in Blelham Tarn, English Lake District, *Hydrobiologia*, **138**, 43-64.

Reynolds, C. S. (1987a). Community organisation in the freshwater plankton., In *Organisation of Communities, Past and Present* (eds J. H R. Gee & P. S. Giller), pp. 297-325. Blackwell Scientific Publications, Oxford.

Reynolds, C. S. (1987b). The response of phytoplankton communities to changing lake environments. *Schweizerische Zeitschrift für Hydrologie*, **49**, 220-236.

Reynolds, C. S. (1987c). Cyanobacterial water blooms. In *Advances in Botanical research, Vol, 13* (ed. J. A. Callow), pp 67-143. Academic Press, London.

Reynolds, C. S. (1989a). The development of ecological concepts in limnology, with particular reference to the last twenty-five years. *Journal of the Limnological Society of Southern Africa*, **14**, 54-59.

Reynolds, C. S. (1989b). Relationships among the biological properties, distribution and regulation of production by planktonic cyanobacteria. *Toxicity Assessment*, **4**, 229-255.

Reynolds, C. S. (1989c). Physical determinants of phytoplankton succession. In *Plankton Ecology* (ed. U. Sommer), pp. 9-56. Springer-Verlag, Berlin.

Reynolds, C. S. (1990). Temporal scales of variability in pelagic environments and the responses of phytoplankton. *Freshwater Biology*, **23**, 25-53.

Reynolds, C. S. (1991). Lake communities: an approach to their management for conversation. In *The Scientific Management of Temperate Communities for Conservation* (eds. I. F. Spellerberg, F. B. Goldsmith & M. G. Morris), 199-225. Blackwell Scientific Publications, Oxford.

Reynolds, C. S. (1992a). Swings and roundabouts: engineering the environment of algal growth. In *Urban Waterside Regeneration* (eds K. N. White, E. G. Bellinger, A. J. Saul, K. Hendry & M. R. Bristow), pp. 000. Simon & Schuster, Chichester.

Reynolds, C. S. (1992b). Dynamics, selection and composition of phytoplankton in relation to vertical structure in lakes. *Ergebnisse der Limnologie*, **35**, 13-31.

Reynolds, C. S. (1992). Scales of disturbance and their role in plankton ecology. *Hydrobiologia* (in press).

Reynolds, C. S. & Bellinger, E. (1992). Patterns of abundance and dominance of the phytoplankton of Rostherne Mere, England: evidence from an 18-year data set. *Aquatic Sciences*, **54**, 10-36.

Reynolds, C. S. & Reynolds, J. B. (1985). The atypical seasonality of phytoplankton in Crose Mere, 1972: an independent test of the hypothesis that variability in the physical environment regulates community dynamics and structure. *British Phycological Journal*, **20**, 227-242.

Reynolds, C. S. & Wiseman, S. W. (1982). Sinking losses of phytoplankton maintained in closed limnetic systems. *Journal of Plankton Research*, **4**, 489-522.

Reynolds, C. S., Wiseman, S. W. & Clarke, M. J. O. (1984). Growth- and loss-rate responses of phytoplankton to intermittent artificial mixing and their potential application to the control of planktonic algal biomass. *Journal of Applied Ecology*, **12**, 11-39.

Reynolds, C. S., Wiseman, S. W., Godfrey, B. M. & Butterwick, C. (1983). Some effects of artificial mixing on the dynamics of phytoplankton populations in large limnetic enclosures. *Journal of Plankton Research*, **5**, 203-234.

Rhee, G.-Y. (1978). Effects on N:P atomic ratios and nitrate limitation on algal growth, cell composition and nitrate uptake. *Limnology and Oceanography*, **23**, 10-25.

Rhee, G.-Y. (1982). Effects of environmental factors on phytoplankton growth. In *Advances in Microbial Ecology, Vol. 6* (ed. K. C. Marshall), pp. 33-74. Plenum Press, New York.

Rhee, G.-Y. & Gotham, I. J. (1980). Optimum N:P ratios and the co-existence of planktonic algae. *Journal of Phycology*, **16**, 486-489.

Riddolls, A. (1985). Aspects of nitrogen fixation in Lough Neagh. II. Competition between *Aphanizomenon flos-aquae, Oscillatoria redekei* and *Oscillatoria agardhii*. *Freshwater Biology*, **15**, 299-306.

Ridley, J. E. (1970). The biology and management of eutrophic reservoirs. *Water Treatment and Examination*, **19**, 374-399.

Robbins, J. A. & Edgington, D. N. (1975). Determination of recent sedimentation rates in Lake Michigan using Pb-210 and Cs-137. *Geochimica et Cosmochimica Acta*, **39**, 285-304.

Rodhe, W. (1948). Environmental requirements of freshwater plankton algae: experimental studies in the ecology of phytoplankton. *Symbolae Botanicae Upsaliensis*, **10**, 5-149.

Sakamoto, M. (1966). Primary production by phytoplankton community in some Japanese lakes and its dependence upon lava depth. *Archiv für Hydrobiologie*, **62**, 1-28.

Sas, H. (1989). *Lake Restoration by Reduction of Nutrient Loading: Expectations, Experiences, Extrapolations.* Academia Verlag Richarz, St Augustin.

Schlinder, D. W. (1977). Evolution of phosphorus limitation in lakes. *Science*, **196**, 260-262.

Sell, D. W., Carney, H. J. & Fahnenstiel, G. L. (1984). Inferring competition between natural phytoplanton populations: the Lake Michigan example re-examined. *Ecology*, **65**, 325-328.

Shapiro, J. (1973). Blue-green algae: why they became dominant. *Science*, **179**, 382-384.

Shapiro, J. (1990a). Current beliefs regarding dominance by blue-greens: the case for the importance of CO_2 and pH. *Verandlungen der Internationale Vereinigung für Theoretische und Angewandte Limnologie*, **24**, 38-54.

Shapiro, J. (1990b). Biomanipulation: the next phase – making it stable. *Hydrobiologia*, **200-201**, 13-27.

Shapiro, J., Forsberg, B., Lamarra, V., Lindmark, G., Lynch, M., Smeltzer, E. & Zoto, G. (1983). *Experiments and experiences in biomanipulation: studies of biological ways to reduce algal abundance and eliminate blue greens (Project Summary EPA-600/53-82 096)* U.S. Environmental Protection Agency, Corvallis.

Shapiro, J., Lamarra, V. & Lynch, M. (1975). Biomanipulation: an ecosystem approach to lake restoration. In *Proceedings of a Symposium on Water Quality Management through Biological Control* (eds P. L. Brezonik & J. L. Fox), pp. 85-89. University of Florida, Gainesville.

Smith, V. H. (1983). Low nitrogen to phosphorus ratios favour dominance by blue-green algae in lake phytoplankton. *Science*, **221**, 669-671.

Sommer, U. (1981). The role of *r*- and *K*-selection in the succession of phytoplankton in Lake Constance. *Acta Oecologia*, **2**, 237-242.

Sommer, U. (1984). The paradox of the plankton: fluctuations of phosphorus availability maintain diversity of phytoplankton in flow-through cultures. *Limnology and Oceanography*, **29**, 633-636.

Sommer, U. (1991). A comparison of the Droop and the Monod models of nutrient-limited growth applied to natural populations of phytoplankton. *Functional Ecology*, **5**, 535-544.

Sommer, U., Gliwicz, Z. M., Lampert, W & Duncan, A. (1986). The PEG-model of seasonal succession of planktonic events in fresh waters. *Archiv für Hydrobiologie*, **106**, 433-471.

Stauffer, R. E. (1991). Environmental factors influencing chlorophyll v nutrient relationships in lakes. *Freshwater Biology*, **26**, 279-295.

Steel, J. A. (1972) The application of fundamental limnological research in water supply system design and management. *Symposia of the Zoological Society of London*, No. **29**, 41-67.

Steel, J. A. (1975). The management of Thames Valley reservoirs. In *The Effects of Storage on Water Quality* (ed. R. E. Youngman), pp. 371-419. Water Research Centre, Medmenham.

Sterner, R. W. (1989). The role of grazers in phytoplankton succession. In *Plankton Ecology* (ed. U. Sommer), pp. 107-170. Springer Verlag, Berlin.

Szilágyi, F., Somlyódy, L., Herodek, S. & Istvánovics, V. (1990). The Kis-Balaton Reservoir system as a means of controlling eutrophication of Lake Balaton, Hungary. In *Guidelines of Lake Management, Vol. 3* (eds. S. E. Jørgensen & H. Löffler), pp. 127-151. ILEC, Otsu.

Talling, J. F. (1957). The phytoplankton population as a compound photosynthetic system. *New Phytologist*, **56**, 29-50.

Talling, J. F. (1992). Comparative seasonal changes and interannual variability and stability in a 26-year record of total phytoplankton biomass in four English lake basins. [Paper in review].

Talling, J. F. & Heaney, S. I. (1988). Long-term changes in some English (Cumbrian) lakes subjected to increased nutrient inputs. In *Algae and Aquatic Environment* (ed. F. E. Round), pp. 1-29. Biopress, Bristol.

Thienemann, A. (1918). Untersuchungen über dir Beziehungen zwischen der Sauerstoffgehält des Wassers und der Zusammensetzung der Fauna in norddeutschen Seen. *Archiv für Hydrobiologie*, **12**, 1-24.

Tilman, D. (1977). Resource competition between planktonic algae: an experimental and theoretical approach. *Ecology*, **58**, 338-348.

Tilman, D. & Kilham, S. S. (1976). Phosphate and silicate growth and uptake kinetics of the diatoms *Asterionella formosa* and *Cyclotella meneghiniana* in batch and semicontinuous culture. *Journal of Phycology*, **12**, 375-383.

Tilman, D., Kilham, S. S. & Kilham, P. (1982). Phytoplankton community ecology: the role of limiting nutrients. *Annual Review of Ecology and Systematics*, **13**, 349-372.

Tilman, D. & Kilham, S. S. & Kilham, P. (1984), A reply to Sell, Carney and Fahnenstield. *Ecology*, **65**, 328-332.

Tolland, H. G. (1977). Destratification/aeration in reservoirs. A literature review of the techniques used for water quality management. *Technical Reports of the Water Research Centre*, TR**50**, 37pp.

Toms, I. P. (1987), Developments in London's water supply system. *Ergebnisse der Limnologie*, **28**, 149-167.

Trimbee, A. M. & Harris, G. P. (1983). Use of time-series analysis to demonstrate advection rates of different variables in a small lake. *Journal of Plankton Research*, **5**, 819-834.

Trimbee, A. M. & Prepas, E. E. (1987). Evaluation of total phosphorus as a predictor of the relative biomass of the blue-green algae with emphasis on Alberta lakes. *Canadian Journal of Fisheries and Aquatic Sciences*, **44**, 1337-1342.

Turpin, D. H. (1988). Physiologicial mechanisms in phytoplankton resource competition. In *Growth and Reproductive Strategies of Freshwater Phytoplankton* (ed. C. D. Sandgren), pp. 316-368. Cambridge University Press, New York.

Vollenweider, R. A. (1965). Calculation models of photosynthesis-depth curves and some implications regarding day rate estimates in primary production measurements. In *Primary Productivity in Aquatic Environments* (ed. C. R. Goldman), pp. 425-457. University of California Press, Berkeley.

Vollenweider, R. A. (1968). *Scientific fundamentals of the eutrophication of lakes and flowing waters with particular reference to nitrogen and phosphorus as factors in eutrophication*. OECD, Paris.

Vollenweider, R. A. (1976). Advances in defining critical load levels for phosphorus in lake europhication. *Memorie dell'Istituto Italiano di Idrobiologia*, **33**, 53-83.

Vollenweider, R. A. & Kerekes, J. (1980). The loading concept as basis for controlling eutrophication philosophy and preliminary results of the OECD programme on eutrophication. *Progress in Water Technology*, **12**, 5-38.

The impact of physical processes on algal growth

Dieter M. Imboden

Swiss Federal Institute of Technology, Swiss Federal Institute for Water Resources and Water Pollution Control, CH-8600, Dübendorf, Switzerland

Mixing and transport processes in surface waters strongly influence the structure of aquatic ecosystems. The impact of mixing on algal growth is species-dependent, affecting the competition among species and acting as a selective factor for the composition of the biocoenose. Were it not for the ever-changing "aquatic weather", the composition of pelagic ecosystems would be relatively simple. Probably just a few optimally adapted algal species would survive in a given water-body.

In contrast to terrestrial ecosystems, in which the spatial heterogeneity is primarily responsible for the abundance of niches, in aquatic systems (especially in the pelagic zone) the niches are provided by the temporal structure of physical processes. The latter are discussed in terms of the relative sizes of physical versus biological time-scales. The relevant time-scales of mixing and transport cover the range between seconds and years. Correspondingly, their influence on growth of algae is based on different mechanisms: rapid changes are relevant for the fast biological processes such as nutrient uptake and photosynthesis, and the slower changes are relevant for the less dynamic processes such as growth, respiration, mineralization, and settling of algal cells. Mixing time-scales are combined with a dynamic model of photosynthesis to demonstrate their influence on algal growth.

Introduction

It is well-known that plants living on land are strongly influenced by climatic factors, such as precipitation, air temperature, solar radiation, wind speed and others. These factors, together with the chemical properties of the soil, are the major factors which determine the competition between different plant species and the composition of the terrestrial ecosystem.

It may be less obvious that in aquatic ecosystems the situation is quite similar: algal competition is controlled by the combined action of the chemistry of the water and the "aquatic weather", i.e. the physical processes in the water column. Obviously, there are some important differences between the terrestrial and aquatic climate. Firstly, precipitation cannot be a selective factor for the aquatic system. Secondly, water currents are not analogues of terrestrial winds, as most organisms living in the pelagial region of aquatic ecosystems are freely floating. Therefore, water currents are not primarily a stress factor, in the way that terrestrial winds affect plants, but the currents are agents of movement which transport the living cells from one region to another, e.g. from light to dark zones and *vice versa*.

The impact of mixing and transport processes on algal growth is species-dependent, affecting the competition among species and acting as a selective factor for the composition of the biocoenose. Were it not for the ever-changing "aquatic weather", the composition of pelagic ecosystems would be relatively simple. Probably just a few optimally adapted algal species would survive in a given water-body.

In contrast to terrestrial ecosystems, in which the *spatial heterogeneity* is primarily responsible for the abundance of niches, in aquatic systems (especially in the pelagic zone) the niches are provided by the *temporal structure* of physical processes. The important characteristic quantity is the relative size of physical versus biological time-scales. The relevant time-scales of mixing and transport cover the range between seconds and years (Imboden 1990). Correspondingly, their influence on growth of algae is based on different mechanisms: rapid changes are relevant for the fast biological processes such as nutrient uptake and photosynthesis, and the slower changes are relevant for the less dynamic processes such as growth, respiration, mineralization, and settling of algal cells.

Figure 1 gives a simplified sketch of the relationship between physical processes in lakes and algal population growth. Obviously, the situation is far too complex to be exhaustibly covered in a short review article. Therefore, two types of mechanisms will be chosen in order to exemplify a situation which actually is extremely complicated and not yet understood in every detail. The first mechanism is related to the long-term balance of biomass and nutrients in the trophogenic (productive) surface layer. Here we are looking at processes controlling the nutrient and algal concentrations which in turn are directly influencing algal growth. We shall call this the *capacity limitation* of growth. The second mechanism is given by the short-term processes of mixing, which causes individual algal cells to continuously change their depth and thus the light intensity to which they are exposed. The corresponding response of photosynthesis to temporal changes of light is called *dynamic limitation* of growth.

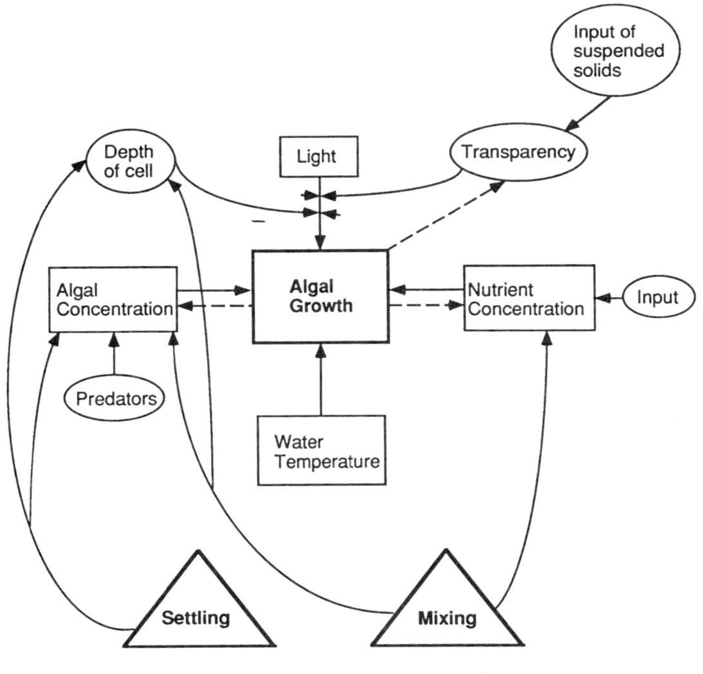

Figure 1. Schematic diagram to show the relations between algal growth and physical processes in lakes represented by the two major transport phenomena, (particle) "settling" and "mixing".

Capacity limitation: Long-term physical processes

Photosynthesis depends on the presence of two types of resources, i.e. light and nutrients. While the former is restricted to the water layers close to the surface, nutrients tend to be depleted in the productive zone, leading to a steady-state in which internal recycling and inputs from outside are balanced by uptake of nutrients by algae. This simple balance represents one important control mechanism for the overall capacity of the biological productivity in the system.

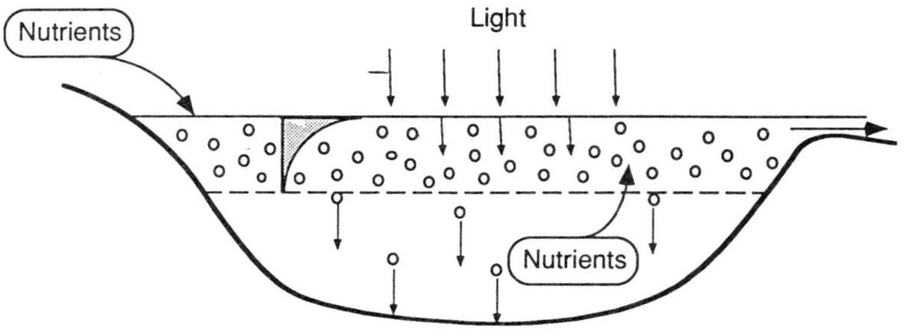

Figure 2. Epilimnic mass balance for nutrients. The nutrient concentration (nitrate, phosphate, etc.) in the trophogenic layer is influenced by external loading, internal diffusion from the hypolimnion, settling of biomass, and loss through the outlet. The depth of the trophogenic layer is controlled by the light extinction in the water column and thus is generally smaller in waters with large biomass concentrations.

Let us discuss the problem of capacity limitation in a quantitative manner by considering a very simple mass balance scheme for nutrients in the trophogenic layer (Fig. 2). For a simple box model (trophogenic layer is completely mixed) the following equations can be formulated:

$$\frac{dN}{dt} = J_e + J_i - E_N - G + R \qquad (1)$$

$$\frac{dB}{dt} = E_B + G - R \qquad (2)$$

where N is the mean nutrient concentration in the trophogenic layer, B is the mean biomass concentration expressed as the concentration of nutrient N incorporated in the biomass, J_e is the external input of nutrient, J_i is the internal input of nutrients by vertical mixing through the thermocline, E_N is loss of nutrient from the layer by transport, E_B is loss of biomass from the layer by transport, R is loss of biomass from the layer by respiration or mineralization, and G is uptake of nutrient by the biomass.

Note that all variables appearing on the right hand side of eqns (1) and (2) are expressed as change of mass per unit time and volume.

In a greatly simplified way nutrient uptake, photosynthesis and algal growth can all be considered as synonyms, at least for long-term considerations for which changes of the stoichiometric composition of the biomass are less important than for short-term processes. If N is the limiting nutrient, algal growth per unit area, ΣG, expressed by the combined Michaelis–Menten/self-saturation equation, is given by (Imboden & Gächter 1978):

$$\Sigma G = hG = \mu \frac{N}{K_N + N} \frac{B}{\varepsilon_0 + \beta B} \qquad (3)$$

where μ is the maximum growth rate, K_N is the Michaelis constant for nutrient N, ε_0 is the light extinction coefficient for water without biomass, β is the biomass-specific increase of extinction coefficient, and h is the depth of mixed layer.

All the other terms shall be expressed by first-order processes, i.e.

$$E_N = k_w N \qquad (4)$$

the loss of nutrient through the outlet, where k_w is the flushing rate of the mixed layer;

$$E_B = (k_w + k_s)\, B \qquad (5)$$

the loss of biomass through the outlet and by settling, where k_s is the settling rate;

$$R = k_r B \qquad (6)$$

the degradation of biomass by respiration or mineralization, where k_r is the respiration/mineralization rate.

Inserting the expressions (3) to (6) into eqns (1) and (2), and solving for steady-state, yields a functional relationship between the productivity per unit area and time, ΣG, and the total nutrient input to the trophogenic layer, $J_e + J_i$. The term ΣG can be decomposed into the "minimum" production, ΣG_o, which holds if no nutrients from the hypolimnion are recycled ($J_i = 0$), and the additional production, ΔG_i, originating from the internal nutrient supply J_i. Thus

$$\Sigma G = \Sigma G_o + \Sigma G_i \qquad (7)$$

Note that in ΣG the much faster epilimnic phosphorus uptake and release is not taken into account but only the net phosphorus uptake which, over longer periods, acts in parallel with carbon uptake and thus with algal growth. An average Redfield factor of 40g C versus 1g P is used to convert ΣG into carbon uptake, ΣG^C. Mixing depth (h) and production rate (ΣG) show significant seasonal variations, with the result that most of the yearly productivity occurs during the stagnation period, i.e. during about eight months. In a simplified way, the annual primary productivity is calculated from the steady-state concentrations N_{ss} and B_{ss}:

$$\Sigma G_o^C = 40\,(T_{St} \times \mu_{St}) \left(\frac{N_{ss}}{k_N + N_{ss}}\right)\left(\frac{B_{ss}}{\varepsilon_o + \beta B_{ss}}\right) \qquad (8)$$

where T_{St} is 240 days per year, the average duration of the stagnation period, and μ_{St} is the mean uptake rate during this period. Average parameter values from Imboden & Gächter (1978) are listed in Table 1. The calculated variation of ΣG_o^C as a function of external loading per unit area and time, $L_P = hJ_e$, is very similar to the lower boundary of measured values (Fig. 3).

Table 1. *Definitions and values of parameters for a simple phosphorus lake-model* (From Imboden & Gächter 1978)

Symbol	Values and Units	Definitions
N	mg m^{-3}	concentration of dissolved nutrient (e.g. phosphorus)
B	mg m^{-3}	concentration of biomass, expressed as particulate nutrient
L_P	mg P m^{-2}yr^{-1}	annual P-input in lake per unit surface area
h	6 m	depth of trophogenic layer
k_W	yr^{-1}	exchange of epilimnic water by river inlets and outlets
k_r	0.07 d^{-1}	rate constant of mineralization of organic material
k_s	0.025 d^{-1}	loss of biomass from epilimnion by sedimentation
G	mg P m^{-3}d^{-1}	P-uptake taken as measure for primary production
μ	1 d^{-1}	maximum P-uptake rate
K_N	1.5 mg P m^{-3}	Michaelis constant for nutrient N
ε_o	0.3 m^{-1}	extinction coefficient of water without biomass
β	0.015 m^{-2} (mg P)$^{-1}$	specific influence of biomass on extinction
ΣG	mg P m^{-2}yr^{-1}	uptake of P per unit time and water column
ΣG^C	mg C m^{-2}yr^{-1}	carbon uptake (assumed Redfield ratio $\Sigma G^C = 40\,\Sigma G$)

The role of internal phosphorus supply becomes more evident by using the so-called P-efficiency, n_p, defined as the average number of times each phosphorus atom takes part in the photosynthetic process before leaving the lake through the outlet or permanent incorporation into the sediments:

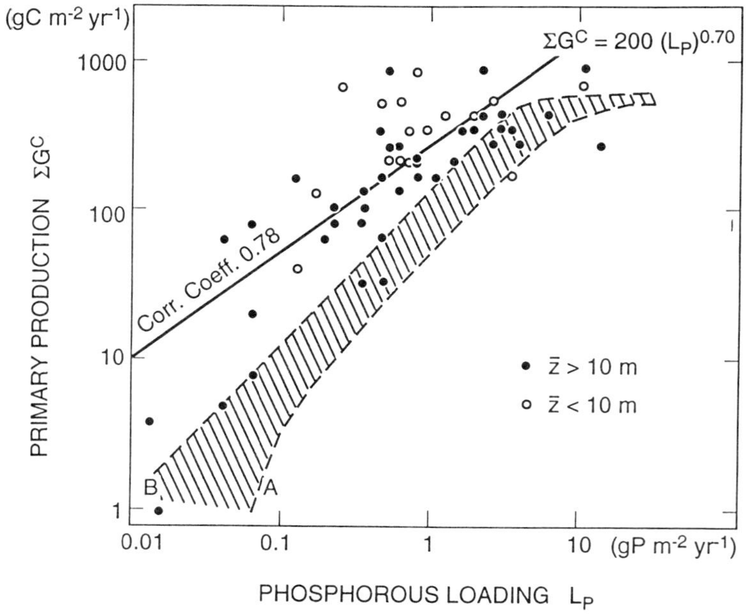

Figure 3. Relationship between external annual P-loading per unit surface area, L_P, and primary productivity ΣG^C, for various European and North-American lakes, taken from the OECD eutrophication program (Vollenweider & Kerekes 1982). The shaded area was calculated with eqn (8) for epilimnic water-residence times ($1/k_w$) between 0.06 years (curve A) and 15 years (curve B). Shallow lakes (mean depth $\bar{z} < 10$m) show a slight tendency towards larger productivities, indicating a possible influence from internal nutrient loading, J_i. From Imboden & Gächter (1979).

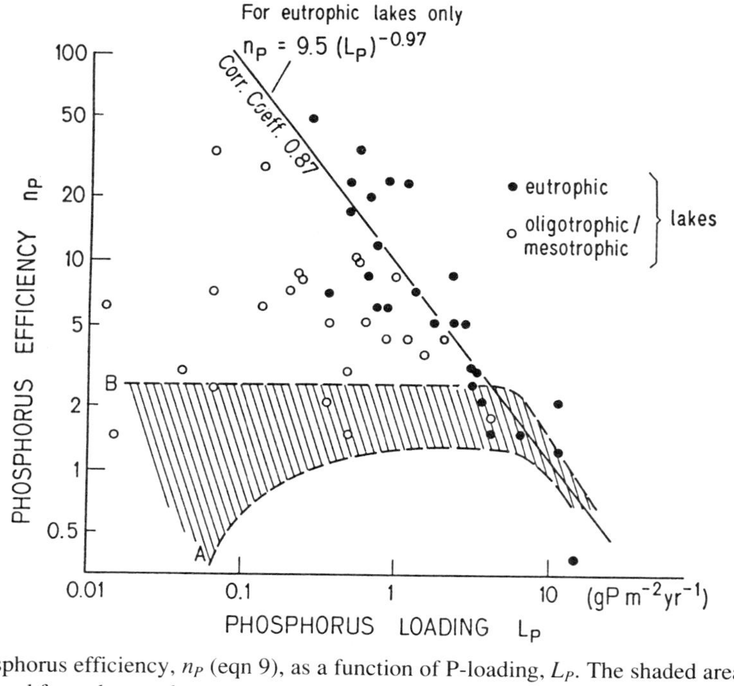

Figure 4. Phosphorus efficiency, n_P (eqn 9), as a function of P-loading, L_P. The shaded area (curves A and B) was calculated from the steady-state model, as explained in Fig. 3. From Imboden & Gächter (1979).

$$n_P = \Sigma G / L_P = \frac{\Sigma G^C}{40\,(L_P)} \qquad (9)$$

In Figure 4, values of measured n_P are plotted (for the same set of lakes as in Fig. 3) and compared to n_P values computed from the steady-state model (eqn 8). For large values of L_P, measured values agree with the steady-state model, indicating that ΣG is controlled by light (self-shadowing), while internal nutrient recirculation does not influence such lakes which are too richly nourished already.

The scattering of the n_P values as a function of L_P is significantly reduced if only eutrophic lakes are considered. The linear regression found between log L_P and log n_P for eutrophic lakes is nearly identical with the relation: ΣP = 380 g C m^{-2} year^{-1}. In other words, primary productivity in these eutrophic lakes seems to be remarkably independent of L_P and has a mean value of about 400 g C m^{-2} year^{-1}.

The important message here is that the difference between the minimum (externally driven) growth as calculated by the model, and the measured growth, reflects the role of internal nutrient supply by mixing. However, mean depth alone cannot explain the magnitude of ΣG_i; probably other factors such as lake morphometry, exposure to wind, and chemical processes occurring at the sediment–water interface are of comparable importance.

Dynamic limitation: short-term physical processes

In this section the following dynamic sequence of causes and effects will be considered:

(Transport processes in the water column) ⟶ (Depth of algal cell) ⟶ (light exposure of cell) ⟶ (algal growth)

In this context our interest will be focused on the vertical motions in the water. Since in most cases, especially during the period of greatest algal productivity, lakes are vertically stratified, the vertical motion of the water is greatly damped by buoyancy effects. Figure 5 shows two kinds of vertical motion: (1), the reversible excursion of isopycnals (surfaces of constant water density ρ_A, ρ_B, etc.) about their equilibrium depths (designated as ρ_A°, ρ_B°, etc.) as caused by internal waves and seiches, and (2), the irreversible mixing of the water across the isopycnals due to turbulent mixing.

Reversible motion

Because of the kinetic energy input by winds and inlets, isopycnals are hardly ever at rest. The long standing waves found in closed basins, called internal seiches (Mortimer 1974), are often only weakly damped and can therefore store wind energy over long time-periods (Fig. 6). Typical vertical amplitudes are in the order of 1 to 2 metres, corresponding to relative variations of light intensity by a factor of 2 to 10. (Note: Take $e^{\varepsilon \Delta z}$ with light extinction ε = 0.3 to 0.5 m^{-1} and Δz = 1 to 2 m). During extreme situations (e.g. storm events) vertical isopycnal displacements can reach 40 m and more in just a few hours (Fig. 7).

The complex temporal variation of isopycnal depth can be analyzed by spectral analysis in order to identify the most important time-scales of motion. As shown in Figure 8, most energy is concentrated in the relatively slow standing waves (seiches). However, at the higher frequency end of the spectrum, which is not sufficiently resolved by the data used to calculate the spectra shown in Figure 8, one can often find distinct maxima (Fig. 9). They represent the so-called stability frequency defined by

$$N = \left(\frac{g}{\rho}\,\frac{dp}{dz}\right)^{1/2} \qquad (10)$$

where g is acceleration of gravity (9.81 m s^{-2}) and dp/dz is the vertical density gradient in the water column.

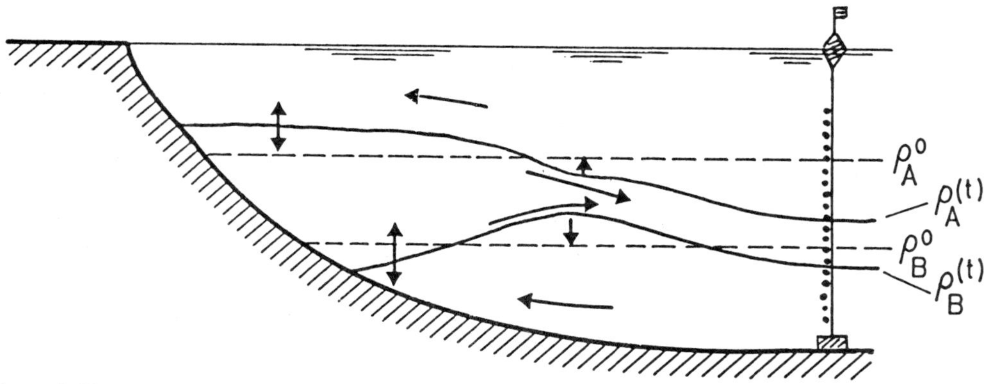

Figure 5. Two types of vertical motion in lakes: a reversible excursion of isopycnals (surfaces of constant water density), ρ_A, ρ_B .. about their equilibrium position, ρ_A°, ρ_B° ... (shown by single arrows), and irreversible mixing of water across isopycnals by turbulence (shown by double arrows).

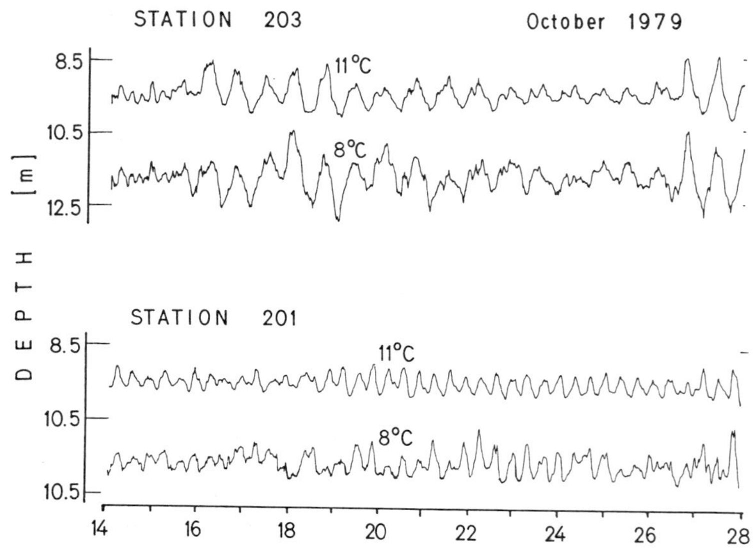

Figure 6. Vertical oscillations of isotherms in Lake Baldegg (Switzerland). Damping is small so that oscillation survives between wind events (e.g. from 19 October to 26 October). From Imboden *et al.* (1983).

The corresponding period $T = 2\pi/N$, depends on the strength of the stratification. In the thermocline during the summer, T is of the order of a few minutes. As will be discussed below, this time-scale is of the same order as the response of the planktonic photosynthesis to changing light intensities.

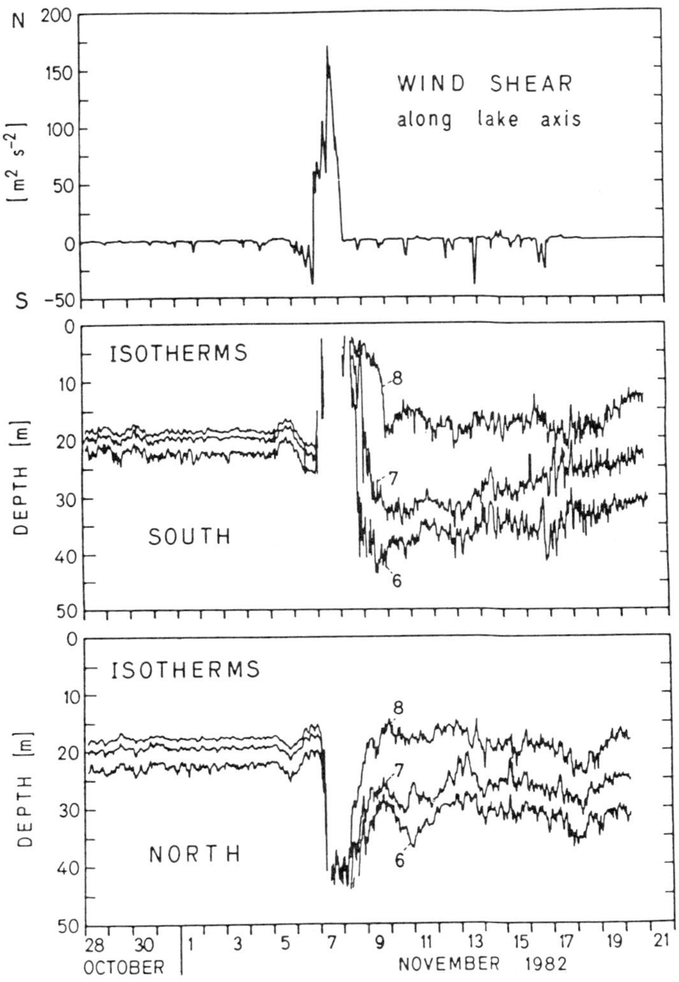

Figure 7. Directed wind shear ($m^2 s^{-2}$) along the main axis of Lake Zug (Switzerland) and isotherm depths (m) at two stations, North and South, demonstrate the influence of a severe storm on 7/8 November 1982. From Imboden *et al.* (1988).

Irreversible motion

Turbulence leads to irreversible mixing between adjacent water layers and to irreversible vertical displacements of the suspended plankton. The dynamics of turbulence in the uppermost layers is driven by two processes, wind mixing and heat exchange. For water temperatures above the density maximum (at about 4°C for fresh water), warming causes the water column to become more stable while cooling leads to convective mixing, even in the absence of any wind. In fact, cooling occurs during most nights of the year (Marti & Imboden 1986), while during most days (except at low air temperature in the winter) the water column gains some heat. The diurnal change of the direction of the heat flux between water and atmosphere causes a diurnal stabilization/mixing rhythm which becomes even more complex by the additional influence from the wind.

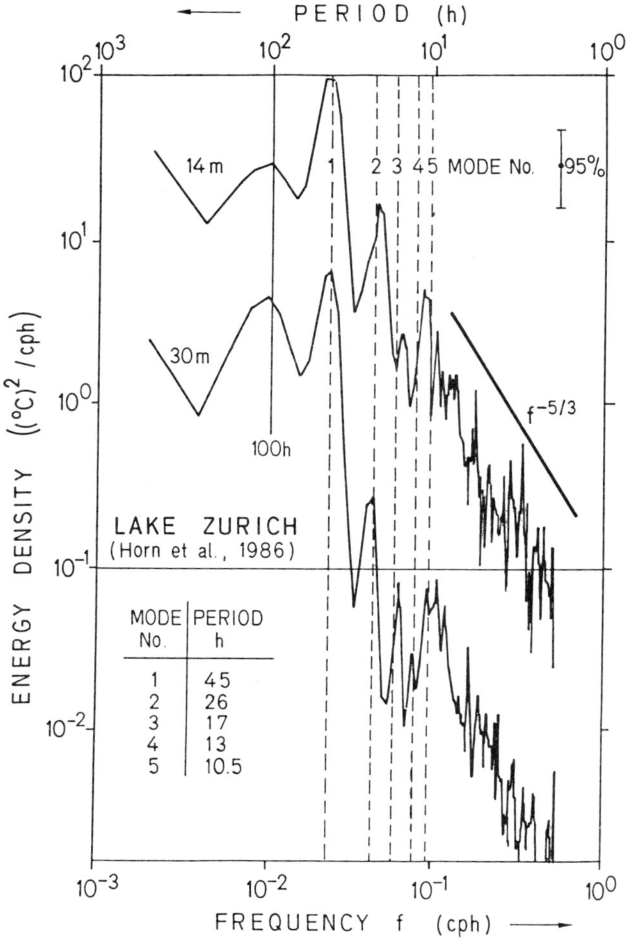

Figure 8. Spectra of fluctuations of hourly mean temperature in Lake Zürich (Switzerland) for a period of 1000 hours. The peaks denoted as mode no. 1 to 5 are standing internal waves (seiches). From Horn *et al.* (1986).

During the day, wind and heat flux usually act against each other: the heat flux suppresses vertical mixing and the wind drives it. The Monin-Obukhov length, L, is a measure for the interplay between wind and heat flux. It defines the depth to which the water would eventually be mixed if heat flux and wind persistently remained constant. As can be seen in Figure 10, L is shallow for small wind speeds and large heat fluxes (e.g. 5 May 1983, Figs. 10, 11). The same heat flux, combined with stronger winds, mixes the water column to greater depths (e.g. 2 May 1983, Figs. 10, 11).

Frequently, the onset of algal blooms is triggered by changes in the vertical mixing patterns. Calm situations are favourable for algal growth, as demonstrated by the occurrence of algal blooms below an ice cover which eliminates most of the kinetic energy input into the water column.

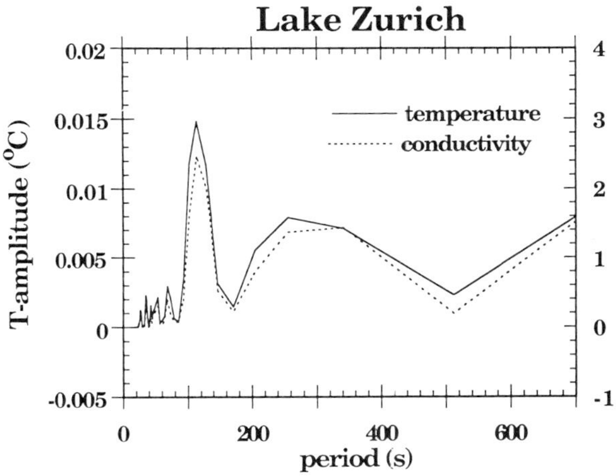

Figure 9. Spectra of fluctuations of temperature (changes in amplitude, °C) and electrical conductivity (µS cm^{-1} at 20°C) measured every second in Lake Zurich (Switzerland). The peak at 100 s represents the stability frequency, a measure for the strength of the vertical density gradient.

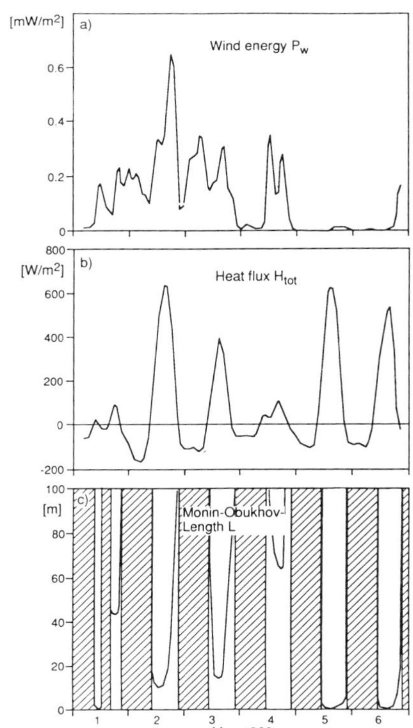

Figure 10. (a) Input of kinetic energy (mW m^{-2}), (b) heat flux (Wm^{-2}) at the water surface (positive if directed into the lake) and (c) the Monin–Obukhov length (m) in Lake Sempach (Switzerland) for 1 May to 6 May 1983. From Marti & Imboden (1986).

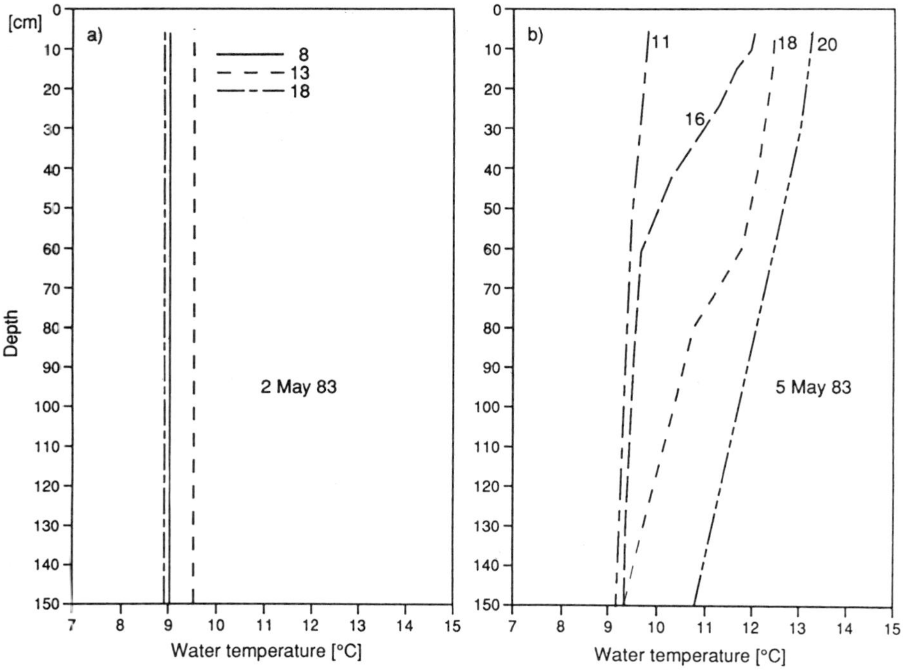

Figure 11. Temporal evolution of the vertical temperature structure in the top 1.5 metres of Lake Sempach (Switzerland) on (a) 2 May 1983 and (b) 5 May 1983. From Marti & Imboden (1986).

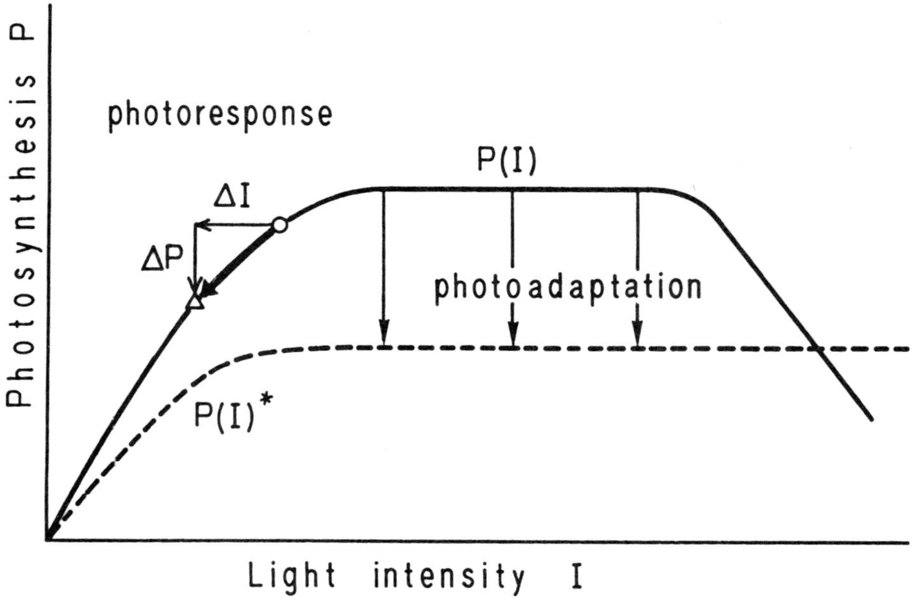

Figure 12. Schematic representation of the relation between rate of photosynthesis, P, and light intensity, I. See text for further explanations.

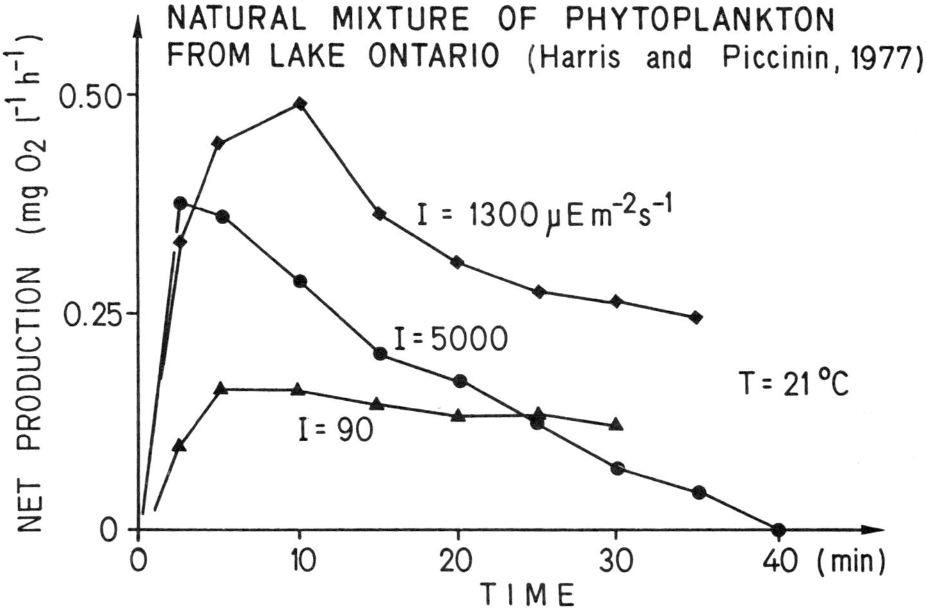

Figure 13. Temporal variation of net photosynthesis at different light intensities for a natural mixture of phytoplankton from Lake Ontario (N. America). In all experiments, the algae were kept in the dark for 5 min and then exposed to different light intensities for periods of 30–40 min at 21°C. Photosynthesis was measured by the rate of oxygen production (mg 0_2 l^{-1} h^{-1}). From Harris & Piccinin (1977).

Photoadaptation and photoresponse

The final link between mixing and algal growth is related to the dynamics of photosynthesis (*P*) under changing light intensity (*I*). Figure 12 explains the difference between photoadaptation, the permanent change of the *P*/*I* curve due to long-term changes of the light, and photoresponse, the change ΔP along a fixed *P*/*I* curve due to a change of ΔI of the light. Of course, this distinction is somewhat artificial. If the time-scale of change in *I* is continuously varied, a gradual change from the (reversible) photoresponse to a more permanent (though not really irreversible) photoadaptation can be observed.

As shown by various experiments (e.g. Harris & Piccinin 1977) the *P*/*I* curve is not static. At low intensities the maximum *P* is only reached after some minutes (Fig. 13) while at larger values of *I*, a significant depletion of *P* is observed after some time (photoinhibition).

Pahl-Wostl & Imboden (1990) have developed a mathematical model to describe the dynamic response of *P* to changing *I*. In Figure 14, this model is applied to phytoplankton suspended in a Langmuir cell (Leibovich 1983). As can be seen, algal growth is favoured by the dynamic response of the plankton to changing light intensity, compared to the exposure of algae to a fixed light intensity.

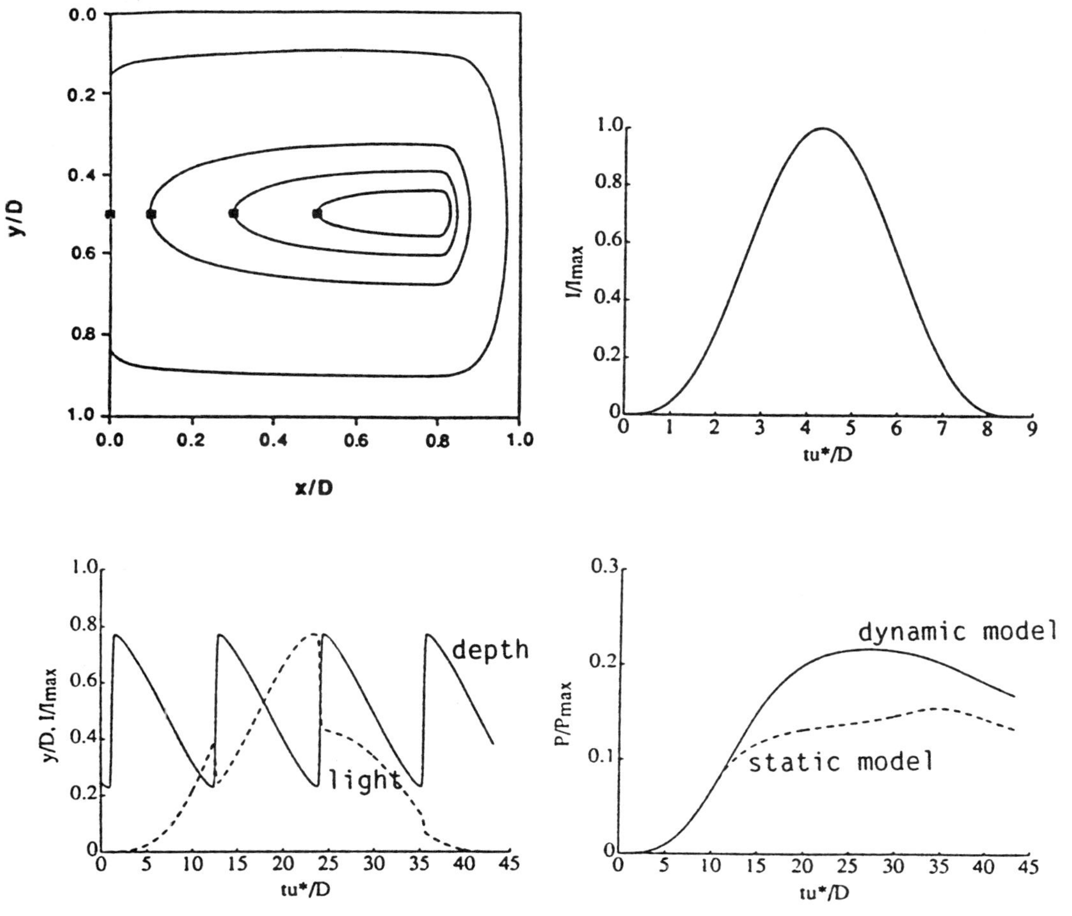

Figure 14. Model responses of phytoplankton suspended in Langmuir cells. Top (left), motion of algae in Langmuir cells while exposed (right) to a diurnal light variation. Below (left), light experienced by an individual cell and (right) the mean photosynthesis of all cells, calculated by a static model and a dynamic model. D is depth of the Langmuir cell; u^* is the maximum surface velocity; tu^*/D is non-dimensional time; I_{max} is the surface maximum light intensity. From Patterson (1991).

Conclusions

Based on theoretical as well as experimental evidence (Lewis *et al.* 1984) it can be concluded that mixing in lakes and oceans must be an important criterion for competing algae. Changes in the mixing pattern must have an influence on the composition of the planktonic community. Although a quantitative understanding of these phenomena does not yet exist, artificial or natural changes of the physical structure of lakes have also to be considered under these aspects.

References

Harris, G. P. & Piccinin, B. B. (1977). Photosynthesis by natural phytoplankton populations. *Archiv für Hydrobiologie*, **80**, 405-457.

Horn, W., Mortimer, C. H. & Schwabe, D. J. (1986). Wind-induced internal seiches in Lake Zürich observed and modelled. *Limnology and Oceanography*, **31**, 1232-1254.

Imboden, D. M. (1990). Mixing and transport in lakes: mechanisms and ecological relevance. In *Large Lakes: Ecological Structure and Function* (eds. M. Tilzer & C. Serruya), pp. 47-80. Springer, Berlin.

Imboden, D. M. & Gächter, R. (1978). A dynamic lake model for trophic state prediction. *Ecological Modelling*, **4**, 77-98.

Imboden, D. M. & Gächter, R. (1979). The impact of physical processes on the trophic state of a lake. In *Biological Aspects of Freshwater Pollution* (ed. O. Ravera), pp. 93-110. Pergamon Press, Oxford.

Imboden, D. M., Lemming, U., Joller, T. & Schurter, M. (1983). Mixing processes in lakes: mechanisms and ecological relevance. *Schweizerische Zeitschrift für Hydrologie*, **45**, 11-44.

Imboden, D. M., Stotz, B. & Wüest, A. (1988). Hypolimnic mixing in a deep alpine lake and the role of a storm event. *Verhandlungen der Internationale Vereinigung für Theoretische und Angewandte Limnologie*, **23**, 67-73.

Leibovich, S. (1983). The form and dynamics of Langmuir circulations. *Annual Reviews of Fluid Mechanics*, **15**, 391-427.

Lewis, M. R., Horne, E. P. W., Cullen, J. J., Oakey, N. S. & Platt, T. (1984). Turbulent motions may control phytoplankton photosynthesis in the upper ocean. *Nature*, **311**, 49-50.

Marti, D. E. & Imboden, D. M. (1986). Thermische Energieflüsse an der Wasseroberfläche: Beispiel Sempachersee. *Zeitschrift für Hydrologie*, **48**, 196-229.

Mortimer, C. H. (1974). Lake hydrodynamics. Mitteilungen Internationale Vereinigung für Theoretische und Angewandte Limnologie, **20**, 124-197.

Pahl-Wostl, C. & Imboden, D. M. (1990). DYPHORA – A dynamic model for the rate of photosynthesis of algae. *Journal of Plankton Research*, **12**, 351-369.

Patterson, J. C. (1991). Modelling the effects of motion on primary production in the mixed layer of lakes. *Aquatic Sciences*, **53**, 218-238.

Vollenweider, R. & Kerekes, J. (1982). *Eutrophication of waters, monitoring assessment, control.* OECD, Paris.

The predicting power of models for eutrophication

GERRIT VAN STRATEN

Systems and Control Group, Department of Agricultural Engineering and Physics, AGROTECHNION, Bomenweg 4, 6703 HD Wageningen, The Netherlands

Restoration of water-bodies from eutrophication has proved to be extremely difficult. Mathematical models have been used extensively to provide guidance for management decisions. The aim of this paper is to elucidate important problems of using models for predicting environmental changes.

First, the necessity for a proper uncertainty assessment of the model, upon calibration, has not been widely recognized. Predictions must not be a single time trajectory; they should be a band, expressing system uncertainty and natural variability. Availability of this information may alter the decision to be taken.

Second, even with well-calibrated models, there is no guarantee they will give correct projections in situations where the model is used to predict the effects of measures designed to bring the system into an entirely different "operating point", as is typically the case in eutrophication abatement. The concept of educated speculation is introduced to partially overcome this difficulty. Lake Veluwe is used as a case to illustrate the point.

Third, as questions become more detailed, such as "what about expected algal composition", there is a greater probability of running into fundamental problems that are associated with predicting the behaviour of complex non-linear systems. Some of these systems show extreme initial condition sensitivity and even, perhaps, chaotic behaviour, and are therefore fundamentally unpredictable.

Introduction

In eutrophication research, mathematical modelling has been extensively used as a powerful tool to study interactions between various biological components of the aquatic ecosystem, and to determine how these interactions are influenced by non-biological factors such as nutrient availability and light. In this context, mathematical models can be viewed as vehicles to organize the thoughts of a researcher. Hypotheses can be tested on data collected from the system as a whole, and the relative importance of various sub-processes can be assessed in a quantitative way. Through modelling the researcher is forced to formulate detailed process descriptions. This alone is a tremendous help in identifying blind spots in existing knowledge. Also, by sensitivity analysis, additional fundamental process research can be directed in the most economical way.

The use of models for research purposes as indicated above can not be advocated enough. On the other hand, quite a large proportion of eutrophication research is motivated by the desire to provide support for practical control measures. Questions to be answered for control, however, seriously differ from those for research, thus putting different requirements to our models. From a tool to understand, they should now become a tool to predict. In view of the vast amounts of money that may be associated with proposed control measures, it is essential that model results are reliable, and, if possible, accurate. The aim of this paper is to study what we can say about the predictive power of models for eutrophication.

Three items will be discussed. First, it is argued that no model results should be presented without a proper analysis of uncertainties arising from model calibration. Both a stochastic and an unknown-but-bounded approach to uncertainty assessment will be demonstrated.

Second, it is argued that even a well-calibrated model does not guarantee a correct prediction when the control measure foreseen will bring the system into a completely different working point, as is often the case in eutrophication abatement. The concept of educated speculation about future parameter values is introduced, as a partial remedy to this problem.

Third, as attention is shifting from predictions about total biomass to the more intricate question of species composition, we run into model equations that are highly non-linear. It will be pointed out that these equations may show chaotic behaviour, or at least extreme initial condition and parameter sensitivity, thus posing fundamental questions of predictability in such systems.

Uncertainty assessment

Prediction uncertainty arises from different sources (e.g. Beck 1987). In this paper we do not consider the uncertainty associated with uncertain future inputs. These effects can be assessed through the generation of proper synthetic input time-series with realistic stochastic properties.

Rather, uncertainties arising from the calibration, i.e. the tuning of the model parameters, are addressed. Although parameters are assumed to have physical meaning, calibration is always necessary, because parameters estimated from measurements in a laboratory situation show large variations, and because model parameters very often refer to lumped state variables. For example, a model specifying diatoms, blue-green algae and green algae requires specific growth rate coefficients for each group, but these simply do not exist as a unique number because they depend upon the actual detailed species composition within each group.

Calibration uncertainties can be accounted for in two ways. If there are sufficient data, a stochastic sub-model may be used to accommodate the remaining, non-explained uncertainty. We explore the potential of the equation error stochastic modelling, which is relatively unknown in the field of eutrophication modelling. If the data are sparse, an alternative is provided by the unknown-but-bounded parameter estimation and uncertainty propagation technique. This will be the subject of the second subsection. The result of either uncertainty analysis will be an uncertainty band around predicted future values of the model's state or output variables.

Equation error stochastic modelling

The usual procedure in calibrating a model is as follows. The model is used to generate output variables. These are then compared to the observations. Subsequently, a parameter search is performed in order to minimize some norm of the difference between model results and observations. Finally, the residual error is plotted as a function of time, and if this still shows clear correlations with itself, or with input data, it is concluded that the model needs revision, because not all dynamics are "explained".

In the research context this procedure seems quite adequate. However, model updating usually requires more state variables, and very often, more input data. When the models must be used for management purposes, there may simply not be time enough for this reductionistic approach. Also, it is not certain that this procedure will indeed lead to better predictions. For instance, replacing a fixed re-aeration term by one dependent upon wind speed will improve the model fit on past data, but will hardly contribute to reducing the future prediction uncertainty, since this is totally dictated by the future wind speeds which will remain unknown.

The procedure outlined above is known as the output error method, i.e. all residual errors between model and data are attributed to the output equations. In equation form:

$$\dot{\boldsymbol{x}} = \boldsymbol{f}(\boldsymbol{x}, \boldsymbol{u}; \boldsymbol{p})$$
$$\boldsymbol{y} = H\boldsymbol{x}$$
$$\boldsymbol{z} = \boldsymbol{y} + \boldsymbol{v}$$

where $\boldsymbol{x}(t)$ is the vector of state variables (a function of time), $\boldsymbol{f}$ are the time derivatives of $\boldsymbol{x}$, a vector valued function of the states $\boldsymbol{x}$ themselves, the input time series $\boldsymbol{u}$ (t), and the constant parameter vector. The vector $\boldsymbol{y}(t)$ represents the model output, here assumed to be some linear function of $\boldsymbol{x}(t)$, expressed in the transformation matrix H (the elements of H may be themselves parameters). The observations $z(t)$ are assumed to be generated by the model but associated with some noise $\boldsymbol{v}(t)$. The idea of the output error method is to find the parameters $\boldsymbol{p}$ that minimize some norm of the residual error sequence $\boldsymbol{v} = \boldsymbol{y} - \boldsymbol{z}$, at the same time requiring that the remaining sequence $\boldsymbol{v}$ is white noise, i.e does not contain any further structural information.

The alternative approach is to consider residual non-white noise as generated by a stochastic process entering the system equation:

$$\dot{\boldsymbol{x}} = \boldsymbol{f}(\boldsymbol{x}, \boldsymbol{u}; \boldsymbol{p}) + \boldsymbol{w}$$
$$\boldsymbol{y} = H\boldsymbol{x}$$
$$\boldsymbol{z} = \boldsymbol{y}$$

(e.g. Ljung 1987). Here, $\boldsymbol{w}$ is a white noise stochastic process. The essential difference with the output error method is that this noise is filtered by the system equations, and therefore appears in the output as non-white noise. In eutrophication modelling this is a much more common situation than white residual errors. The key here is that non-white output error does not necessarily lead to the conclusion that a model update is needed. Rather, the addition of a stochastic part to the equation may be quite as adequate without the need to introduce further state variables or inputs.

As an example consider the diurnal variation in dissolved oxygen in a lake due to photosynthesis. A minimum information, yet reasonable, model for the average over the water column is

$$\frac{dC}{dt} = qI + k_r(C_s - C) - R$$

where c is the dissolved oxygen concentrations, C_s is the saturation concentration (a function of temperature), I is the incident solar radiation, q is a photosynthetic dissolved oxygen production parameter, k_r is the re-aeration coefficient, and R is the total respiration. Thus, in this model the state variable is C, the inputs are C_s (through the temperature) and I, and the parameters are q, k_r and R.

Figure 1 shows the incident light and the associated diurnal pattern as measured in Loosdrecht Lakes. The data are 20-minute averages over sensor output sampled every minute. The dashed line indicates the pattern simulated when using parameters estimated by the output error method. The data are presented as deviations from the mean. Details of the calculation procedure can be found in Van Straten & Kouwenhoven (1991).

After a short period related to the dying out of the initial condition, the model fits the data fairly well. However, it is also clear that the real data are more shaky and irregular than the model. The peaks and valleys of the model are much more smooth and of less amplitude than those observed in the field. Figure 2 shows the auto-correlation of the prediction error. The signal is highly correlated and far from white noise, suggesting that the model is not covering the full dynamics present in the data. The discrepancy between model and data also becomes apparent if the model is simulated while adding a synthetically generated output error of the same variance as identified in the parameter estimation stage. This is shown in Figure 3. Cleary, the rapid changes in dissolved oxygen in the simulation are not appropriate mimics of

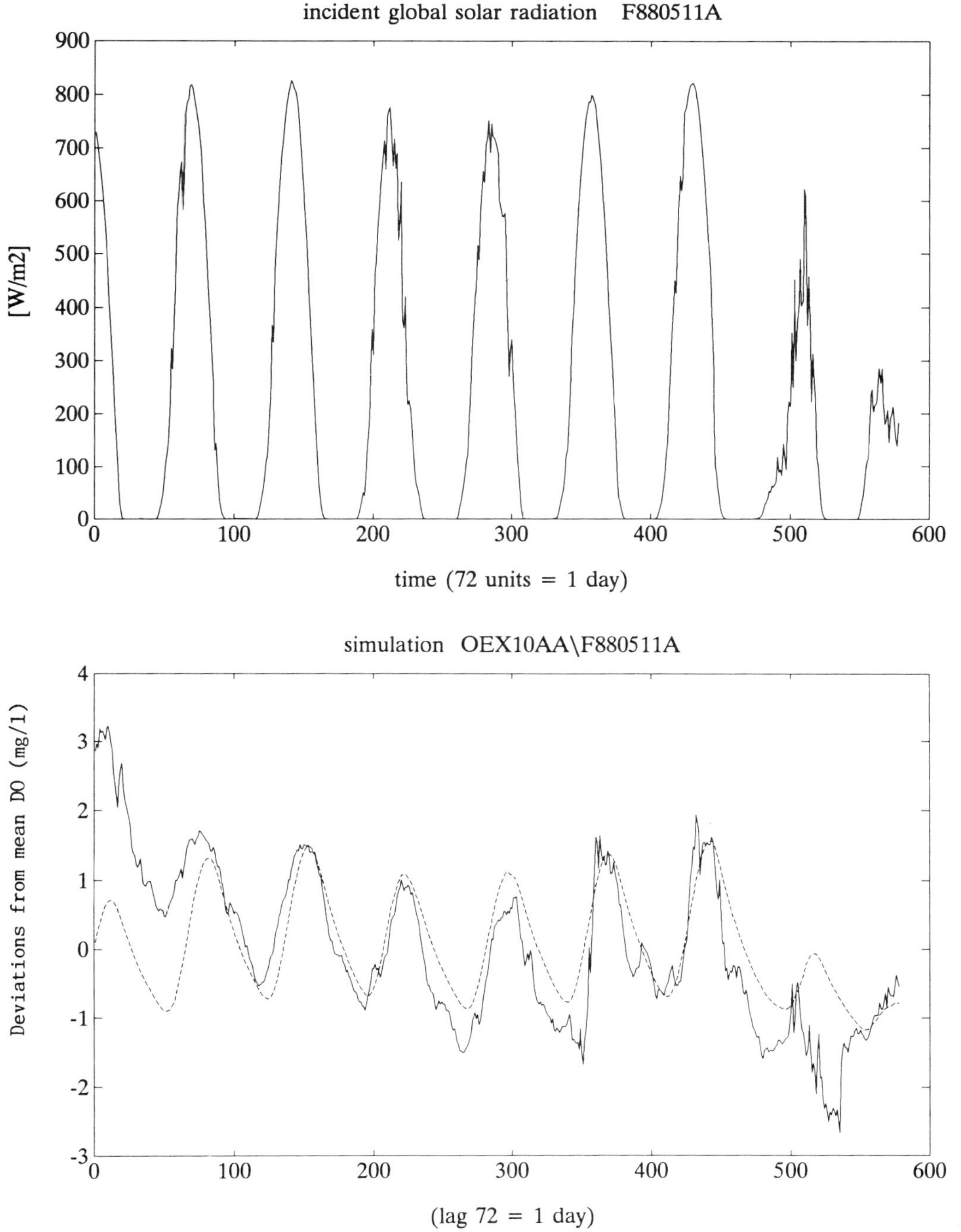

Figure 1. Solar radiation input (upper) and dissolved oxygen concentration (DO), measured (lower, solid line) and simulated with output error parameter estimates (lower, dashed line). DO (mg l^{-1}) is shown as deviations from the mean concentration (approximately equal to saturation in this case), plotted against time.

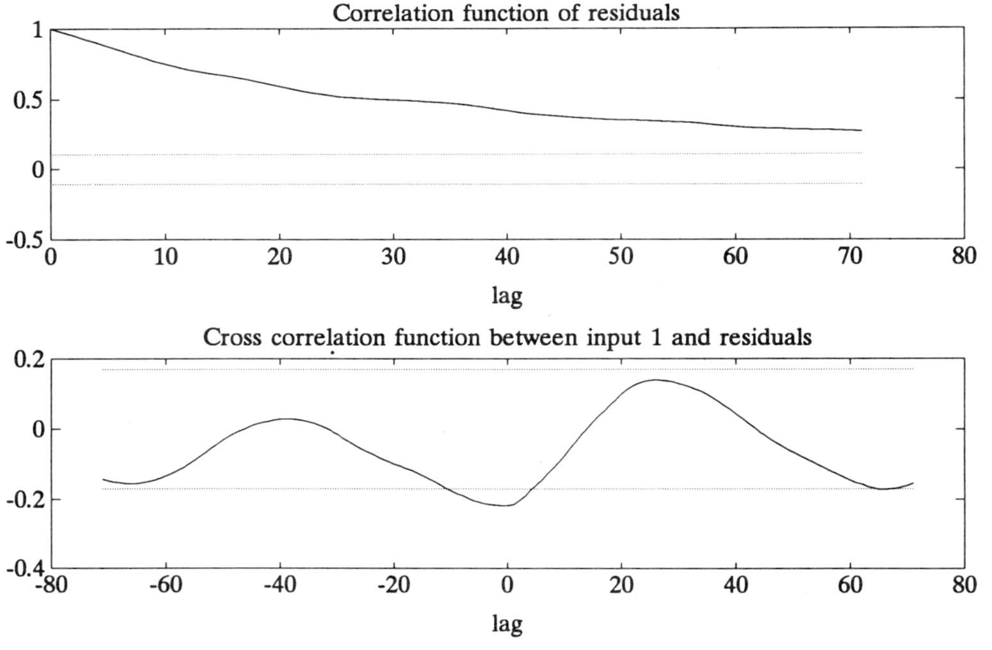

Figure 2. Correlation of residual error, i.e. difference between one-step-ahead prediction and measured value.

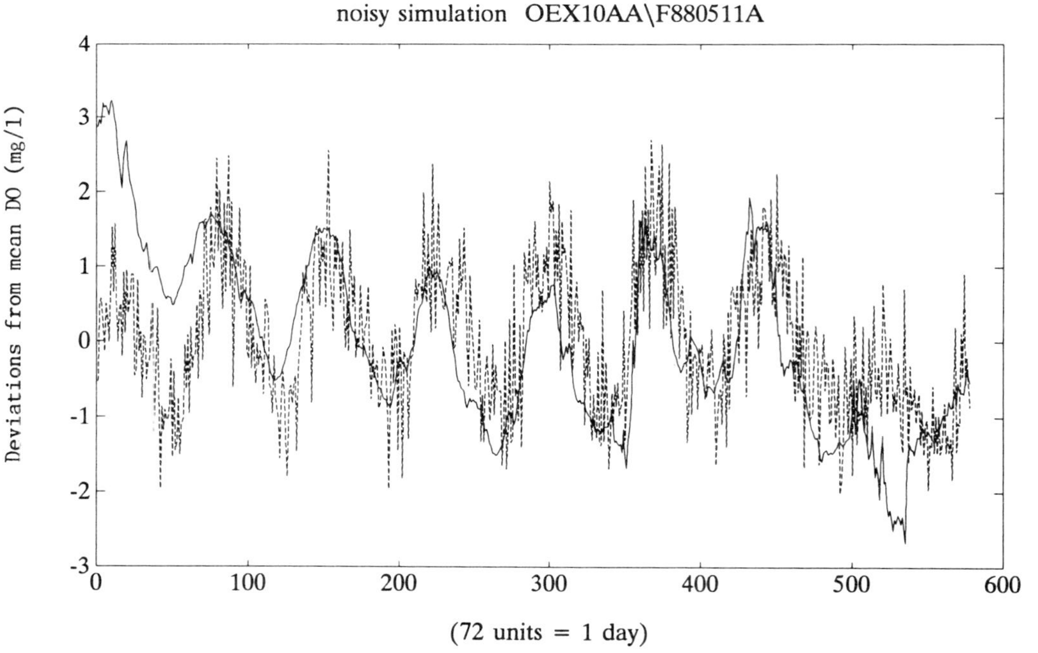

Figure 3. Sample noisy simulation with output error model (broken line) and deviations from the measured mean dissolved oxygen concentration (solid line; as in Fig. 1.)

what happens in the field.

The usual procedure is now to reject the model and to look for other explanatory factors. In this case one might think of variations in re-aeration rate due to wind, or horizontal transport of patches of algae by wind-induced currents. A much more complicated model would be needed. As outlined above an alternative, however, would be first to look for other models for the stochastic part of the system's behaviour, for instance in terms of an equation error model:

$$\frac{dC}{dt} = qI + k_r(C_s - C) - R + w$$

Here, the additional term w represents a stochastic systems noise term. Of course introducing equation noise results in a stochastic differential equation. The parameter estimation was solved in this case by transforming the model into a linear discrete time version, resulting in an auto-regressive with exogenous variable (ARX) model. Solutions for this problem are readily available, e.g. in the System Identification Toolbox of MATLAB. In the general non-linear case, finding the parameters would require the solution of an extended Kalman filter.

The simulation result without noise is presented in Figure 4. At first sight there is not much difference from the output error method. However, when we now simulate the model again while adding a realisation of the noise to the equation, we obtain the pattern of Figure 5. The coincidence in appearance and characteristics of the real data is striking. Sharp and sudden changes in dissolved oxygen, as in the output error model, no longer occur, and the equation error model gives a much more "look-a-like" behaviour. This is also confirmed by looking at the correlation function of the one-step-ahead prediction error which, indeed, is practically white (Fig. 6). Note that with a stochastic model we require that the one-step-ahead prediction starting from the last observation is as close as possible to the next observation. We cannot expect the free-running stochastic model to fit the data, because we can only simulate with a

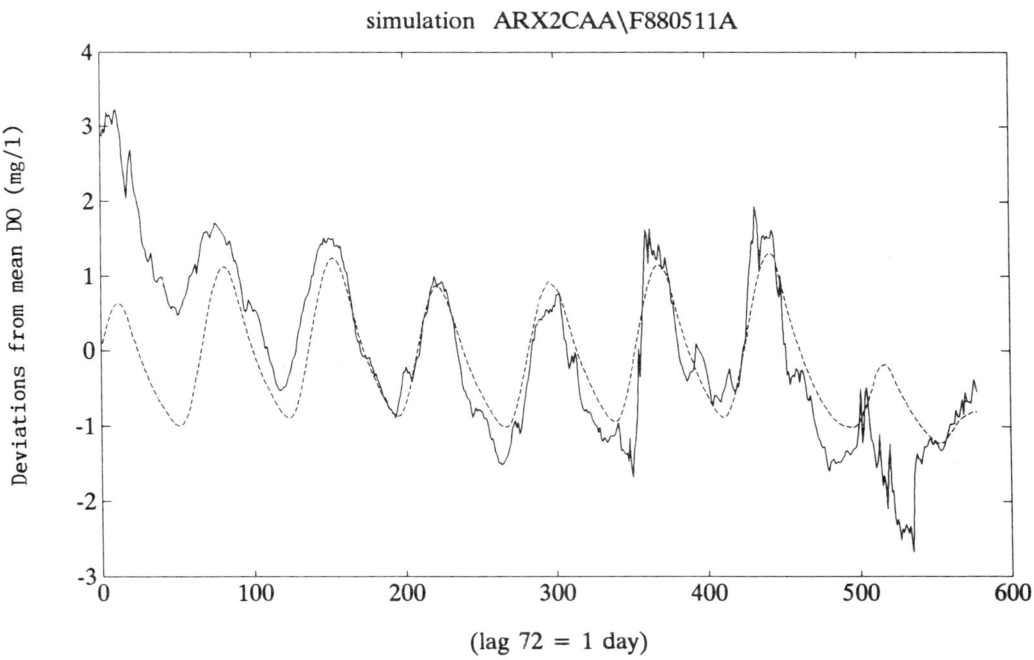

Figure 4. Simulation with equation error model (ARX parameters) (dashed line) and deviations from the mean dissolved oxygen concentration (solid line; as in Fig. 1).

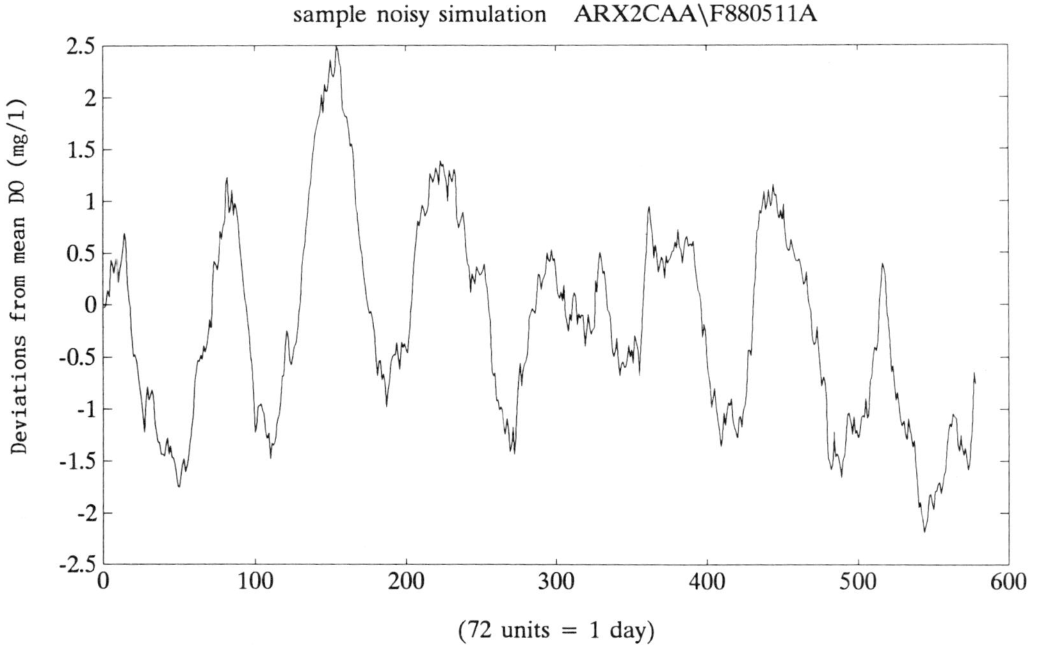

Figure 5. Sample noisy simulation with equation error model, plotted as deviations from the mean dissolved oxygen concentration (as in Fig. 1).

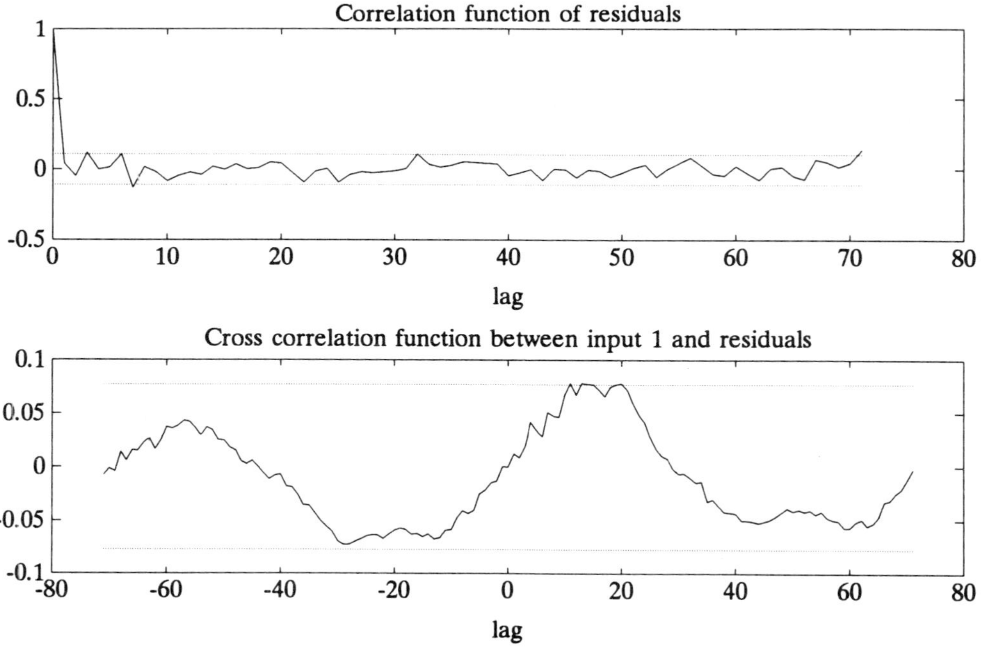

Figure 6. Correlation of residual error for equation error model.

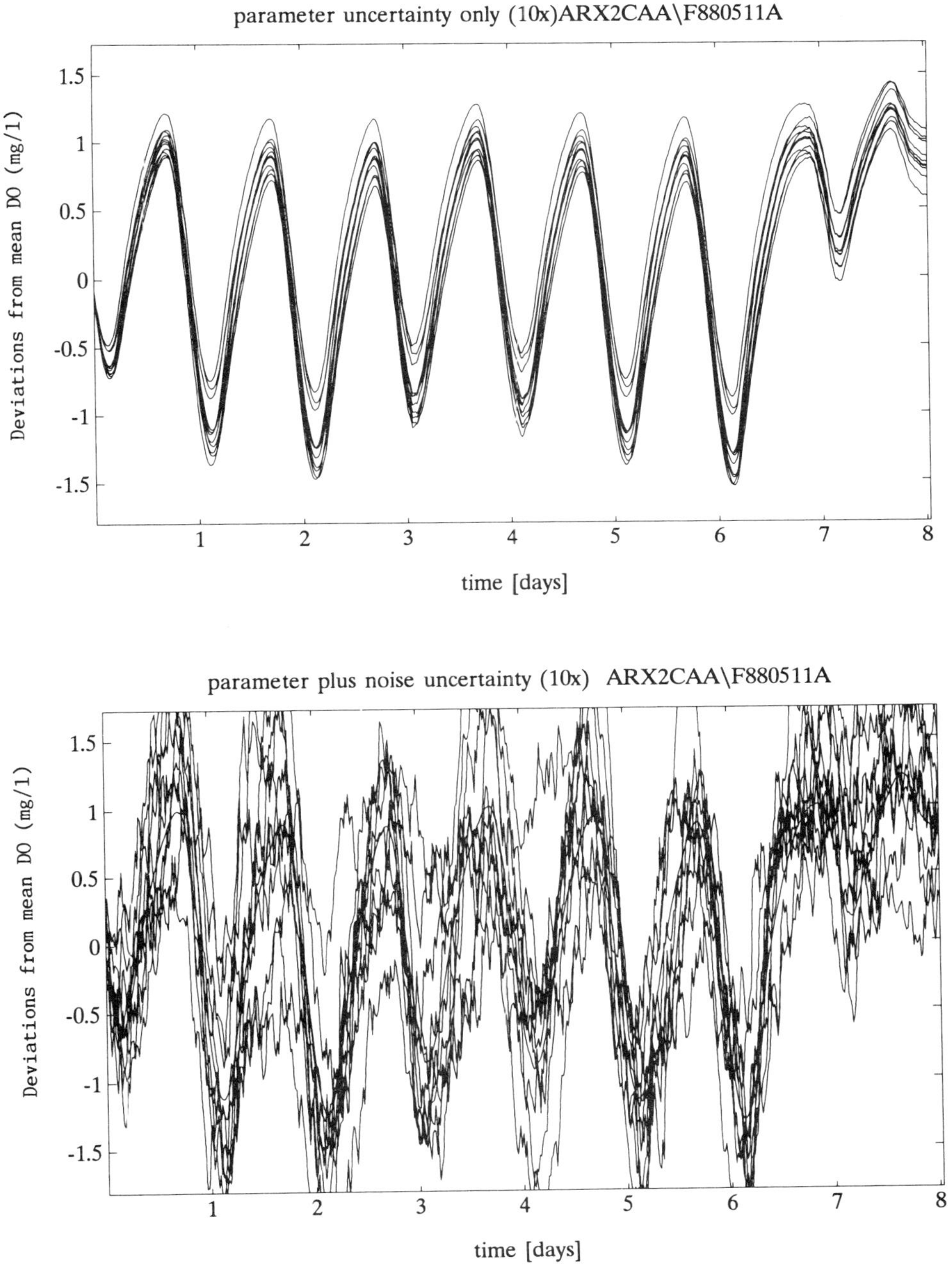

Figure 7. Prediction uncertainty with equation error model due to parameter uncertainty only (upper) and including equation noise (lower). The plots show deviations from the mean dissolved oxygen concentration (as in Fig. 1).

particular realisation of the noise that will always be different from what happened in the field. What we can do, however, is to repeat the simulation several times in order to show the prediction uncertainty range. Figure 7 (upper) shows the range due to parameter uncertainty only (small in this case), and Figure 7 (lower) shows the additional uncertainty due to the systems noise. Although the uncertainty is in the order of magnitude of about 1.5 mg l^{-1} the distinction between day and night is still clearly visible. Also, the firm statement can be made that there is hardly a chance that dissolved oxygen will drop more than 2 mg l^{-1} below its average value in this system.

Unknown-but-bounded uncertainty assessment

In cases where data have not been sampled frequently enough to allow for a properly validated stochastic model, an alternative is provided by postulating an unknown-but-bounded model. In contrast to stochastic modelling the only assumption to be made is that the uncertainty is bounded. Several authors have pioneered this method of uncertainty modelling (cf. the review by Beck (1987)). A procedure based on a Monte Carlo approach has been developed by Keesman (1989), and was applied to eutrophication modelling by Van Straten & Keesman (1991). The basic idea is as follows.

First, a bounded area of uncertainties is associated to the observed time series of the data. This is called the behaviour definition. The bounds are based upon known or estimated measurement and sampling noise, and estimates on spatial and temporal variations of which it is known that they are not described – intentionally – by the designed model. Also, a set of feasible ranges for the parameters of the model is defined, based on physical knowledge.

Next, random parameter vectors are taken from the feasible range by Monte Carlo selection, and the model is run for each vector. During the simulation the results are checked against the defined behaviour range. If the model runs outside the behaviour range, then the associated parameter vector is classified as a non-behaviour-giving vector. Thus, after many runs, two sets are obtained within the feasible parameter range: a behaviour and a non-behaviour set.

Since the only initial assumption was that the error was bounded, all parameters in the behaviour set are equivalent, and equally possible. Therefore, when the model is run to make predictions, all these parameters are used to simulate the future systems behaviour. The result is quite naturally a bounded uncertainty range of the model predictions, rather than a single trajectory. Several details, e.g. tricks to improve the efficiency of this procedure, can be found in Keesman (1989).

An example of this procedure is worked out by Van Straten & Keesman (1991) for modelling eutrophication in Lake Veluwe, a shallow lake in the central parts of the Netherlands. Heavy loads of phosphorus in this lake had created very high chlorophyll-*a* concentrations, and an almost permanent dominance of the cyanobacteria *Oscillatoria aghardii*. During the late 1970s the loads were gradually reduced, without much success. In order to enforce the process of restoration, the water authority decided to start flushing the lake, by modifying the water discharge regime of the adjacent polder. Consequently, since the start of flushing in late 1979, the water replenishment rate was increased from approximately 1-2 times per year to about 4 times a year with peaks of 7-8 times per year in winter, by pumping in water low in phosphorus and high in calcium.

The data up till the start of flushing were used to develop models to describe chlorophyll-*a*, ortho-P and total phosphorus. Unknown parameters were estimated with the classical least-squares technique, with the result shown in Figure 8. Two models were developed, one with a simple exchange term with the sediment, and the other with a more elaborate sediment sub-model. Fairly standard equations were used. The case was an almost ideal test to see how these models behave when used for making predictions, because post-flushing data were also

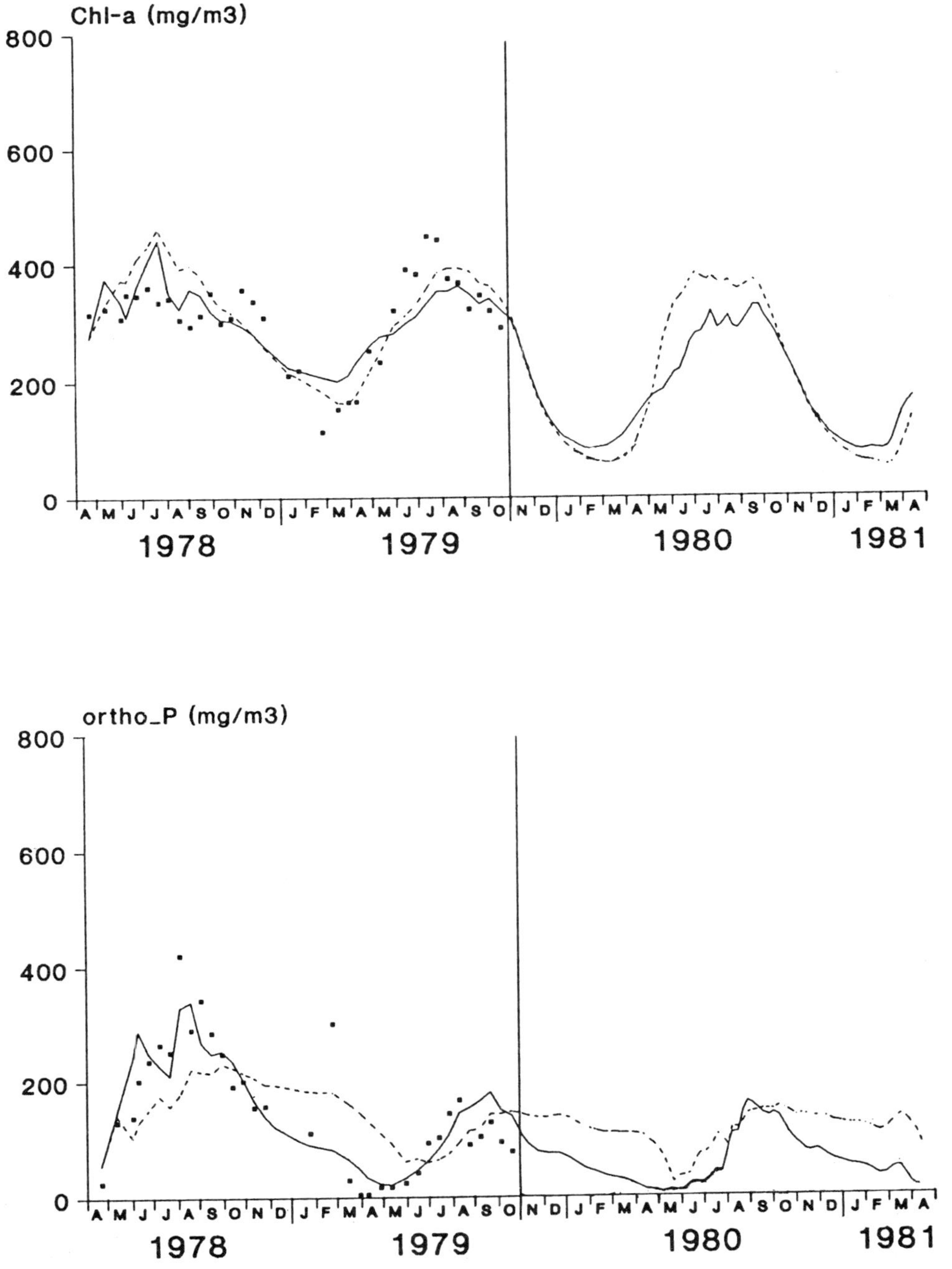

Figure 8. Least-squares fit to measured data (dots) for chlorophyll-*a* (upper) and orthophosphate-P (lower) in Lake Veluwe, using only the pre-flushing data (1978-1979) fitted to Model I (dashed lines) and Model II (solid lines). Predictions are extrapolated into 1980-1981. Measured data for the latter period are shown in Fig. 9.

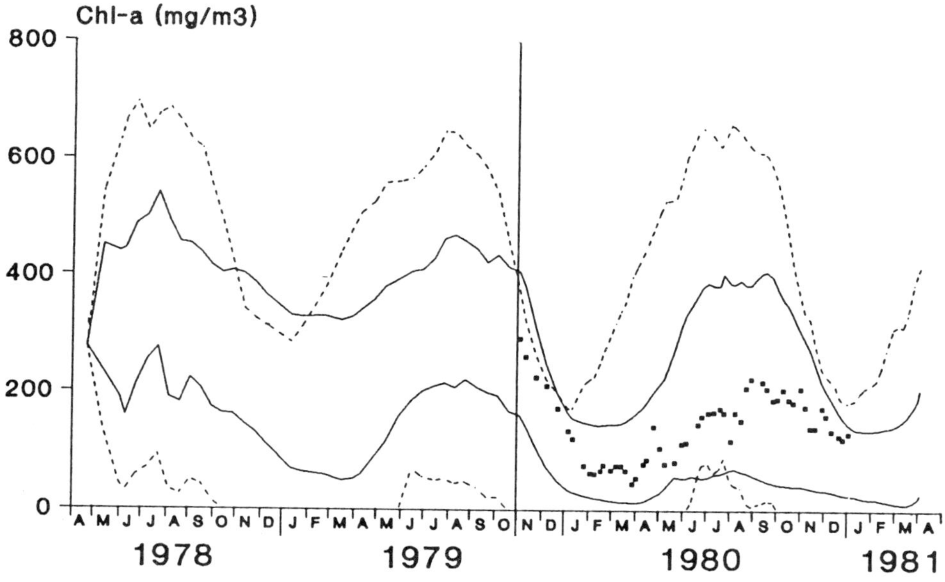

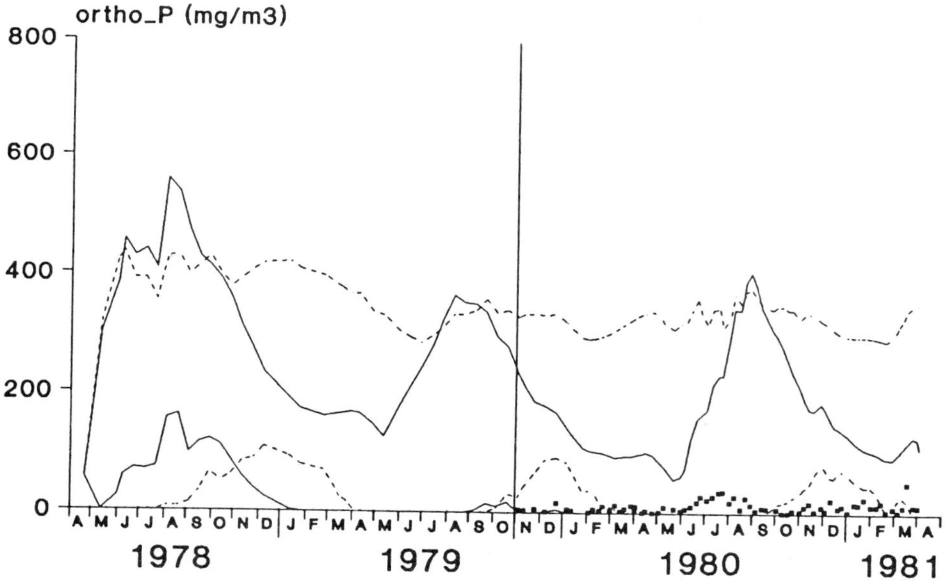

Figure 9. Unknown-but-bounded ranges for Model I (dashed lines) and Model II (solid lines). Estimations are based on measured data for pre-flushing period in 1978-1979 (see Fig. 8). Prediction uncertainty for the post-flushing period is also shown against the measured data (dots) for chlorophyll-*a* (upper) and orthophosphate (PO_4–P; lower).

available. Comparing the predictions with actual observations show that the models did not make very accurate predictions, despite a seemingly good fit in the pre-flushing period.

To assess the uncertainties, an unknown-but-bounded approach using the same data was undertaken as well. This is shown in Figure 9. Larger uncertainty areas were obtained for Model I, because the initially set "acceptable" uncertainty band of 300 mg m^{-3} chlorophyll-*a* was not sufficient. This shows that uncertainties can be reduced by incorporating more details in the models, at the expense of more state variables and more data (for Model II, sediment dissolved oxygen consumption data were used as well).

The calibration uncertainties propagate into the prediction period, as shown in Figure 9 as well. By assessing uncertainties, it becomes apparent that the seemingly high precision of the deterministic predictions in Figure 8 are not justified. The interpretation of the bands given in Figure 9 is that algal peaks will go down, but that there is still a possibility of fairly large values even after flushing. On the other hand, the plots show that there is, in fact, also a possibility of very low values, not existing before the flushing was effected. So, despite large uncertainties, a manager might have decided that flushing was worth a try, a conclusion not at all apparent when using the best-fit parameters of Figure 8 without uncertainty assessment.

Educated speculation

The example of Lake Veluwe can also serve as an illustration of a problem of considerable interest in predicting environmental change in general (Beck 1991). After all, when implementing measures for eutrophication reduction we intentionally bring the system into another operating point. How can we be sure that we are dealing with the same system as before measures were taken? In the Lake Veluwe case, for instance, flushing introduced large amounts of calcium-rich water. It is quite conceivable that this had a definite effect upon the phosphorus exchange process with the sediment. The reduction of dissolved oxygen consumption due to lower settling of detritus material was taken care of in the model, but not the possible change of adsorptive capacity. Yet, having the model and knowing its assumptions does allow for speculation about non-modelled parameter changes brought about by control, which without a model would not have been possible.

In Figure 10, the results shown from Model I by reducing the apparent phosphorus equilibrium concentration in the sediment to 25% of its original value, as an educated guess of the effect of the larger calcium content of the flushing water. We see that even with the poor Model I this educated speculation leads to a predicted effect of flushing.

Principle problems of prediction in non-linear systems

As long as the questions remain rather crude, such as those related to phosphorus reduction measures, it is still possible to obtain reasonable answers since, after all, the mass balance just determines an upper limit to the biomass that can be sustained at any one moment. However, if we move into the much more intricate question of species composition or the inclusion of higher elements of the food web, things start to be more complicated. The equations used are highly non-linear. As a consequence, such models may exhibit strong initial sensitivity behaviour, catastrophic transitions between pseudo-equilibrium states, and even full chaotic behaviour. Moreover, the cyclic nature makes a proper calibration very difficult. The crux of this short third section of the paper is to make modellers aware of these intrinsic difficulties or even impossibilities of predicting the behaviour of cyclic, non-linear systems, and to stimulate further research in this area.

Nice examples of the effects of non-linearities in ecosystems models are given by Scheffer (1990). The practical interest lies in the options for active biological control. Some of Scheffer's results suggest that the shallow system consisting of algae, almost no zooplankton,

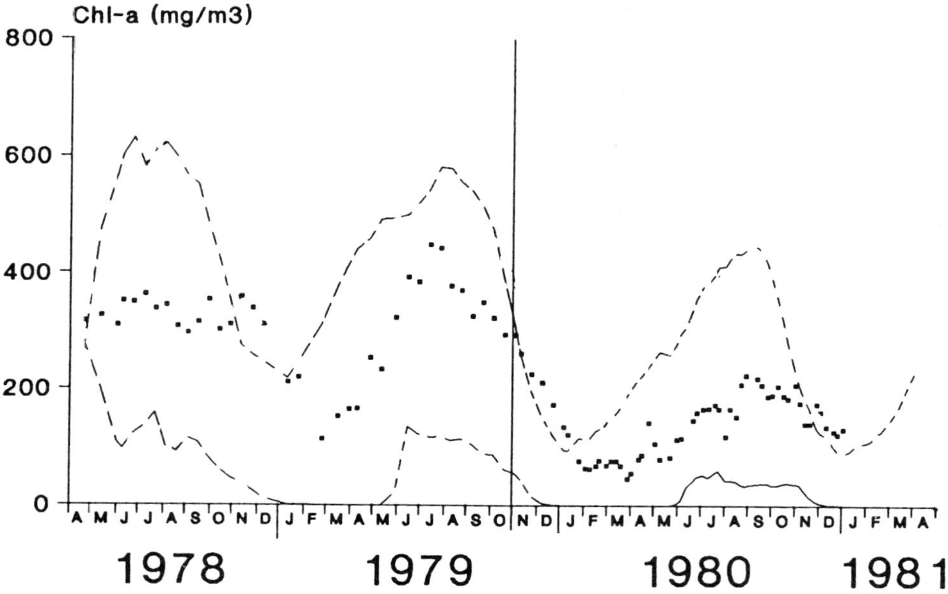

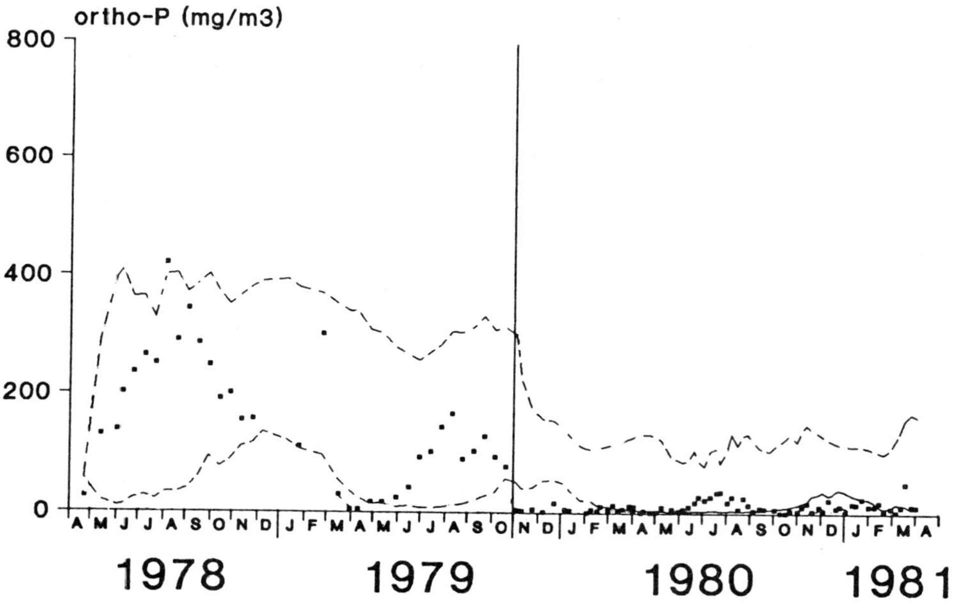

Figure 10. Educated speculation using Model I (see the text). A reduction by 75% of the equilibrium P-concentration due to flushing is assumed. Measured data (dots) are shown for chlorophyll-*a* (upper) and orthophosphate (PO_4–P; lower) for both the pre-flushing and post-flushing periods.

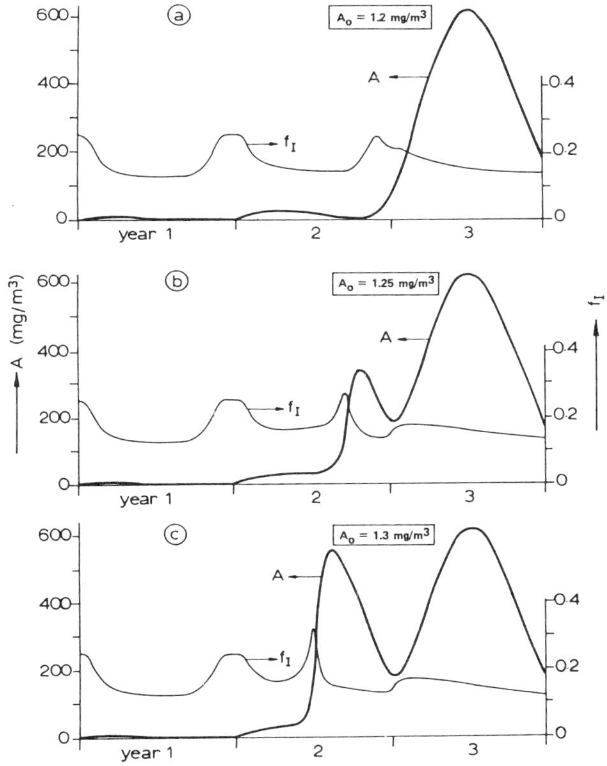

Figure 11. Demonstration of extreme initial condition sensitivity in a simple self-shading model for cyanobacteria in a lake. Solid line A shows algal concentration in equivalent chlorophyll-*a* units (left-hand scale). Solid line f_1 shows the light dependency factor of the vertically and daily averaged growth rate (low in summer due to inhibition). The three plots (a – c) depict the behaviour at slightly different initial concentrations of algae (A_0).

large amounts of white fish, no predatory fish, and a lack of water plants which might serve as shelter, due to the high turbidity brought about by the algae, is the natural end station of a strongly eutrophic situation, and moreover shows a remarkable robustness against load reductions. The results can be qualitatively used to promote fish control as an additional control measure, but it is extremely difficult to quantify the effects, because the dynamics are very sensitive to parameter values, and to the assumed form of the non-linearities.

A very simple model of algal blooms, assuming a light inhibition curve, as is typical for cyanobacteria, with just one single state variable, already exhibits extreme initial condition sensitivity, as is shown in Figure 11 (Van Straten 1986). Very small changes in concentrations in winter-time have a dramatic effect upon the occurrence of algae even three years later. With more species of algae, deterministic chaotic behaviour can easily be demonstrated, and since initial conditions cannot be measured with infinite accuracy, we are facing a fundamental problem of prediction in such systems. The prediction horizon is intrinsically limited, much as in predicting the weather, and if research shows that this is truly the case, then new avenues are needed to reliably use models for ecosystems management.

Conclusions

Model uncertainties should be and can be assessed if model predictions are to be used for management. If sufficient data are available stochastic methods are well suited. Non-white residual error can sometimes conveniently be remedied by supplementing the deterministic model with an equation error type stochastic model, rather than with an output error model. If successful, this procedure is a serious alternative to cumbersome deterministic model updates with larger data demands, more costs and longer development times, which is particularly relevant if the models have to be used for management rather than research.

When data are scarce and of poor quality, the set-theoretic unknown-but-bounded approach offers an attractive alternative to stochastic modelling in order to assess uncertainties. A proper and honest uncertainty assessment upon calibration may well reveal the true – sometimes limited – significance of model forecasts, despite the tendency of these to be on the pessimistic side.

Even when there are large uncertainties, modelling offers advantages over qualitative statements when it comes to making predictions about the systems behaviour under changed management conditions. This is because engineering judgement as well as learning from the lessons of experience can be translated into educated speculation about parameter responses to future measures.

Finally, we should be aware of fundamental problems in predicting detailed time patterns of complicated ecosystems. Due to essential non-linearities we can expect chaotic behaviour and initial condition sensitivity. More research is needed to confront and reconciliate modelling with biological observation and experience based on pattern recognition and extrapolation techniques.

References

Beck, M. B. (1987). Water quality modelling: a review of the analysis of uncertainty. *Water Resources Research*, **23**, 1393-1442.

Beck, M. B. (1991). Forecasting environmental change. *Journal of Forecasting*, **10**, 3-19.

Keesman, K. J. (1989). A set-membership approach to the identification and prediction of ill-defined systems: application to a water quality system. *PhD dissertation, University of Twente, Enschede, The Netherlands.*

Ljung, L. (1987). *System Identificaiton; Theory for the User.* Prentice Hall, New Jersey.

Scheffer, M. (1990). Simple models as useful tools for ecologists. *PhD thesis, University of Utrecht, The Netherlands.*

Van Straten, G. (1986). Identification, uncertainty assessment and prediction in lake eutrophication. *PhD thesis, University of Twente, Enschede, The Nertherlands.*

Van Straten, G. & Keesman, K. J. (1991). Uncertainty propagation and speculation in projective forecasts of environmental change: a lake-eutrophication example. *Journal of Forecasting*, **10**, 163-190.

Van Straten, G. & Kouwenhoven, J. P. M. (1991). Identification of a mechanistic model for the diurnal dissolved oxygen pattern in lakes with discrete time series analysis. *Water Science and Technology*, **24**, (6), 17-23.

Structural dynamic eutrophication models

S. E JØRGENSEN

DFH, Institut A, MILJØKEMI (Environmental Chemistry), Universitetsparken 2, 2100 Copenhagen Ø, Denmark

This article discusses problems of modelling the seasonal succession of algal species in lakes and reservoirs, and the adaptive selection of certain groups of algae in response to changes in the inputs and relative concentrations of nutrients and other environmental variables. A new generation of quantitative models is being developed which attempts to translate some important biological properties of species (survival, variation, inheritance, reproductive rates and population growth) into predictions about the survival of the fittest, where "fitness" is measured or estimated in thermodynamic terms. The concept of "exergy" and its calculation is explored to examine maximal exergy as a measure of fitness in ecosystems, and its use for calculating changes in species composition by means of structural dynamic models. These models accomodate short-term changes in parameters that affect the adaptive responses (species selection) of algae.

Introduction

Eutrophication models have been developed (Jørgensen 1986) that are able to predict changes in the level of eutrophication due to changes in the "forcing functions", i.e. the input of nutrients to the lake in question. The application of eutrophication models has, however, also demonstrated the shortcomings of current models: due to rapid changes in the composition of algal species it is hardly possible to make predictions about seasonal variations and when the spring and summer blooms will occur. This is regrettable because predictions of seasonal and annual changes in species composition are often more important in lake management than the actual level of eutrophication.

If we follow a general modelling procedure (see, for instance, Jørgensen 1988) we obtain a model describing the processes in the focal ecosystem, but the parameters represent the properties of the state variables (components) as these exist in the ecosystem during the period of examination. They are not necessarily valid for another period of time because we know that ecosystems are able to regulate, modify and change them, if needed, as a response to changes in the prevailing conditions. The latter are determined by the forcing functions and the particular interrelationships existing between the state variables of each ecosystem. Our present models have rigid structures and a fixed set of parameters, so that alterations or replacements of the components are not possible. We need, however, to introduce parameters (properties) that can alter according to changes in the forcing functions affecting an ecosystem and the general set of conditions for the state variables (components) in that ecosystem.

Models that can account for a change in species composition and the ability of the species to alter their properties (i.e. the biological components of the models can adapt to the prevailing conditions imposed on the species), are sometimes called structural dynamic models, to indicate that they are able to accommodate structural changes in the ecosystem. They may also be called the next generation of ecological models, to underline the fact that they are radically different from previous modelling approaches and can do more: they describe changes in species composition.

It could be argued that the natural ability of ecosystems to replace present species with other, better-fitted species, can be modelled by constructing models which encompass all of the actual species known to occur for the entire period of time which the model attempts to cover. However, this approach has two essential disadvantages. Firstly, the model becomes very complex, as for each trophic level it will contain many state variables. Thus the model will also contain many more parameters that have to be calibrated and validated, introducing a large uncertainty in the model and making its application very case-specific (Nors Nielsen 1992a,b). In addition, the model will still be rigid and not possess the property of real ecosystems, where the parameters may alter even if there is no change in species composition.

Later sections of this paper present some results obtained by applying the thermodynamic concept "exergy", as an optimization principle for current changes in the parameters of an ecosystem. Exergy has been used in some kinds of algorithm as a goal function. It expresses the biogeochemical energy (or free energy, or energy able to do mechanical work) of the system, or the distance from thermodynamic equilibrium. First, however, some basic ideas concerning the responses of organisms to changes in their environment are considered, as these have to be incorporated into models of ecosystems and are especially important for modelling the seasonal, short-term dynamics of plankton in lakes and reservoirs.

Ecological structure and species composition

Ecosystems are "soft systems" in the sense that they are able to meet changes in external factors, or impacts with many regulatory processes, on different levels. The results are that only minor changes are observed in the function of the ecosystem. This means that the state variables – but not necessarily the biotic species – are maintained almost unchanged, in spite of changes in external factors.

During recent years there has been wide discussion (Odum 1983; Straskraba 1980) on ways of describing these regulatory processes, particularly those on the ecosystem level, i.e. the changes in ecological structure and the species composition.

Neo-Darwinian theory, expanded to include (1) co-evolution, (2) transfer of knowledge (information) from parents to offspring, (3) the ability of organisms to regulate their environment and thereby the selection pressure on them, (4) D-genes, and modernized according to Brooks *et al.* (1989) and Wiley (1988), describes competition among species. It states that the species which are best fitted to prevailing conditions in the ecosystem will survive. This formulation may be interpreted as a tautology and the following is preferable: life is a matter of survival and growth. Given a particular set of conditions, determined by the external and internal functional properties of the ecosystem, the question is: which of the available species (and there are more available species than are needed) have the combinations of properties required to give the best guarantee for survival and growth? Those species, or rather the particular combination of species, which have the best properties may be denoted the fittest and will be selected. In other words, Darwin's theory may be used to describe changes in ecological structure and the composition of species, but with the present formulation it can not be directly applied quantitatively in ecological modelling.

The problem of describing changes in ecological structure and species composition, using quantitative methods developed by translating the "survival of the fittest" into thermodynamic terms, will be discussed below. In this context it is presented as a hypothesis which attempts to unite the theories of Monod (1970) and Prigogine & Stengers (1979) by explaining ecosystem development and evolution in terms of the concept of exergy.

Hypotheses on survival and growth of organisms under the conditions currently prevailing in an ecosytem

All biological species in an ecosystem are confronted with the question: how is it possible to grow, or even survive, under the prevailing conditions? The latter are considered to be *all* factors that influence the species, i.e. all external and internal factors, including those originating from other species.

These factors are dynamic; the conditions are steadily changing and there are always many species waiting in the wings, ready to take over if they are better fitted to the emerging conditions than the species dominating in the current prevailing conditions.

A wide spectrum of species, representing different combinations of properties, is available to occupy the ecosystem. The question is: which of these species are best able to survive and grow under the present conditions and which species are best able to survive and grow under new conditions one time-step later, two time-steps later, and so on? The "necessity", in Monod's sense, is given by the prevailing conditions; in order to survive a species must have genes and a phenotype (i.e. properties) which match the prevailing conditions. But the natural

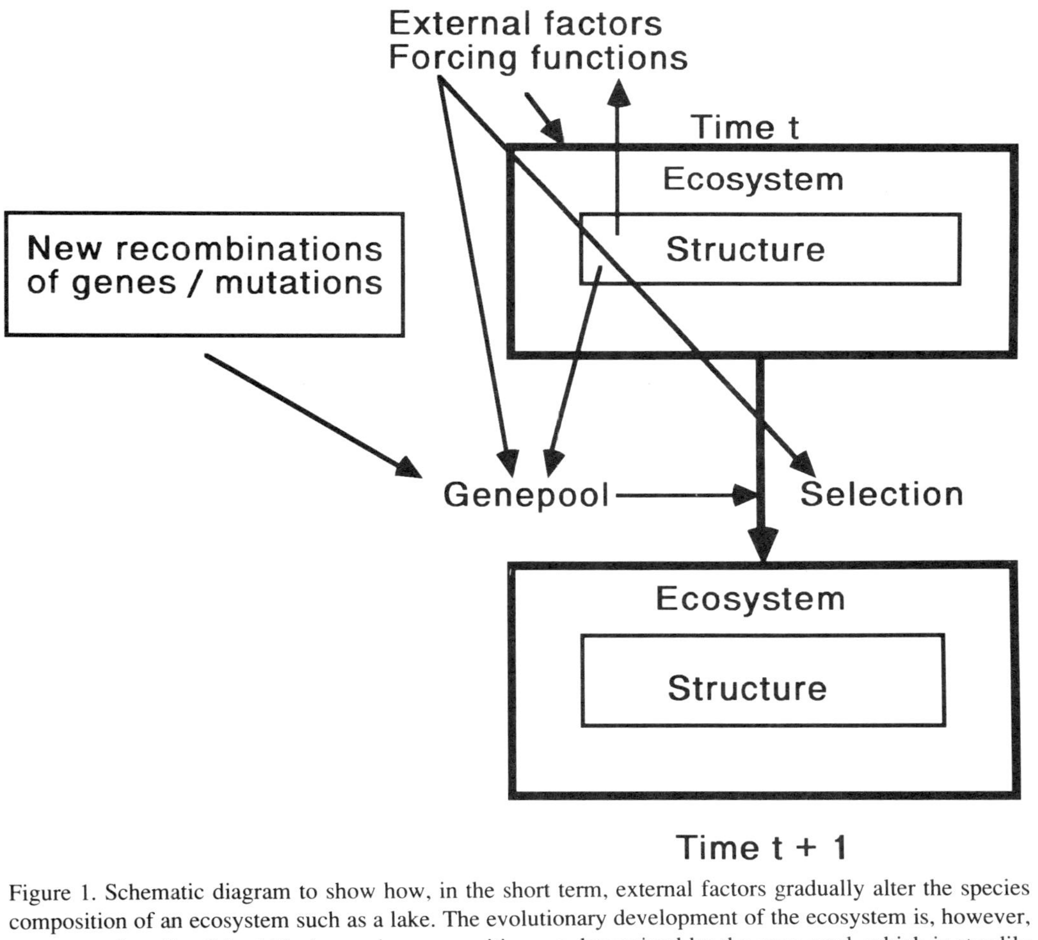

Figure 1. Schematic diagram to show how, in the short term, external factors gradually alter the species composition of an ecosystem such as a lake. The evolutionary development of the ecosystem is, however, more complex. Possible shifts in species composition are determined by the gene pool, which is steadily changed due to mutations and new sexual recombinations of genes; these are tested for fitness, by natural selection. Arrows leading from "structure" to "external factors" and "selection" account for the possibility that a species can modify its own environment and thereby its own selection pressure, and an arrow from "structure" to "gene pool" accounts for the possibility that species can to a certain extent change their own gene pools.

external factors, and the genetic pool available for testing by natural selection, may change by chance. Gradually, new mutations and sexual recombinations emerge and progressively provide new material to be tested against the question: which species are best fitted under the conditions prevailing *now* ?

These ideas are illustrated in Figure 1. The external factors steadily alter; some slowly, others relatively quickly and partly at random, e.g. meteorological factors. The species present in the ecosystem are selected from the pool of species available and are represented by the genetic pool which alters by chance, slowly but surely. "Ecological development" is the process of changes with time in nature, caused by the dynamics of external factors, giving the ecosystem sufficient time to react. The ideas behind Figure 1 are analogous with those of Wiley (1988), Brooks *et al.* (1988), Ulanowicz (1980) and Wicken (1988). These authors claim that the following features are characteristic for the evolution of ecosystems from a thermodynamic point of view.

(1). Dollo's Law of an irreversible evolution – in geological time, the same species will never reappear – is valid, because the same situation will not occur again with the same forcing functions and state variables.

(2). The entire ecosystem is evolving because everything is linked to everything else, underpinning the concept of ascendency.

(3). The past history of the ecosystem is important, because it determines which genes are available to find the best solution to the problem of survival. The history also determines the initial conditions, which are of great importance for the development of the system – compare also with the indirect effect (Patten 1991).

(4). Natural selection is based upon given conditions in the entire ecosystem and is separated from the evolution of the genetic pool. On the other hand, evolution is related to the genetic pool, and is the result of the relationship between the dynamics of the external factors and the dynamics of the genetic pool. The external factors are steadily changing the conditions for survival, and the genetic pool is steadily producing new solutions to the problem of survival.

(5). The most complex ecosystem does not necessarily give the best answer to the problem of survival. Therefore, perhaps, we can not find any relationship between complexity and stability. It does not exist, as was emphasized by May (1981).

Natural selection in ecosystems

Darwin's theory of natural selection presumes that populations consist of individuals which have the following properties.

(i). High reproductive rates: on average, individuals produce more offspring than are needed to simply replace themselves when they die.

(ii). Inheritance: progeny resemble their parents more than they resemble randomly-chosen individuals in the population.

(iii). Variation: individuals vary in heritable traits which influence reproduction and survival, i.e. fitness.

All three properties are part of the ideas presented here. High reproductive rates are needed to get relatively rapid changes (seasonal and annual) in the species composition, caused by rapid changes in external factors. Variability is represented in the short-term and long-term changes in the genetic pool, and inheritance is necessary for the fitness test to take effect. The species are continuously tested against the prevailing conditions (external as well as internal factors) and the better they are fitted, the better they are able to maintain their biomass and also to grow. Indeed, the specific rate of population growth may be used as a measure of fitness (e.g. see Stenseth 1986). But the property of fitness must be inheritable to have any effect on the species composition and the ecological structure of the ecosystem.

Man-made changes in external factors, i.e. anthropogenic pollution, has created new problems because new genes fitted to these changes do not develop overnight, although resistance to insecticides, for example, sometimes develops surprisingly fast. Most natural changes have occurred many times previously and the genetic pool is, therefore, prepared to meet the majority of natural changes but not many of the man-made changes, because these are new and untested in the ecosystem.

Survival and growth in terms of exergy

Darwin's theory may be translated into thermodynamic terms, with exergy as the basic concept. Survival implies maintenance of the biomass, and growth means an increase of biomass. The construction of biomass requires exergy, which is transferable to support other exergy-demanding (i.e. energy-demanding) processes. Therefore survival and growth can be measured by the thermodynamic concept, exergy, which is *the amount of free energy relative to the environment*. Thus Darwin's theory may be reformulated in thermodynamic terms as follows:

> The prevailing conditions of an ecosystem are steadily changing and the system will continuously select the species that can contribute most to the maintenance of, or even growth of, the exergy of the system.

Ecosystems are open systems, and receive an inflow of solar energy. This carries low entropy, while the radiation from the ecosystem carries high entropy. If the power of the solar radiation is W and the average temperature of the ecosystem is T_0, then the exergy gain per unit of time, $\Delta\Sigma x$, is (Eriksen *et al.* 1976):

$$\Delta\Sigma x = T_0 \, [W \, (1/T_2 - 1/T_1)] \qquad (1)$$

where T_2 is the temperature of the environment (ecosystem) and T_1 is the temperature of the sun. This flow of exergy can be used to construct and maintain structures that are a long way from equilibrium.

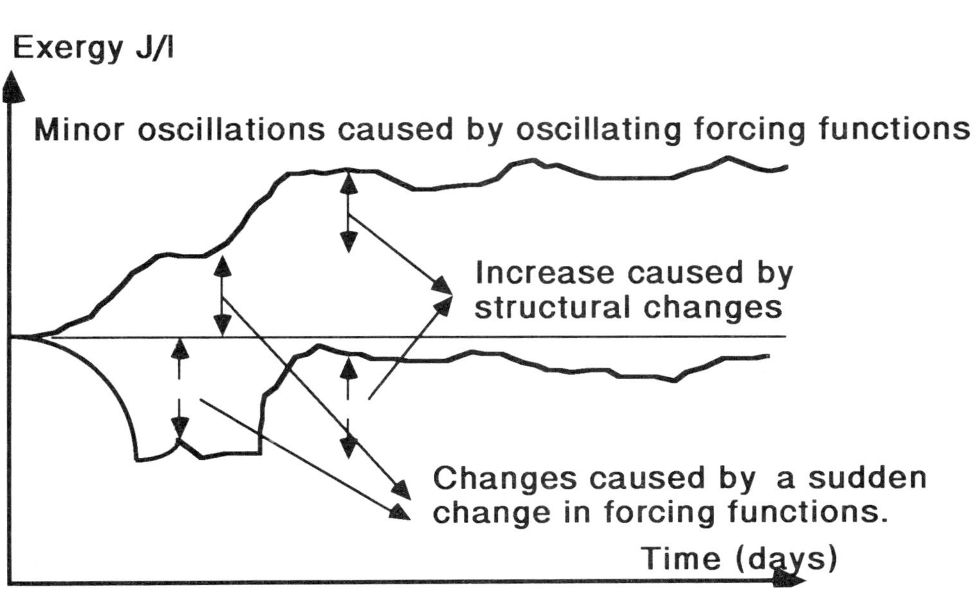

Figure 2. Exergy response to increased and decreased nutrient concentrations.

Thermodynamic translation of Darwin's theory also requires that populations must have the above-mentioned properties of reproduction, inheritance and variation. For selection to occur in favour of species that contribute most to the exergy of a system, under the prevailing conditions, there must be enough individuals with differing properties so that selection can take place. Reproduction and variation must be high so that once a change has occurred due to better fitness, it can be conveyed rapidly to the next generation, at least in the case of algal populations in a lake or reservoir. Furthermore, it should be noted that changes in exergy are not necessarily always positive or zero in value (Fig. 2), depending on changes in the resources of the ecosystem. Nevertheless, general theory supposes that ecosystems attempt to reach the highest possible exergy level under a given set of circumstances, with the available genetic pool ready for this attempt. This is illustrated schematically in Figure 2, which shows the response to changed forcing functions in a eutrophication model.

Calculating exergy

It is not possible to measure exergy but it can be computed, if the composition of the ecosystem is known. From thermodynamics, the following equation is valid for the components of an ecosystem (Mejer & Jørgensen 1979):

$$Ex = RT \sum_{i=1}^{i=n} [C_i \times ln\,(C_i/C_{eq.i}) - (C_i - C_{eq.i})] \qquad (2)$$

where R is the gas constant, T is the temperature (degrees Kelvin) of the environment, C_i represents the ith component expressed in a suitable unit (e.g. milligrams of phytoplankton per litre of lake water), $C_{eq.i}$ is the concentration of the ith component at thermodynamic equilibrium (which can be found in Morowitz (1968)), and n is the number of components. $C_{eq.i}$ is a very small concentration of organic components, corresponding to the very small probability of forming a complex organic compound in an inorganic soup (at thermodynamic equilibrium). Morowitz (1968) has calculated this probability and found that for proteins, carbohydrates and fats the corresponding concentration is about 10^{-60} mg l^{-1}. This figure may be used for mono-cellular species, but at higher trophic levels the probability of cells working together in a more complex organism is several magnitudes lower. Therefore it is relevant to let $C_{eq.i}$ be two magnitudes lower for zooplankton and four magnitudes lower for fish.

For more complex compounds such as hormones and genes, the value will be even smaller and, therefore, even if they are present in very small concentrations, they also contribute significantly to the exergy of the ecosystem.

The term $[C_i \times \ln\,(C_i/C_{eq.i})]$ corresponds to the theoretical minimum work (free energy) required to be done on the ecosystem in order to change its concentration relative to the reference level, and the term $(C_i/C_{eq.i})$ is the work done by the constant concentration's surroundings. Jørgensen *et al.* (1992) show that exergy is always non-negative, independent of whether the system is more or less concentrated than the reference state. In practice the system is more concentrated than the surroundings and the second term $(C_{eq.i})$ becomes negligible, except for the inorganic components, where the equilibrium concentration corresponds to the total concentration.

Equation (2) is valid for systems with an inorganic net inflow and passive outflow. It can be derived from basic thermodynamic equations; e.g. see Mejer & Jørgensen (1979). It is important to emphasize that all computations of exergy have the following shortcomings. They are based on a model and therefore will not be more correct than the model. Hence the results are more appropriate for examining relative differences in exergy by comparing an ecosystem under different conditions. Additionally, like all calculations in thermodynamics, the computations of exergy are based on approximations and assumptions.

The theory behind the application of exergy may be correct – but should of course be considered a hypothesis at this stage – but the practical application of the theory on real ecosystems will suffer from the above-mentioned shortcomings.

Nevertheless the exergy of an ecosystem can be illustrated by a simple example. Consider a pond containing only one species of algae. The phosphorus concentration of the algae is 1 mg l^{-1} or 0.6 mg l^{-1} orthophosphate. The temperature is 300K. If we are concerned only with the exergy contribution from the phosphorus compounds, Ex_p, then from equation (2) we find:

$$Ex_p = 8.31 * 300\,[(1/31 \ln 10^{-60}) - (1/31 - 10^{-60}/31) + (0.61/31 \times \ln 0.6/1.6) - (0.6/31 - 1.6/31)] = 9214 \text{ mJ } l^{-1} \quad (3)$$

Similarly, it is possible to calculate the exergy contribution for nitrogen and *all* other compounds, including the contribution originating from the more complex compounds such as hormones and genes. The theoretical definition of exergy implies that we must find the concentrations of all components, calculate the contribution of every component, and add all these contributions to get the total exergy. In practice, however, we can not find the concentrations of all the components, but if we are concerned only with the changes caused by phosphorus, for example, it is sufficient to make the computations shown above to see the *relative changes* in exergy.

Calculations of exergy are often carried out in relation to ecological models, in which case the results will give the exergy of the model ecosystem and not that of a real ecosystem. However, if the model is used in two different situations, i.e. two different sets of external factors are imposed on the model ecosystem, the computations of exergy will give some useful indications of the differences that are likely to occur in a real ecosystem, provided that the model incorporates the essential features of the ecosystem in relation to changes in external factors. Notice that exergy is defined as the free energy of the system relative to the environment and, if the latter is defined as the inorganic soup present on earth 4 billion years ago, the baseline for exergy is the low concentrations of biological components present in this original soup.

Jizhong *et al.* (1992) use exergy computations to produce exergy balances for plants and animals. They distinguish between exergy and anexergy, which is the non-exergetic part of the energy. Jizhong *et al.* compute the exergy efficiency as the ratio between the output exergy + exergy change of the system, and exergy input. They propose to use exergy efficiency as an objective function for optimizing an ecosystem, i.e. the organisms that have a higher efficiency in the use of exergy will be selected for.

Shieh & Fan (1982) have suggested that exergy contents in structurally complicated materials can be estimated as follows, in kcal per kg under standard conditions (1 atmos. pressure and 298.15K):

$$8177.79\,[C] + 5.25\,[N] + 27892.63\,[H] + 4364.33\,[S] - 3173.66\,[O] + 5763.41\,[F] + 2810\,[Cl] + 1204.3\,[Br] + 692.5\,[I] - (298.15\,S_{ash}\,W_{ash}) + 0.15\,[O] \times (7837.667\,[C] + 33888.889\,[H] + 3828.75\,[S] - 4236.10\,[O] + 4447.37\,[F] + 1790.9\,[Cl] + 681.97\,[Br] + 334.86\,[I] \quad (4)$$

In equation (4), [C], [H] etc. refers to the respective elements in terms of kg per kg of structurally complicated material, S_{ash} refers to the specific entropy of ash, assumed to have a value of 0.17152 kcal per kg ash °K, and W_{ash} refers to the weight of ash expressed as kg per kg structurally complicated material. If the composition of a particular organic material is known, equation (4) is straightforward to use. However, calculations of this kind do not take into consideration the exergy stored as information. The application of exergy calculations therefore requires that a model, containing all of the focal information for the specific problem, is used simultaneously. As already emphasized, the results are dependent on the accuracy of the model as well as the accuracy of the computations of exergy.

Structural dynamic models of ecosystems

The aim of the new generation of models presented here is to find a new set of parameters, limited for practical reasons to the most crucial (= sensitive) parameters, which are better fitted for the prevailing conditions of the ecosystem. "Fitted" is defined in the Darwinian sense, as the ability of the species to survive and grow, which may be measured by use of exergy (Mejer & Jørgensen 1979; Jørgensen 1981, 1986; Jørgensen & Mejer 1977, 1979). Figure 3 shows the proposed modelling procedure which has been applied in the cases presented below.

Exergy has previously been tested as a "goal function" for ecosystem development, for example by Jørgensen & Mejer (1979), Jørgensen (1986) and Herendeen (1989). However, in all these cases the model applied did not include the "elasticity" of the system, obtained by

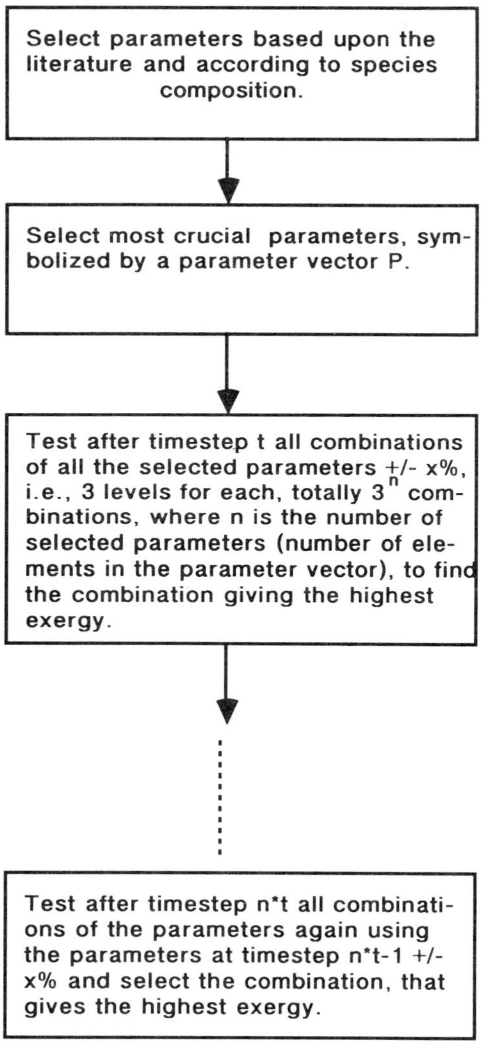

Figure 3. The procedure used to develop structural dynamic models.

using variable parameters, and therefore the models did not reflect the properties of real ecosystems. A realistic test of the exergy principle requires the application of *variable* parameters, used only in the few cases referred to here.

Case 1: algal growth as a variable parameter

In this case the growth of algae was used as the only variable parameter (Jørgensen 1986). This gave a significantly improved validation of the model, which encouraged further investigation of the possibilities for developing and applying such new modelling approaches. Values for the maximum growth rate (μmax), and the respiration rate (F, set equal to 0.15 μmax), were changed in the model, relative to values for growth rates previously found by experimental calibration (μc). Thus:

$$\mu max = F \times \mu c \qquad (5)$$

The model was run for a series of F values and three levels of phosphorus (and nitrogen) input. As shown in Figure 4, maximum algal growth, giving maximum exergy, decreases when the phosphorus concentration increases, in accord with general ecological observations. When nutrients are scarce, competition between phytoplankton species is dependent on their rates of nutrient uptake. Smaller species have faster uptake rates due to a larger specific surface area and they grow more rapidly. On the other hand, high nutrient concentrations will not favour small species because interspecific competition is based on avoidance of grazing by zooplankton, which favours the larger algae. The results shown in Figure 4 were used to improve the prognosis published in Jørgensen *et al.* (1981, 1986), by introducing selective alteration of the parameters according to the procedure summarized in Figure 3. The standard deviation between model and measurement was significantly reduced, compared to the original prognosis where fixed parameters were applied.

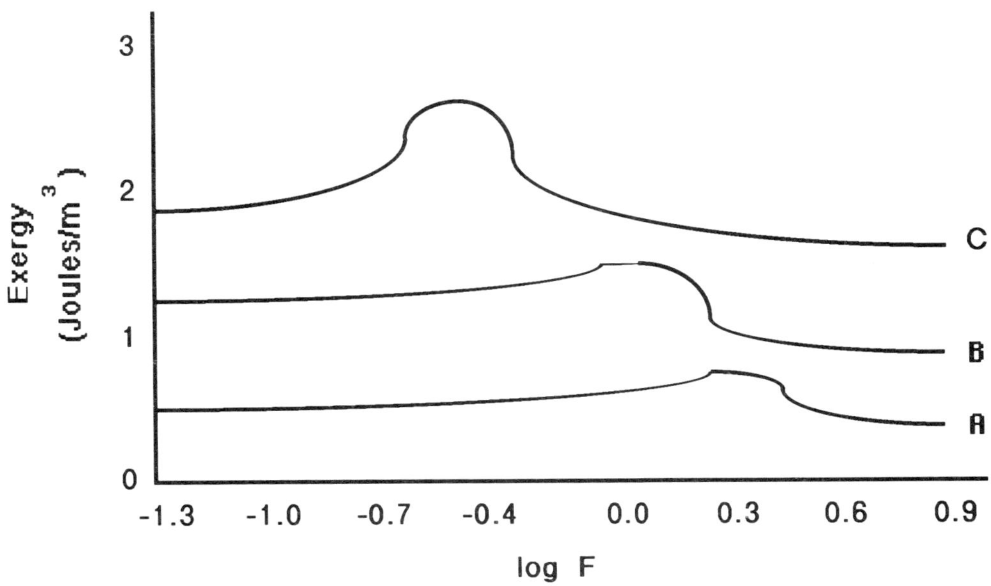

Figure 4. Exergy (joules m^{-3}) plotted against F at different levels (A-C) of phosphorus in a lake, where F is algal respiration rate relative to the maximum growth rate (equation (5), see the text: Case 1). Maximum exergy occurs at different F values for different P-loadings. A represents an oligotrophic lake corresponding to nutrient inputs of 0.04 mg P l^{-1} and 0.3 mg N l^{-1}, B corresponds to nutrient inputs of 1 mg P l^{-1} and 8 mg N l^{-1}, and C corresponds to a hypereutrophic lake where the inputs are 2 mg P l^{-1} and 16 mg N l^{-1}.

Case 2: exergy of half-saturation constants for uptake of silica and phosphorus

This second case focuses on differences in the half-saturation constants for phosphorus and silica, which may play an important role in the selection of different diatoms (see Tilman & Kilham 1976). If the P/Si ratio is low, the diatoms with a low half-saturation constant for phosphorus are selected, while a high Si/P ratio is selective for species with a low half-saturation constant for silica. If the P/Si ratio becomes very low, diatoms may be replaced by green algae.

An examination with a simple single-species model (see Fig. 5) explores the relationship between the Si/P ratio and the half-saturation constants for silica and phosphorus uptake, corresponding to the highest internal exergy value for the system. The results are shown in Figure 6, where the exergy at steady state is plotted against Si/P ratios for two species of algae with different half-saturation constants. The results are completely in accord with chemostat results obtained by Tilman & Kilham (1976), including the shift between the two species at an Si/P ratio of 2 (Fig. 6). These results therefore provide strong support for the maximum exergy principle and for a wider use of this principle in structural dynamic modelling.

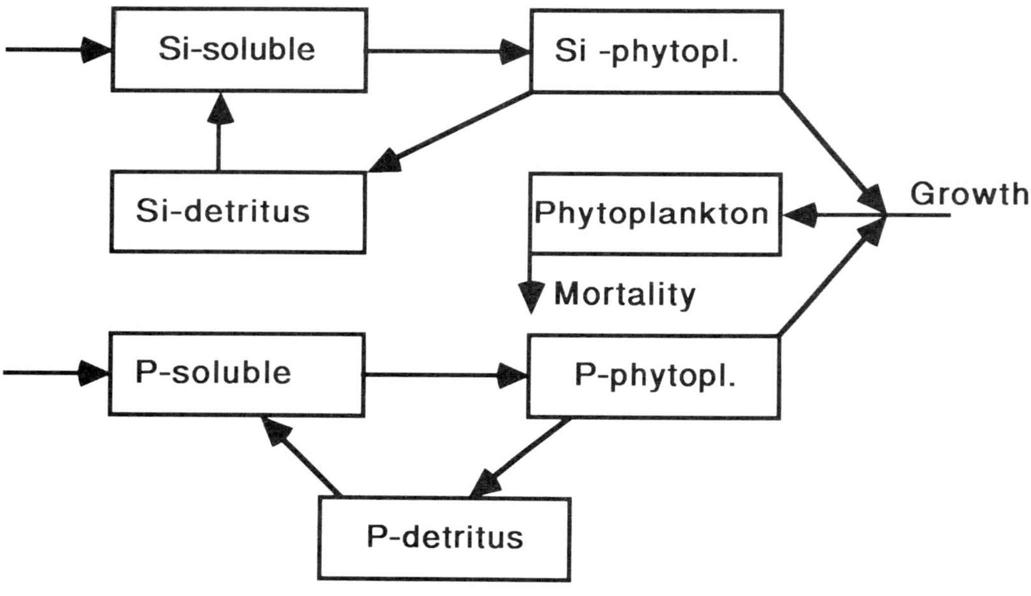

Figure 5. A single-species model (Case 2, see text) applied to show that the results obtained by Tilman & Kilman (1976) are consistent with the maximum exergy principle or the Ecological Law of Thermodynamics. The model has seven state variables. The cycles of silica and phosphorus are independent of each other, while their concentrations in the algae (phytoplankton) determine algal growth.

Case 3: exergy of a model for diatoms and blue-green algae

Nitrogen-fixing species of algae become increasingly predominant as the ratios of nitrogen to phosphorus progressively decrease and phosphorus concentrations increase. A two-species model has been used to simulate these observations. The results are shown in Figure 7, where the internal exergy for diatoms and blue-green algae is plotted against the logarithm of the concentration ratios of nitrogen to phosphorus in the ecosystem, whilst silica is held constant at a relatively high concentration. Exergy was calculated for varying nitrogen and fixed

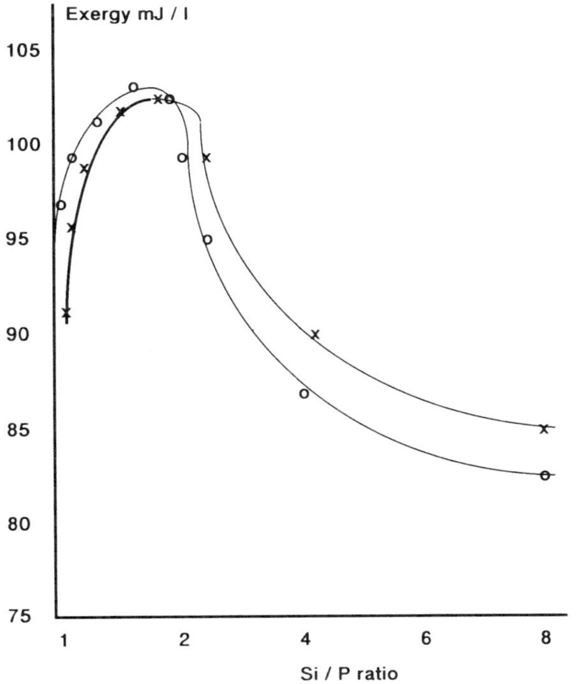

Figure 6. Exergy (mJ l^{-1}) plotted against the ration of silica: phosphorus for two species of algae. X, algae with half-saturation constants of 0.003 mg P l^{-1} and 0.5 mg Si l^{-1}. O, algae with half-saturation constants of 0.1 mg P l^{-1} and 0.1 mg Si l^{-1}.

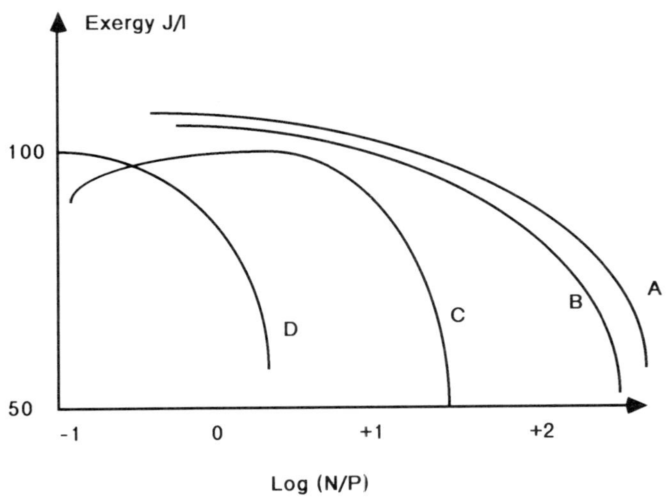

Figure 7. Internal exergy (joules l^{-1}) for diatoms and blue-green algae plotted against the logarithm of concentration ratios for nitrogen:phosphorus, when silica is kept constant at a relatively high concentration. A represents diatoms and B represents blue-green algae at a concentration of 0.02 mg P l^{-1}. C represents diatoms and D represents blue-green algae at a concentration of 1.0 mg P l^{-1}.

phosphorus concentrations of 1.0 mg P l^{-1} (curves A and B, Fig. 7) and 0.02 mg P l^{-1} (curves C and D, Fig. 7). Blue-green algae give the highest exergy at low N/P ratios when P = 1.0 mg l^{-1}, while the diatoms give the highest exergy at all other situations. This accords with the appearance of blue-green nitrogen-fixing algae in eutrophic lakes with high phosphorus concentrations.

Case 4: exergy of a model for diatoms and green algae

Diatoms are dominant in many tropical lakes at the high rates of precipitation which usually occur during the summer. This is surprising because diatoms usually have a high settling rate and, as the retention time under these conditions is often low, diatoms should be relatively poor competitors. However, the very heavy rains stir up the photosynthetic layers, thereby reducing the tendency of diatoms to settle out and providing an opportunity for them to move, from time to time, into layers that offer the most favourable light conditions.

Table 1 shows the results of a two-species model for the exergy of diatoms and green algae. Under given circumstances of temperature, retention time and ratios of silica to phosphorus, the dominance of diatoms gives the highest internal exergy for a tropical ecosystem (reservoir) in the summer rainy season, while a distribution between the two classes of algae gives the highest exergy in the winter dry period. These theoretical results agree with an actual case reported by T. Tundisi (*pers. comm.*, 1988)

Table 1. *The ratios of diatoms to green algae at maximum exergy, calculated for summer and winter conditions in a tropical reservoir.*
(Q = flow, V = volume; Q/V = relative retention time).

Season	Simulation characteristics	Ratio
Summer	Very low settling rate Q/V = 0.1 No light limitations	∞
Winter	Normal settling rate Q/V = 0.02 Light limitations	6:4

Case 5: exergy of a model for nitrogen and phosphorus turnover

Another examination of structural changes in an ecosystem is based on a model similar to the one shown schematically in Figure 5, except that nitrogen and phosphorus were considered instead of silica and phosphorus. All rates were given average values (growth, mortality and mineralization) for the nitrogen and phosphorus cycles. All rates were multiplied by the same factor in each of the two cycles and the inputs of nitrogen and phosphorus were varied. The factors giving the highest exergy were found, and the ratios between the corresponding nitrogen and phosphorus turnover rates were calculated. The results are summarized in Figure 8, where the turnover ratios are plotted against the N/P concentration ratios, and they are completely in accord with Vollenweider (1975). The latter considered it a paradox that the higher the ratio of N/P, the higher was the turnover rate of nitrogen to phosphorus, as it may be argued that relatively high nitrogen concentrations make it unneceesary to have a rapid turnover of nitrogen.

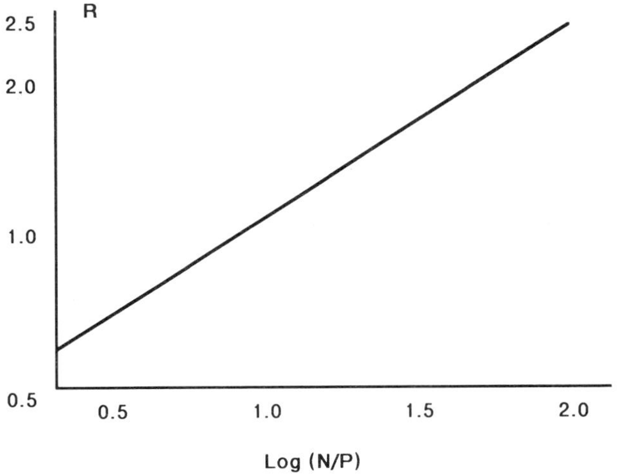

Figure 8. The ratio of nitrogen:phosphorus turnover (R) at maximum exergy, plotted against the logarithm of concentration ratios for nitrogen:phosphorus. Note that both scales are logarithmic, making the relationship almost linear.

Conclusions

The five cases referred to above all show that the principle of maximum exergy agrees with the observed selection and seasonal succession of algal species in natural ecosystems. Therefore it would be an obvious step to use the principle to predict the short-term natural selection and sequence of events that will occur under given circumstances, i.e. to develop structural dynamic models, as was attempted in the first case above. The results obtained so far are promising for a wider application, but far more experience is needed before a general application of this approach to structural dynamic modelling can be recommended.

References

Brooks, D. R., Collier, J., Maurer, B. A., Smith, J. D. H. & Wiley, E. O. (1989). Entropy and information in evolving biological systems. *Biology and Philosophy*, **4**, 407-432.

Brooks, D. R., Cumming, D. D. & LeBlond, P. H. (1988). Dollo's law and the second law of thermodynamics: analogy or extension? *In Entropy, Information and Evolution: New Perspectives on Physical and Biological Evolution* (eds B. H. Weber, D. J. Depew & J. D. Smith), pp. 189-224. MIT Press, Cambridge, Massachusetts.

Eriksson, B., Eriksson, K. E. & Wall, G. (1976). *Basic Thermodynamics of Energy Conversions and Energy Use.* Institute of Theoretical Physics, Göteborg, Sweden.

Herendeen, R. (1989). Energy intensity, residence time, exergy, and ascendency in dynamic ecosystems. *Ecological Modelling*, **48**, 19-44.

Jizhong, Z., Shijun, M. & Hinman, G. W. (1992). Ecological exergy analysis: a new method for ecological energetics research. *Ecological Modelling*, in press.

Jørgensen, S. E. (1982). A holistic approach to ecological modelling by application of thermodynamics. In *Systems and Energy* (eds W. Mitsch *et al.*), pp. 61-72. Ann Arbor.

Jørgensen, S. E. (1986). Structural dynamic model. *Ecological Modelling*, **31**, 1-9.

Jørgensen, S. E. (1988). *Fundamentals of Ecological Modelling*. Elsevier, Amsterdam.

Jørgensen, S. E. & Meyer, H. F. (1977). Ecological buffer capacity. *Ecological Modelling*, **3**, 39-61.

Jørgensen, S. E. & Meyer, H. F. (1979). A holistic approach to ecological modelling. *Ecological Modelling*, **7**, 169-189.

Jørgensen, S. E., Mejer, H. F. & Friis, M. (1978). Examination of a lake model. *Ecological Modelling*, **4**, 253-279.

Jørgensen, S. E., Nors-Nielsen, S. & Jørgensen, L. A. (1991). *Handbook of Ecological Parameters and Ecotoxicology*. Elsevier, Amsterdam.

Jørgensen, S. E., Patten, B. & Straskraba, M. (1992). Ecosystem emerging. *Ecological Modelling*, **60**, in press.

Jørgensen, S. E., Jørgensen, L. A., Kamp-Nielsen, L. & Mejer, H. F. (1981). Parameter estimation in eutrophication modelling. *Ecological Modelling*, **13**, 111-129.

Jørgensen, S. E., Kamp-Nielsen, L., Christensen, T., Windolf-Nielsen, J. & Westergaard, B. (1986). Validation of a prognosis based upon a eutrophication model. *Ecological Modelling*, **32**, 165-182.

May, R M. (Editor) (1981). *Theoretical Ecology: Principles and Applications*. 2nd Edition. Blackwell Scientific Publications, Oxford.

Mejer, H. F. & Jørgensen, S. E. (1979). Energy and ecological buffer capacity. In *State-of-the-Art of Ecological Modelling. Environmental Sciences and Applications, 7th Proc. Conf. Ecological Modelling, 28th August – 2nd September 1978* (ed. S. E. Jørgensen), pp. 829-846. International Society for Ecological Modelling, Copenhagen.

Monod, J. (1972). *Chance and Necessity*. Random House, New York.

Morowitz, H. J. (1968). *Energy Flow in Biology*. Academic Press, New York.

Nors-Nielsen, S. (1992a). Modelling the structural dynamical changes in a Danish shallow lake. *Ecological Modelling* (in press).

Nors-Nielsen, S. (1992b). Optimization of exergy in a structural dynamical model. *Ecological Modelling*, (in press).

Odum, H. T. (1983). *Systems Ecology*. Wiley Interscience, New York.

Patten, B. C. (1991). Network ecology: indirect determination of the life-environment relationship in ecosystems. In *Theoretical Studies of Ecosystems: The Network Perspective* (eds M. Higashi & T. P. Burns), pp. 288-351. Cambridge University Press.

Prigogine, I. & Stengers, I. (1979). *La Nouvelle Alliance*. Gallimard, Paris.

Shieh, J. H. & Fan, L. T. (1982). Estimation of energy (enthalpy) and energy (availability) contents in structurally complicated materials. *Energy Resources*, **6**, 1-46.

Stenseth, N. C. (1986). Darwinian evolution in ecosystems: a survey of some ideas and difficulties together with some possible solutions. In *Complexity, Language, and Life: Mathematical Approaches* (eds J. L. Casti & A. Karlqvist), pp. 105-129.

Straskraba, M. (1980). The effects of physical variables on freshwater production: analyses based on models. In *The Functioning of Freshwater Ecoystems* (eds E. D. Le Cren & R. H. McConnell), pp. 13-31.

Tilman, D. & Kilham, S. S. (1976). Phosphate and silicate growth and uptake kinetics of the diatoms *Asterionella formosa* and *Cyclotella meninghiniana* in batch and semi-continuous culture. *Journal of Phycology*, **12**, 375-383.

Ulanowicz, R. E. (1980). An hypothesis on the development of natural communities. *Ecological Modelling*, **85**, 223-245.

Vollenweider, R. A. (1975). Input-output models with special reference to the phosphorus loading concepts in Limnology. *Schweizerische Zeitschrift für Hydrologie*, **37**, 53-84.

Wicken, J. S. (1988). Thermodynamics, evolution, and emergence: ingredients for a new synthesis. In *Entropy, Information and Evolution: New Perspectives on Physical and Biological Evolution* (eds B. H. Weber, D. J. Depew & J. D. Smith), pp. 139-172. MIT Press, Cambridge, Massachusetts.

Wiley, E. O. (1988). Entropy and evolution. In *Entropy, Information and Evolution: New Perspectives on Physical and Biological Evolution* (eds B. H. Weber, D. J. Depew & J. D. Smith), pp. 173-188. MIT Press, Cambridge, Massachusetts.

The scope for biomanipulation for improving water quality

B. Moss

Department of Environmental and Evolutionary Biology, University of Liverpool, Derby Building, P.O. Box 147, Liverpool L69 3BX, England

Biomanipulation is a form of biological engineering in which organisms are selectively removed or encouraged to alleviate the symptoms of eutrophication. Most examples involve fish and grazer zooplankton though mussels have also been used. The technique involves continuous management in many deeper lakes and is not a substitute for nutrient control. In some lakes, alterations to the lake environment have given longer-term positive effects. And in some shallow lakes, biomanipulation may be essential, alongside nutrient control, in re-establishing former aquatic-plant-dominated ecosystems which have been lost through severe eutrophication.

The emergence of biomanipulation techniques emphasises that lake systems are not simply chemical reactors which respond simply to engineered chemical changes, but very complex and still very imperfectly understood ecosystems which require a yet profounder understanding before they can be restored with certainty.

Introduction

In the summer of 1990, my postgraduate student, Laurence Carvalho and I were puzzled by events in a small and shallow lake, Little Mere in Cheshire. The water was extremely clear and there was a large development of aquatic plants, yet the algal growth potential of the water was huge. Effluent from a small sewage treatment works entered the lake, which had concentrations of available inorganic phosphate-phosphorus and ammonium-nitrogen each of the order of several milligrams per litre. The only algae of any consequence in the water were a few large colonies of *Volvox*. The resolution of this paradox came when we noticed large populations of a large-bodied grazer, *Daphnia magna*, in the zooplankton. The animals were unusually bright red with haemoglobin and the water, even in the middle of the day, had very low oxygen saturation values and concentrations less than $1 mgl^{-1}$. Subsequently we established that there were almost no fish in the lake despite normal coarse fish communities in Mere Mere, a larger lake which lies immediately upstream of Little Mere, and the stocking of Little Mere with fish by residents whose houses border the lake. The quality of the sewage effluent was such as to deoxygenate the water sufficiently to kill fish and hence remove major predation pressure on the *Daphnia*. Grazing was then sufficient to prevent the development of the large algal crops the water was otherwise capable of supporting. It was a further challenge to the idea that algal crop size is determined largely by "bottom–up" (in the foodchain) mechanisms and predictable simply from nutrient supply. This is undoubtedly true where prediction of the potential crop is concerned (Cranfield & Bachmann 1981; Dillon & Rigler 1974) but subject to modification by other biological processes when it is the actual crop that is of interest (Shapiro *et al.* 1975).

The concept of biomanipulation

There is now a variety of examples demonstrating the importance of grazing by the larger-bodied species of water-fleas (Cladocera) in determining algal crop sizes in fresh waters (e.g.

Pace 1984) and by mussels and other bivalve molluscs in saline lagoons (Officer *et al.* 1982). This has led to the concept of biomanipulation (Shapiro *et al.* 1975; Shapiro 1990; Gophen 1990) which is definable as the manipulation of the food webs of aquatic ecosystems in such a way as to increase the numbers of grazers on algae. At present this largely concerns the phytoplankton but conceivably could include the periphyton which covers aquatic plants and other surfaces also. Biomanipulation might be used in concert with nutrient control as an additional measure or possibly as a sole technique in the improvement of water quality from the point of view of its algal content. Techniques used so far have involved the complete removal of fish and maintenance of near zero fish stocks (van Donk *et al.* 1989); removal and then restocking with a fish community which gives greater precedence to the zooplanktivorous fishes' own predators (Shapiro & Wright 1984); direct stocking of grazers (examples include only molluscs) (Reeders & Bij de Vaate 1990); and major inadvertent restructuring of the ecosystem through changes at the tops of the food webs. The latter favour grazing without the need for continuous management by readjustment of fish communities, which inevitably otherwise drift back towards some sort of natural equilibrium which may be greatly different from that desired by the lake manager to minimise algal growth. A particularly good example is that of Lake Washington, which is described below.

Five key questions

The key questions for water supply management are as follows: (i), Does biomanipulation really reduce algal crops significantly in the sorts of lakes and reservoirs used for water supply? (ii), Can biomanipulation be used as a technique on its own or only as an adjunct to nutrient control? (iii), Of the available approaches, which are likely to be most valuable and what are their limitations in terms of practicality? (iv), Are there circumstances where biomanipulation may be essential in reducing algal crops? (v), Notwithstanding the possibilities of its use in water-supply management, are there other roles for biomanipulation?

Question (i): Does biomanipulation work?

The case of Lake Washington in the USA, itself the classic instance of algal reduction by nutrient control, is also an excellent example of the positive benefits of biomanipulation in a very large and deep lake, and an answer to question (i). Lake Washington (Edmondson 1991) became eutrophic by the 1950s as a result of sewage effluent discharge. It had substantial growths of *Oscillatoria rubescens*, one of the blue-green algae, and its transparency was much reduced from that in previous decades. A programme of diversion of sewage effluent to the nearby sea was begun and during the 1960s the clarity of the lake steadily improved until, with complete diversion, it was believed that the algal crops could be reduced no further. However, the clarity of the water continued to increase and this was correlated with an increase in the populations of several *Daphnia* species in the zooplankton (Edmondson & Litt 1982). No measures to reduce fish stocks had been taken and fish were plentiful in the lake. Numbers of an invertebrate predator on the *Daphnia*, a mysid shrimp (*Neomysis mercedis*) had, however, concurrently declined (Murtaugh 1981). Why this should have happened is less clear but investigations showed that the main predator on the mysid was a fish, the long-fin smelt (*Sprinichus thaleichthys*); this had built up its population following inadvertent improvement to its spawning habitat by engineering works in one of the main inflow rivers to the lake.

A change three steps remote from the algae had thus resulted in reduction of their numbers very significantly. This was in a lake that had been much improved previously by nutrient control. Are such dramatic effects possible in lakes without such control?

Question (ii): can biomanipulation be used as a substitute for nutrient control?

An example is that of Brundall Broad in Norfolk (Leah, Moss & Forrest 1978) which, though much shallower than most British reservoirs, is not dissimilar from those used in the Netherlands. Brundall Broad was divided into two by dams and one part of it was isolated from inflow from the nutrient-rich River Yare, which carries effluent from the Norwich sewage treatment works immediately upstream. The other part of the lake was left open to the river. In the first year of investigation both parts developed large diatom crops in the phytoplankton and it was found that leaks had developed in the dams which, on the high tide, allowed river water to enter the supposedly isolated section. It was not possible to repair the leaks so we were surprised when, in the subsequent year, the water in the "isolated" part cleared and lost most of its algae, whilst the control section, open to the river, remained turbid with algae. A large population of *Daphnia longispina* had developed in the "isolated" part and fish were nearly absent. There were large stocks of zooplanktivorous fish in the control section. But why had the "isolated" part lost its fish? A fish-eating cormorant (*Phalacrocorax carbo*) with a damaged wing had been permanently confined to the "isolated" broad during the previous winter and had attracted other cormorants. They fished out the "isolated" broad and fish had been unable to move in from the river to replace the stock. It was a good example of inadvertent biomanipulation and improvement of water quality in a shallow lake in the absence of any nutrient control.

Of course, in a subsequent year when the resident bird died the system would naturally revert to its former state, as indeed it did, but the example does show that biomanipulation can be effective without nutrient control (question ii). Two factors probably were crucial to its success in this instance. First, there was a high replacement rate of the water in the broad because of the twice-daily tidal-flushing, and secondly this maintained a diatom-dominated flora in the phytoplankton. In the particular riverine lake system of the Norfolk Broadland there is a strong relationship between diatom populations and conversely high flushing, and blue-green algal populations and reduced flushing, given high nutrient concentrations (Moss & Balls 1989). Diatoms are eminently edible by daphnids, blue-green algae are not. Experiments in netting enclosures in Broadland lakes have demonstrated that *Daphnia* populations cannot be sustained when filamentous blue-green algae come to dominate the phytoplankton, but do not preclude the daphnids if the algal community is otherwise dominated by diatoms (Moss *et al.* 1991; Irvine *et al.* 1992).

Question (iii): what are the most useful approaches to biomanipulation and what are their limitations?

Of the available approaches to deliberate biomanipulations, some are more usable than others. It is not feasible to grow a culture of suitable *Daphnia* and release it into a lake. The animals will very rapidly be eaten by any normal fish stock, and the levels of concentrations needed to give immediate high grazing impact in the lake (some tens to hundreds of animals per litre) are not much below those in which the animals could be cultured and viably transported. For even a small reservoir of (say) 250 000m^3, enormous numbers of tankerloads of culture would be needed. *Daphnia* is not the new copper sulphate!

Mussels have been deliberately stocked and the zebra mussel (*Dreissena polymorpha*) has been studied in freshwater lakes in the Netherlands (Reeders *et al.* 1989). Large numbers are needed and they must have a hard substratum or be placed in racks which are costly on a large scale and clutter the water-body. Marine mussels (*Mytilus edulis*) have the happy habit of colonising suspended vertical ropes and even natural settlements on the dock walls have successfully cleared enclosed docks in the Mersey Estuary (Allen & Hawkins in press). Unfortunately they die in fresh water.

Daphnia is a ubiquitous genus in temperate fresh waters and if fish predation is decreased, the *Daphnia* populations rapidly increase to effective sizes through their own very fecund parthenogenic reproduction. Doubling times are only a few days at summer temperatures. Ways of decreasing fish predation include total removal, partial removal or stocking of fish predators and, potentially, modification of the fishes' behaviour.

Of course not all fish species eat zooplankters. But of the British fish fauna a great many species at some time in their growth and development do depend on the zooplankton. Predominant among these are the cyprinid fish like roach (*Rutilus rutilus*) and bream (*Abramis brama*), and percids such as *Perca fluviatilis*, which are associated with fertile lowland waters where algae are likely to pose problems for water suppliers. Their young feed on the smallest zooplankters, the rotifers, soon after hatching but rapidly move to the cladocerans and other Crustacea when they are a few centimetres long. After their first year, or second year if growth is slow and they remain small, the fish need to take larger food particles, generally derived from the invertebrates of the bottom or plant beds, or, in some species, other fish. However, because almost all British fish have a reproductive strategy which involves production of large numbers of young, each with a low probability of survival in the long term, a huge number of small zooplanktivorous fish is present in the water from late spring and throughout the summer. Where growth is slow this cohort may be joined by successive ones to give continuous heavy impact on the zooplankton. It is not easy to remove these fish efficiently and selectively by known fishing techniques. Small fish are difficult to net out and respond least well to electrofishing. The complete fish stock can be removed from a small lake but only by very intensive fishing over a long period using staff with an intimate knowledge of the water-body and the habits and movements of the fish within it. This is very expensive in labour. It is very doubtful if all the fish can be taken from a large water-body using manual methods. It might be possible to remove many but if only a few adults are left these will ensure a heavy recruitment the following year, when the reduced stock will present reduced competition for, and greater early survival of, the young-of-the-year. A compromise is the use of nets to remove as many fish as possible and then to use a fish poison like rotenone to kill the remainder. Although this is an acceptable technique in some countries it is unlikely to be welcomed in Great Britain on a large scale.

Even the most intensive fish-removal operation is not complete. Some fish will always escape and will very quickly build up the population again. Fishing has to be repeated at least annually to maintain the stocks low enough to have a major impact on cladoceran survival. Where it can be done the effects are dramatic, as at Cockshoot Broad in Norfolk, where this technique has been used to create clear water for aquatic plant recolonisation (B. Moss, unpublished data).

The difficulties posed by fishing-out large bodies of water have been avoided in some lakes by the addition of fish predators (Benndorf *et al.* 1984, 1988; Scavia *et al.* 1986) or alteration of the ratio of existing predators and their prey. The latter requires continuous management as the fish community will drift back to whatever ratio represents a natural equilibrium, and the former poses problems of what fish to use and what undesirable side effects might be created. The literature is replete with the disasters caused by the introductions of exotic species, some of them fish (Zaret & Paine 1973; Barel *et al.* 1985).

In the USA, the fish fauna is rich enough to allow a degree of experimentation using native species, for example the large-mouthed bass (*Micropterus salmoides*), walleye (*Stizostedion vitreum*), and the several Pacific salmon species. However, it may be just as undesirable to transfer a species from one side of a continent to another as from continent to continent. Good results were obtained by Shapiro & Wright (1984) by adjusting the predator–prey ratio in a

Minnesotan lake. Stocking with predators seems also to have had impacts on the alewife (*Alosa pseudoharengus*, a zooplanktivorous fish even when adult) population of some of the St Lawrence Great Lakes, with consequent increases in zooplankton and reductions in algae (Scavia *et al.* 1986).

In Britain there are problems, however. The British fish fauna is singularly lacking in predators that thrive in eutrophic conditions and those that do, like pike (*Esox lucius*), do not hunt among the open-water shoals of young coarse fish. Pike are marginal-lurking predators, and stocking with additional pike is likely to lead only to pike mortality. Brown trout (*Salmo trutta*) are unlikely to survive under the sorts of conditions that are created by large algal growths and there might be dangers in extending the range of the exotic rainbow trout (*Salmo gairdneri* = *Oncorhynchus mykis*) any further. It is already regarded as an undesirable species among native communities. The perchpike or zander (*Stizostedion luciopercum*) is not now a native British fish though it may have been, prior to the glaciation, and has already been introduced to one part of eastern England. It is an open-water feeder on small fish and might merit further attention. It is unlikely, however, that a more extensive introduction of it would be countenanced by English Nature and other informed conservation opinion. The options for predator introduction are thus not great, and for natural lakes used for drinking-water supply there would be considerable opposition to complete fish removal or even selective removal, for almost all are of conservation importance.

For artificial drinking-water reservoirs, however, which are already severely managed water-bodies, biomanipulation could offer a reasonably cheap way of reducing algal growths. Fish removal would have to be near complete and continually attended to and the reservoir conditions would probably need to favour development of algae other than blue-green algae. There is much controversy about whether or not *Daphnia* can eat particular species of blue-green algae (de Bernardi & Giussani 1990; Gliwicz 1990). In general these algae seem to be poor food unable to sustain *Daphnia* (Arnold 1971) or not ingestible (Gliwicz 1980) or perhaps chemically inhibitory (Infante & Abella 1985). The genus *Oscillatoria* seems to be particularly unavailable (Moss *et al.* 1991) though even this is disputed (Dawidowicz 1990). Smaller blue-green algae may be eaten or at least they are not inhibitory if alternative edible algae are available. It is possible also that in a reservoir which develops blue-green algae under present conditions the development of *Daphnia* early in the season, under biomanipulated conditions, could influence the seasonal periodicity of the algae in favour of other groups.

The reason why fish removal needs to be near complete is that at present no absolute stocking density can be given which corresponds with a particular desired reduction in phytoplankton crop (Benndorf 1990). Doubtless, with the continued accumulation of data, equations will be developed for guidance. However, coarse fish recruitment is often so vigorous that the population will rapidly be replaced even if only a very small stock is left. Any improvement in the year following a partial removal will thus probably be confined to that year or not be seen at all (van Donk *et al.* 1990).

A yet underdeveloped aspect of biomanipulation is that of altering the fish behaviour within a lake without removing the fish. Thus if zooplanktivorous fish can be confined to the edges of the lake, and kept out of the middle, predation on the zooplankton might be greatly reduced. Alarmed fish produce substances that scare away others. A synthetic source of such a substance held in the middle of a lake might be effective (M. Gliwicz, *pers. comm.*) but how effective it might be in ultimately reducing algal crops is not known.

Problems are always best solved by attending to their causes rather than by alleviating their symptoms, and the proper solution to that of large algal growths is nutrient control. Biomanipulation is a symptom treatment in most cases which might be used when the possibilities for reduction of nutrient loading from the catchment have been exhausted. It involves continuous management and this is not an alien concept in the management of artificial reservoirs. Where natural lakes are being used for water supply it is just as undesirable as the managed water levels which leave an unsightly and sterile zone at the lake edge. There are however instances where biomanipulation is not only useful but essential in restoring water quality (question iv) and environmental quality (question v) to a lake. These cases are of shallow lakes, formerly dominated by the aquatic plants of their extensive littoral zones, but in which eutrophication has led to replacement of the plants by dense phytoplankton crops. Such lakes may be used for water supply and are often of considerable fisheries, conservation and amenity importance (de Nie 1988).

Question (iv) & (v):– where biomanipulation may be essential for restoration of water quality or environmental quality

The process by which aquatic plants are lost during eutrophication is a complex one involving much more than the simple shading-out by the algae that is usually invoked as a mechanism. The change does not take place linearly as nutrient loadings increase. In the early stages there is a replacement of short-growing plant species by taller ranker-growing ones (Moss 1988, 1989) but these latter are then able to withstand considerable further increases in nutrient loading without being replaced by phytoplankton (Balls *et al.* 1989). The plant-dominated system appears to be buffered against such change by a number of mechanisms (Irvine *et al.* 1989; Moss 1990). These include the secretion of substances which inhibit algal growth, and the luxury uptake of nutrients. The most effective mechanism, however, is the harbouring of large communities of grazer Cladocera within the plant beds. These animals drift out at night and graze in the adjacent open water. In daylight they are taken by fish but their numbers can easily be replaced from the reserves held in the plant beds. With such a system of refuges, a stable coexistence of zooplankton, fish and plants can be maintained with clear water and very low phytoplankton crops (Timms & Moss 1984).

Plants probably disappear from such systems when something happens to disturb these buffer mechanisms. This might be the removal of the plants themselves by manual clearing, drawdown of the water level, herbicides, or grazing by an introduced herbivore like pinioned geese, muskrat or coypu (Moss 1991). Alternatively it might be destruction of the cladoceran grazer community by pesticide run-off (Stansfield *et al.* 1989). The disturbance need only be short-term, for at increased nutrient loadings it allows the system to switch to phytoplankton dominance, when a new set of buffer mechanisms comes into play which prevents a switch back to the former state. Thus the phytoplankters begin their growth earlier and can shade-out the young plant propagules at the bottom of the water column early in the summer; the open-water environment lacks refuges for the cladocerans, and fish can rapidly remove the larger species and individuals that are the most effective potential grazers. There thus exist two alternative states over a band of nutrient loadings, and nutrient control alone is unlikely to shift the system from one state to another (Irvine *et al.* 1989; Scheffer 1990). Biomanipulation, on the other hand, may be necessary to do so.

In the Norfolk Broadland (Moss 1983), many lakes have lost their plants and, together with them, their roles as habitat for invertebrates like dragonflies, as food for birds, and as stabilizers of sediment and protectors against bank erosion by tides, waves and boats, as well as their intrinsic and conservation importance. It is thus highly desirable to restore the plant communities. Nutrient control, by phosphorus-stripping at the sewage treatment works, has proved ineffective alone. It has led to a significant reduction in the total algal crop (Phillips & Chilvers 1991) but not to a clearing of the water sufficient to allow aquatic plants to grow. This has been because of the operation of the buffers mentioned above and to the still high availability of nutrients from agriculture and release from sediments within the lake.

In one instance, Cockshoot Broad, isolation of the lake from the river did give a marked improvement at first (Moss *et al.* 1986) and plants recolonised a sheltered section of the lake known as Cockshoot Dyke. There they have formed a permanent community. In the more exposed open Broad, plants colonised more slowly and then declined again as phytoplankton crops built up from 1985 onwards (B. Moss, unpublished data). A progressive decline in the size of both the *Daphnia* population and in the body-sizes of its constituent animals suggested that the early restoration had involved both nutrient control and biomanipulation. The fish stock in the lake had not been high previously, for the lake was then very shallow. Pumping out of sediment had probably scared what fish there were back to the river and damming had prevented the return of most of them. In the early years there was thus only a small fish stock, *Daphnia* populations were very high and the water cleared. In the dyke, plants were quick to colonise and the stabilising buffer mechanisms were early re-established. Wave disturbance and bird grazing slowed the colonisation of the main Broad so that when, by 1984-85, fish stocks had built up again from the small inoculum left in the Broad and *Daphnia* populations were decimated, there was not enough plant structure available to allow establishment of the refuge mechanism and the plants succumbed to the increasing algal competition. Since 1988 the fish have been removed each winter, the water has cleared again and the plants are slowly returning. Biomanipulation has thus proved essential in this system.

Despite the success of biomanipulation in Cockshoot Broad and the eventual intention to allow the fish community to re-establish once a sufficiency of plant refugia have re-established, there is still local opposition to even temporary fish removal, largely from anglers. Though this arises from misconception and ignorance and can be dealt with, there are still problems in the use of temporary total fish removal in the Broadland system. The riverine and tidal nature of the area mean that it is not possible to isolate areas once the fish have been removed. This is because there are rights of navigation on tidal waters which prevent connections with the river from being blocked. Our solution to this has been to experiment with artificial refuges designed to build up the *Daphnia* population in the presence of fish so that the water may clear and thus the natural plant refuges return.

None of a series of refuges based on polypropylene rope, suspended fruit-cage netting, or bundles of alder twigs, proved satisfactory for either cost or operational reasons or both (Irvine *et al.* 1990). Our current design (Moss 1990) involves isolation of a 1-ha area by a fence permeable to water and algae but not to fish or their larvae. Fish have been removed from within the fenced area in which it is expected that plants will be able to establish in the clear water. Once a vigorous plant bed has been formed, the refuge enclosure will be progressively extended until a substantial part of the lake is occupied by protected plant beds. We will then remove all the fences at a stage when the natural buffer mechanisms should be able to stabilise the system without further intervention. The water of the lake in question has had its phosphorus content reduced by precipitation at the upstream sewage treatment works and has a diatom-dominated phytoplankton. The refuge enclosures are permeable to water through mesh screens, to maintain a flushing regime which will continue to support edible diatoms and not the blue-green algae which the water chemistry would support in a less well flushed situation. Broads in the downstream part of Broadland which have low flushing rates and extensive crops of filamentous blue-green algae, particularly *Oscillatoria* spp. (Moss & Balls 1989), seem less likely to be restorable by these techniques without prior very severe nutrient control.

Conclusions

Biomanipulation is thus not a palliative to solve eutrophication problems at minimal expense. It is not recommended in large natural lakes but is a potentially useful tool in artificial reservoirs where there is already a high degree of intervention and where water retention times are low. It is most likely to succeed in the UK through total removal of fish and is thus not compatible with amenity fisheries on reservoirs. In these cases it could replace nutrient control or permit controls of low rigour so long as the nutrient and flushing regimes do not favour blue-green algae.

Biomanipulation is likely to be essential for the improvement of water quality in shallow lakes used for water supply where the nature of the usually very fertile catchment limits the degree to which nutrients can be controlled or where internal loading of phosphorus from the sediments cannot be stopped. It is similarly essential for the restoration of the conservation and amenity values of such shallow lakes where once extensive plant communities have been lost.

References

Allen, J. R. & Hawkins, S. J. (in press). Can biological filtration be used to improve water quality? Studies in the Albert Dock complex, Liverpool. *Journal of the Science of the Total Environment.*

Arnold, D. E. (1971). Ingestion, assimilation and reproduction by *Daphnia pulex* fed seven species of blue green algae. *Limnology and Oceanography*, **10**, 906-920.

Balls, H. R., Moss, B. & Irvine, K. (1989). The loss of submerged plants with eutrophication. 1. Experimental design, water chemistry, aquatic plant and phytoplankton biomass in experiments carried out in ponds in the Norfolk Broadland. *Freshwater Biology*, **22**, 71-87.

Barel, C. D. N., Dorit, R., Greenwood, P. H., Fryer, G., Hughes, N., Jackson, P. B. N., Kawanabe, H., Lowe-McConnell, R., Witte, F. & Yamaoka, K. (1985). Destruction of fisheries in Africa's lakes. *Nature*, **315**, 19-20.

Benndorf, J. (1990). Conditions for effective biomanipulation; conclusions derived from whole lake experiments in Europe. *Hydrobiologia*, **200/201**, 187-203.

Benndorf, J., Kneschke, H., Kossatz, K. & Penz, E. (1984). Manipulation of the pelagic food web by stocking with predacious fishes. *Internationale Revue der Gesamten Hydrobiologie*, **69**, 407-428.

Benndorf, J., Schultz, H., Benndorf, A., Unger, R., Penz, E., Kreschke, H., Kossatz, K., Hornig, U., Kruspe, R. & Reichel, S. (1988). Food-web manipulation by enhancement of piscivorous fish stocks: long term effects in the hypereutrophic Bautzen reservoir. *Limnologica (Berlin)*, **19**, 97-110.

de Bernardi, R. & Giussani, G. (1990). Are blue-green algae a suitable food for zooplankton? An overview. *Hydrobiologia*, **200/201**, 29-41.

Canfield, D. E. Jr, & Bachmann, R. W. (1981). Prediction of total phosphorus concentration, chlorophyll *a* and secchi depths in natural and artifical lakes. *Canadian Journal of Fisheries and Aquatic Sciences*, **38**, 414-423.

Dawidowicz, P. (1990). The effect of *Daphnia* on filament length of blue green algae. *Hydrobiologia*, **191**, 265-268.

Dillon, P. J. & Rigler, F. H. (1974). The phosphorus-chlorophyll relationship in lakes. *Limnology and Oceanography*, **19**, 767-773.

van Donk, E., Gulati, R. D. & Grimm, M. P. (1989). Food-web manipulation of Lake Zwemlust: positive and negative effects during the first two years. *Hydrobiological Bulletin*, **23**, 19-35.

van Donk, E., Grimm, M. P., Gulati, R. D., Heuts, P. G. M., de Kloet, W. A., & van Liere, L. (1990). First attempt to apply whole lake food web manipulation on a large scale in the Netherlands. *Hydrobiologia*, **200/201**, 291-301.

Edmondson, W. T. (1991). *The Uses of Ecology. Lake Washington and Beyond.* University of Washington Press, Seattle.

Edmondson, W. T. & Litt, A. H. (1982). *Daphnia* in Lake Washington. *Limnology and Oceanography*, **27**, 272-293.

Gliwicz, Z. M. (1980). Filtering ratio, food size selectivity, and feeding rates in cladocerans – another aspect of interspecific competition in filter feeding zooplankton. *American Society of Limnology and Oceanography Special Symposium*, **3**, 282-291.

Gliwicz, Z. M. (1990). Why do cladocerans fail to control algal blooms? *Hydrobiologia*, **200/201**, 83-97.

Gophen, M. (1990). Biomanipulation: retrospective and future development. *Hydrobiologia*, **200/201**, 1-11.

Infante, A. & Abella, S. E. D. (1985). Inhibition of *Daphnia* by *Oscillatoria* in Lake Washington. *Limnology and Oceanography*, **30**, 1046-1052.

Irvine, K., Moss, B. & Balls, H. (1989). The loss of submerged plants with eutrophication. II. Relationships between fish and zooplankton in a set of experimental ponds and conclusions. *Freshwater Biology*, **22**, 89-107.

Irvine, K., Moss, B. & Stansfield, J. (1990). The potential of artificial refugia for maintaining a community of large bodied Cladocera against fish predation in a shallow eutrophic lake. *Hydrobiologia*, **200/201**, 379-389.

Irvine, K., Stansfield, J. H. & Moss, B. (1991). The use of enclosures to demonstrate the enhancement of *Daphnia* populations when isolated from fish predation in a shallow eutrophic lake. *Memoria del Istituto Italiano di Idrobiologica*, **48**, 325-344.

Leah, R. T., Moss, B. & Forrest, D. E. (1980). The role of predation in causing major changes in the limnology of a hypereutrophic lake. *Internationale Revue der Gesamten Hydrobiologie*, **65**, 223-247.

Moss, B. (1983). The Norfolk Broadland: experiments in the restoration of a complex wetland. *Biological Reviews*, **58**, 521-561.

Moss, B. (1988). The palaeolimnology of Hoveton Great Broad, Norfolk. Clues to the spoiling and restoration of Broadland. In *The Exploitation of Wetlands* (eds P. Murphy & C. Friend), pp. 163-191, British Archaeological Reports, British Series 186.

Moss, B. (1989). Water pollution and the management of ecosystems: a case study of science and scientist. In *Toward a More Exact Ecology* (eds P. J. Grubb & J. H. Whittaker), pp. 401-422, Thirtieth Symposium of the British Ecological Society, Blackwell Scientific, Oxford.

Moss, B. (1990). Engineering and biological approaches to the restoration from eutrophication of shallow lakes in which aquatic plant communities are important components. *Hydrobiologia*, **200-201**, 367-377.

Moss, B. (1991). The role of nutrients in determining the structure of lake ecosystems and implications for the restoration of submerged plant communities to lakes which have lost them. In *Danish Research Programme on Nitrogen, Phosphorus & Organic Matter (NPO)*. pp. 75-85, *International Conference Contributions by Invited International experts*. Helsingor, Denmark.

Moss, B. & Balls, H. R. (1989). Phytoplankton distribution in a temperate floodplain lake and river system. II. Seasonal changes in the phytoplankton community and their control by hydrology and nutrient availability. *Journal of Plankton Research*, **11**, 841-862.

Moss, B., Balls, H. R., Irvine, K. & Stansfield, J. (1986). Restoration of two lowland lakes by isolation from nutrient rich water sources with and without removal of sediment. *Journal of Applied Ecology*, **23**, 391-414.

Moss, B., Stansfield, J. H. and Irvine, K. (1991). Development of daphnid communities in diatom dominated and cyanophyte dominated lakes and their relevance to lake restoration by biomanipulation. *Journal of Applied Ecology*, **28**, 586-602.

Murtaugh, P. A. (1981). Selective predation by *Neomysis mercedis* in Lake Washington. *Ecology*, **62**, 894-900.

de Nie H. (1988). The decrease in aquatic vegetation in Europe and its consequences for fish populations. *EIFAC/CECPI Occasional Papers*, **19**, 1-52.

Officer, C. B., Smayda, T. J. & Mann, R. (1982). Benthic filter feeding: a natural eutrophication control. *Marine Ecology Progress Series*, **9**, 203-210.

Pace, M. L. (1984). Zooplankton community structure but not biomass influences the phosphorus-chlorophyll-*a* relationship. *Canadian Journal of Fisheries and Aquatic Sciences*, **41**, 1089-1096.

Phillips, G. L. & Chilvers, A. (1991). *The Control of Phosphorus in the Catchments of the Rivers Ant & Bure*. 5th Annual Report, Anglian Region, National Rivers Authority.

Reeders, H. H. & Bij van der Vaate, A. (1990). Zebra mussels (*Dreissena polymorpha*): a new perspective for water quality management. *Hydrobiologia*, **200/201**, 437-450.

Reeders, H. H., Bij van der Vaate, A. & Slim, F. J. (1989). The filtration rate of *Dreissena polymorpha* (Bivalvia) in three Dutch lakes with reference to biological water quality management. *Freshwater Biology*, **22**, 133-141.

Scavia, D., Fahnenstiel, G. L., Evans, M. S., Jude, D. J. & Lehman, J. T. (1986). Influence of salmonid predation and weather on long term water quality trends in Lake Michigan. *Canadian Journal of Fisheries and Aquatic Sciences*, **43**, 435-443.

Scheffer, M. (1990). Multiplicity of stable states in freshwater systems. *Hydrobiologia*, **200/201**, 475-486.

Shapiro, J. H. (1990). Biomanipulation: the next phase – making it stable. *Hydrobiologia*, **200/201**, 13-27.

Shapiro, J. H., Lamarra, V. & Lynch, M. (1975). Biomanipulation: an ecosystem approach to lake restoration. In *Proceedings of a Symposium on Water Quality Management through Biological Control* (eds P. L. Brezonik & J. L. Fox), pp. 65-69. University of Florida, Gainsville.

Shapiro, J. H. & Wright, D. I. (1984). Lake restoration by biomanipulation: Round Lake, Minnesota, the first two years. *Freshwater Biology*, **14**, 371-383.

Stansfield, J. H., Moss, B. & Irvine, K. (1989). The loss of submerged plants with eutrophication. III. Potential role of organochlorine pesticides: a palaeoecological study. *Freshwater Biology*, **22**, 109-132.

Timms, R. M. & Moss, B. (1984). Prevention of growth of potentially dense phytoplankton populations by zooplankton grazing in the presence of zooplanktivorous fish in a shallow wetland ecosystem. *Limnology and Oceanography*, **29**, 472-486.

Zaret, T. & Paine, R. T. (1973). Species introduction in a tropical lake. *Science, NY*, **218**, 444-445.

The control of the indirect effects of biomanipulation

JÜRGEN BENNDORF
Institute of Hydrobiology, Dresden University of Technology, Mommsenstrasse 13, D-0-8027 Dresden, Germany

Results from long-term investigations on biomanipulation show that indirect effects are at least as important as direct effects are for the stability of biomanipulation. Three types of indirect effects can be distinguished: (1) a change in quantity or quality of the resource base, (2) behavioural change of the prey, and (3) development of anti-predator traits. Although indirect effects of type (2), (e.g. a change in the pattern of vertical migration of zooplankton), and type (3), (e.g. development of helmets and neck teeth in *Daphnia*), are important mechanisms, the most essential indirect effects regarding biomanipulation belong to type (1). An example of the latter will be demonstrated: the complex of indirect effects of enhanced grazing by large herbivores on the phosphorus metabolism of the lake.

It is concluded that control of the indirect effects is absolutely necessary to stabilize biomanipulation measures, but this is much more difficult than the control of direct effects and needs deeper insights into the structuring mechanisms of food webs. Proper management of fish stocks, in combination with the control of phosphorus load and/or the physical conditions, seems to be the most promising way of controlling the indirect effects of biomanipulation.

Introduction

Biomanipulation is regarded as a new tool for water management (Gulati *et al.* 1990). The development of this new tool was possible because water quality depends to a large extent on the structure and function of aquatic food webs. In turn these food webs are mainly controlled by two complex factors: (1) resource limitation ("bottom–up control") and (2) predation ("top–down control" or biomanipulation).

Hitherto, much of the research performed on biomanipulation was focused on the direct effects of predation and consumption (Fig. 1). Some studies also took into consideration the resulting feedback effects which must be regarded as direct and indirect bottom–up effects, caused by the top–down induced change of the resource base (Fig. 1). In recent years it has become increasingly obvious that the complexity of top–down mechanisms, and hence the reliability of biomanipulation as a water-management tool, cannot be understood without taking into consideration the numerous indirect effects (Kerfoot & Sih 1987; Benndorf 1988; Carpenter 1988; Lammens *et al.* 1990). The purpose of this paper consists in demonstrating that indirect effects could be at least as important for the stability of biomanipulation as direct effects, and this insight leads to essential consequences regarding the principal control strategy in water quality management.

Types of indirect effects of biomanipulation

Three types of indirect effects can be distinguished: trophic linkage, behavioural, and morphological (Fig. 2).

The typical character of a trophic linkage indirect effect consists in changing the abundance of a population (A in Fig. 2) which is not directly connected with the "key" population (C in Fig. 2), by influencing the *abundance (or concentration)* of an intermediate population (or

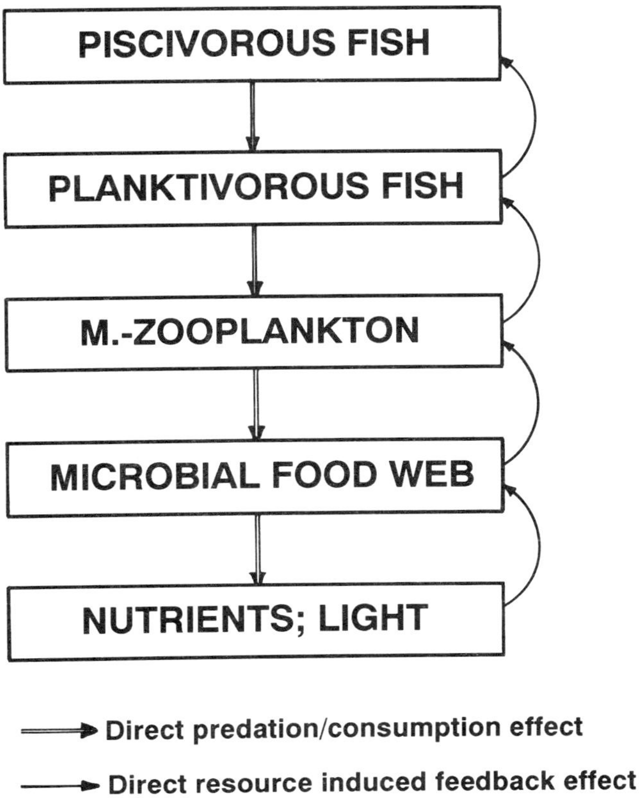

Figure 1. Schematic representation of a simple food chain.

compartment), (B in Fig. 2). Three numerous and typical cases of trophic linkage indirect effects are demonstrated in Figure 3.

Behavioural indirect effects are characterised by a change of the *behaviour* of an intermediate population. That change is a direct effect caused by the "key" population (C in Fig. 2) and leads to a change in abundance of another population which is not directly coupled with the "key" population. A typical example was demonstrated by Werner *et al.* (1983). Zooplanktivorous fish (bluegill) exerted a high predation pressure on large-sized *Daphnia* in a shallow lake densely colonized with macrophytes. After the introduction of a piscivorous fish (large-mouth bass), the bluegills took refuge in the stands of macrophytes and, consequently, their predation pressure on *Daphnia* decreased considerably, though there was almost no change in the abundance of the bluegill population.

If a "key" population changes the *morphology or the chemical composition* of an intermediate population in such a way that another (third) population is also affected, this represents a morphological indirect effect. A typical example was observed in the long-term biomanipulation experiment in Gräfenhain (Köhler *et al.* 1989; Benndorf 1990). Some years after the complete elimination of planktivorous fish, the abundance of *Chaoborus flavicans* larvae increased dramatically. Following this increase, small- and medium-sized *Daphnia pulex* produced neck teeth which protect them against predation by *Chaoborus* (Fig. 4). The observed results were a stable and abundant population of *Daphnia pulex*, and a low concentration of all grazable particles in the lakewater.

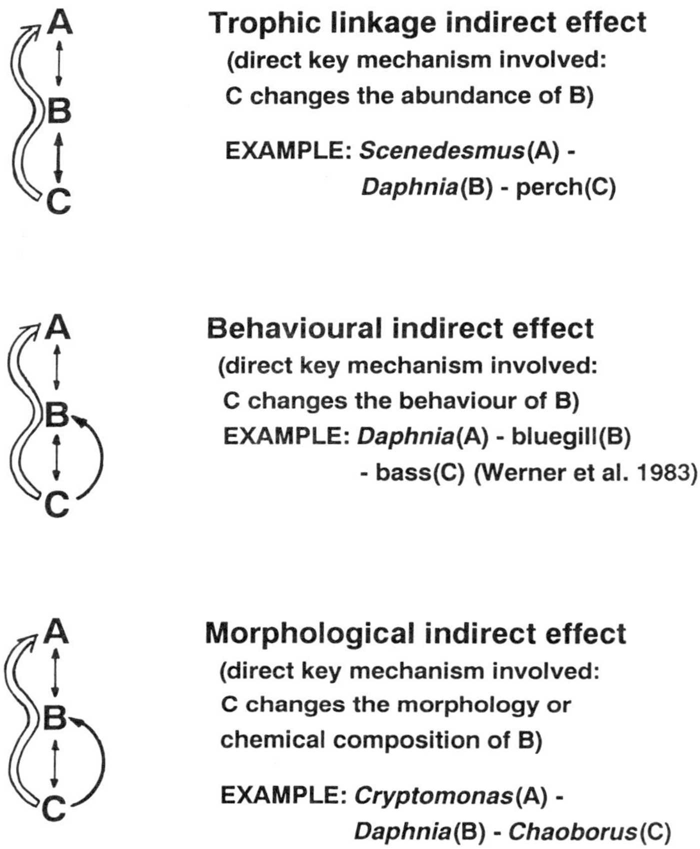

Figure 2. Types of indirect effects. Thick arrows: indirect effects; thin arrows: direct effects (modified after Miller & Kerfoot 1987).

The role of indirect effects in food webs

The enormous complexity of interactions in aquatic food webs is not only caused by the high diversity of types of indirect effects. Another important reason is that each trophic level is occupied by numerous species and many indirect effects exist between these species. An attempt to illustrate this fact is made in Figure 5 by assuming the simplest case, where each trophic level is occupied by only two species. We have to realise, for instance, that a piscivorous fish (PiF_l) may be affected indirectly by another piscivore (PiF_n) if the latter reduces a planktivorous fish ($PlaF_l$) which happens to be a preferred prey of PiF_l. At the same time, the population of PiF_n could also exert a positive indirect influence on population $PlaF_l$ by reducing a planktivorous competitor ($PlaF_n$). Similar indirect top–down effects also exist between all species of the other (lower) trophic levels (Fig. 5A).

Similarly, as in the simple food chain shown in Figure 1, each top–down effect automatically causes a feedback response. In almost all cases this feedback response consists of a direct effect and several indirect effects (Fig. 5B). If, for instance, a microbial population, Mi_l (e.g. a

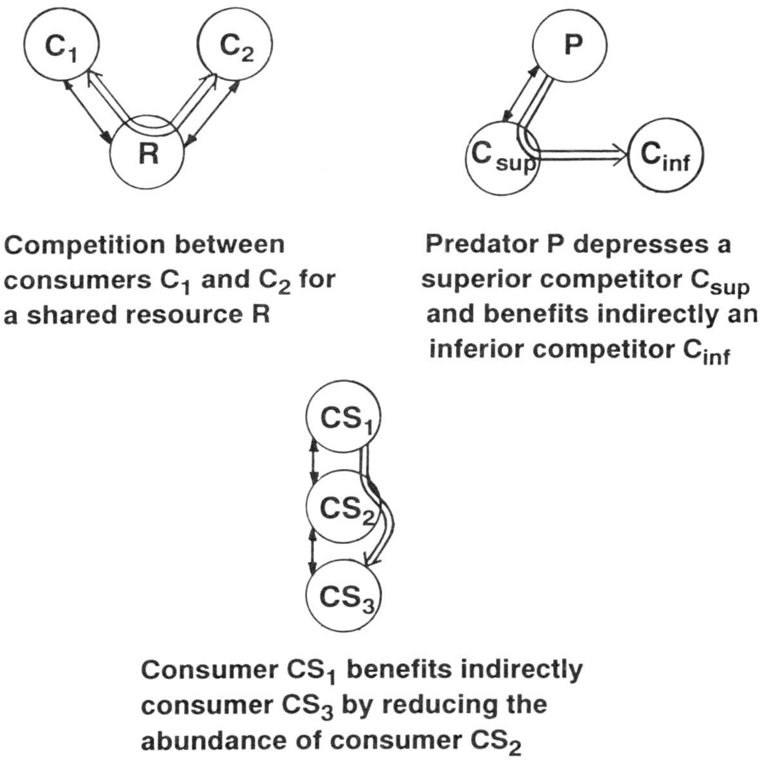

Figure 3. Principal types of trophic linkage indirect effects. Arrows as in Fig. 2 (modified after Crowder *et al.* 1988).

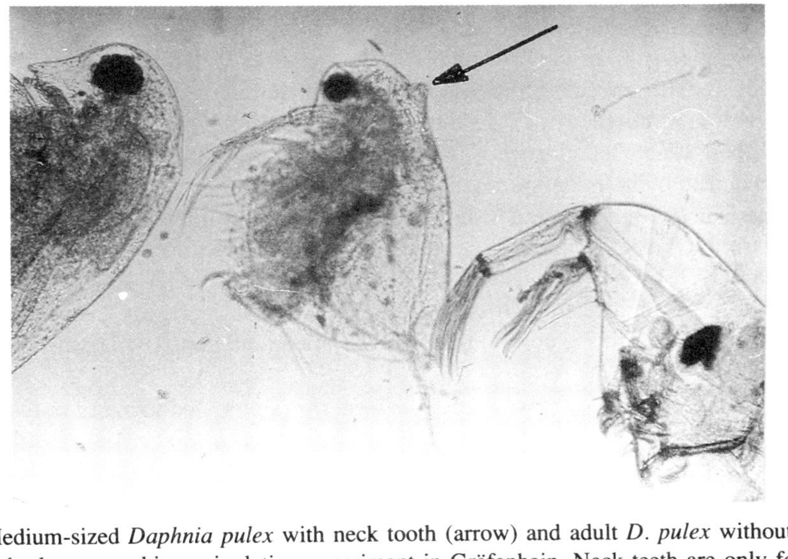

Figure 4. Medium-sized *Daphnia pulex* with neck tooth (arrow) and adult *D. pulex* without neck tooth (left), from the long-term biomanipulation experiment in Gräfenhain. Neck teeth are only formed if the abundance of the predator *Chaoborus flavicans* (right) is high (photograph taken by U. Hornig).

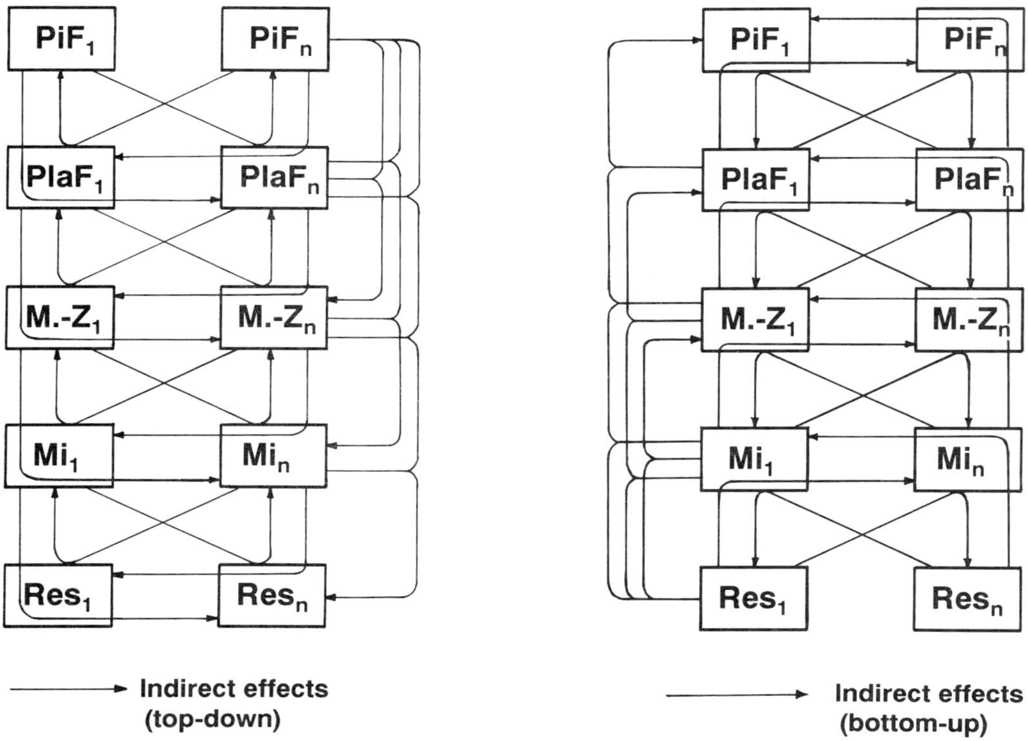

Figure 5. Schematic representation of indirect effects operating top–down (A, left) and bottom–up (B, right) in a simple food web with only two populations on each trophic level. PiF: piscivorous fish, PlaF: planktivorous fish, M.–Z: mesozooplankton/macrozooplankton, Mi:microbial food web (phytoplankton, bacteria, protozoans), Res: resources (nutrients, energy).

grazable microalga), has been reduced as a consequence of top–down mechanisms shown in Figure 1 and Figure 5A, this reduced amount of Mi_l may indirectly affect a mesozooplankton population, $M.–Z_n$ (e.g. a rotifer), through the action of another population, $M.–Z_n$ (e.g. a *Daphnia* species). The inferior competitor ($M.–Z_n$, the rotifer) can not compete successfully with the superior competitor ($M.–Z_i$, the *Daphnia* sp.) under conditions of strong food limitation (Gilbert 1985; Benndorf *et al.* 1988). In similar ways, the other species of all trophic levels are influenced by indirect bottom–up (feedback) effects.

As a comparison of Figures 1 and 5 shows, the number of couplings (i.e. interactions) within and between trophic levels is much greater for indirect effects than it is for direct effects. Thus, it seems already from an *a priori* viewpoint, we should expect that the structuring processes of food webs will be dominated by indirect effects. If this view is correct, then biomanipulation could be much more difficult to control than was assumed (and hoped) originally.

An example: complexity of the indirect effects that enhanced grazing may have on phosphorus concentrations in lakewater

It is well known that in many cases top–down effects are strong at the top of the food web, but in some cases they weaken near the bottom of the trophic pyramid. This "damping phenomenon" is especially pronounced in the more eutrophic water-bodies whereas it seems to be absent under oligotrophic conditions (McQueen *et al.* 1986, Benndorf 1987). There are many indications that this different behaviour of eutrophic and oligotrophic waters is caused by a rather complicated network of indirect effects (Fig. 6).

Enhanced grazing by large and efficient filtrators (e.g. *Daphnia*) exerts indirect effects, some of which cause an increase and others a decrease of the phosphorus concentration in the production zone of a lake or reservoir. Three indirect effects can cause an increase (Fig. 6, left) and five can cause a decrease of phosphorus in the euphotic zone (Fig. 6, right).

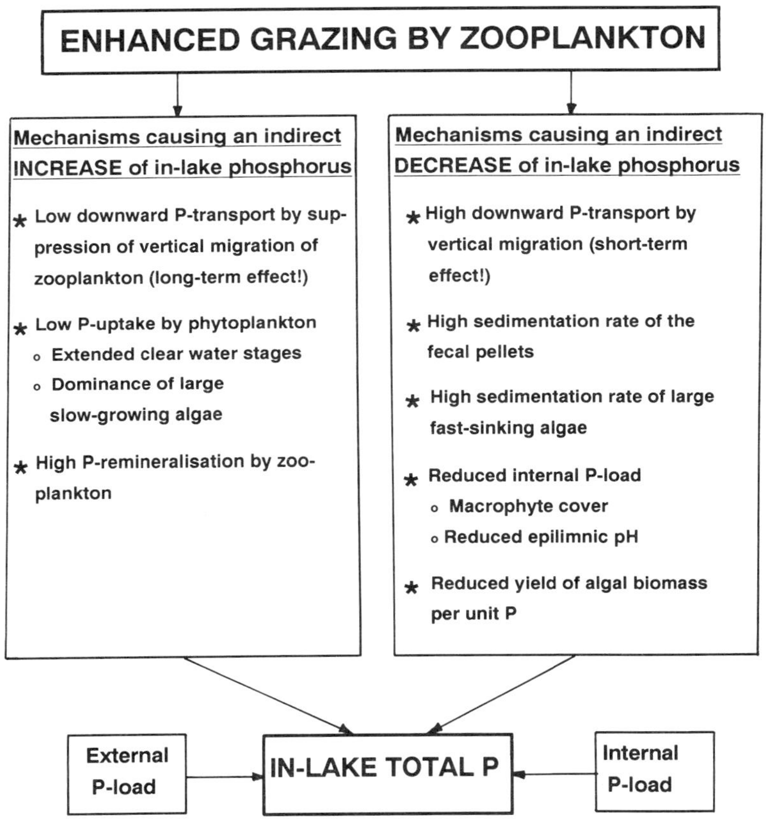

Figure 6. The indirect effects of biomanipulation (enhanced grazing) on the phosphorus concentration in a water-body (modified after Benndorf & Miersch 1991).

Indirect effects that increase the phosphorus concentration in lakewater

(i). The first indirect effect causing an increase in phosphorus concentration is a predator-avoidance mechanism, where the transport of phosphorus from surface waters to deep waters by migrating zooplankters is reduced. Diel vertical migration of the zooplankton, moving to deeper waters in daytime and rising to surface waters at night, must be regarded as a predator-avoidance mechanism which can be shifted by a change in predation pressure. But, as the mechanism is at least partly fixed genetically, a shift in the daily migrating behaviour appears only in a time-scale of months or even years after the change (reduction) in fish predation pressure (Gliwicz 1986; Köhler *et al.* 1989). Therefore, when zooplanktivorous fish are reduced as a management measure, zooplankters will continue their vertical migrations for one or two (or more) following years and thereby continue to transport phosphorus down to the hypolimniom and the sediments. But sooner or later (e.g. after 5 years according to Köhler *et al.* 1989) migrating zooplankters will be outcompeted by non-migrating individuals, which are always present due to large genetic variability, because it is much more advantageous to remain in the warmer and food-rich upper layers of the lake throughout the day when there is no longer a risk of predation.

The indirect consequences of zooplankters remaining in the upper layers are a shift to larger, inedible algae (Lampert 1987; Köhler *et al.* 1989) and a reduced downward transport of phosphorus by zooplankton. This predator-avoidance mechanism shows that a positive short-term effect of food-web manipulation can be counterbalanced by the long-term response of the food web.

(ii). The second indirect effect occurs when phosphorus concentrations remain relatively high because algal uptake is reduced by the low biomass of algae present during zooplankton-induced clearwater stages, and there is a dominance of large, slow-growing algal cells and colonies which cannot be eaten by zooplankters (Sommer *et al.* 1985; Lampert *et al.* 1986).

(iii). The third indirect effect occurs when phosphorus concentrations remain relatively high because the faeces of zooplankters have a high content of mineralized phosphorus. Although the specific excretion of phosphorus (i.e. per unit mass of zooplankton) decreases as the body-size of the zooplankters increases (Esjmont-Karabin 1984; Peters 1987), the overall excretion of phosphorus should increase as a consequence of biomanipulation if there is a considerable increase in total zooplankton biomass.

Indirect effects that decrease the phosphorus concentration in lakewater

(1). An enhanced export of phosphorus from the epilimnion into the hypolimnion and sediments occurs when the zooplankton contains numerous large, vertically-migrating *Daphnia*. These zooplankters feed at night in the upper layers and stay in deeper waters during the day, where a greater proportion of phosphorus is released by excretion (Wright & Shapiro 1984; Dini *et al.* 1987).

(2). The phosphorus-rich faeces of zooplankters have higher sedimentation rates than many small organisms (Benndorf 1979; Carpenter *et al.* 1986; Uehlinger & Bloesch 1987). This settling-out effect is particularly pronounced in oligotrophic–mesotrophic waters where large diaptomids dominate; their faecal pellets sink much faster than the "fluffy", easily disintegrating faeces of *Daphnia*. But even these latter faeces sink faster than the small single cells of algae.

(3). Intensive filtration by zooplankton removes almost all of the small, edible algae, but the large inedible species or strains remain. If the inedible ones are green algae or diatoms, then a higher sedimentation rate of algae-bound phosphorus is to be expected, compared with an algal community dominated by small (slow-sinking) algae.

(4). A reduced internal loading of phosphorus can follow a reduction of algal biomass for two reasons: the development of a macrophyte cover due to an improved underwater light climate (Jagtman *et al.* 1988; Mitchell *et al.* 1988), and phosphorus release from epilimnic sediments is reduced due to lower phytoplankton production and hence a lower pH.

(5). There can be a reduced yield of algal biomass per unit of phosphorus, as a consequence of grazing-induced increases in the population growth rates of phosphorus-limited algae (Olsen 1988).

Indirect effects of grazing in relation to phosphorus loadings

It must be realized that indirect effects which *increase* the phosphorus concentration of lakewater, i.e. (i) to (iii) above, are *independent* of the external and internal phosphorus loadings to the lake. But indirect effects (2) to (5), which *decrease* the phosphorus concentration, are *strongly inhibited* by high external and/or internal phosphorus loadings. The actual concentration of phosphorus in the lakewater is the result of all these indirect effects on the circulation of phosphorus, as well as the external and internal phosphorus loadings to the lake (Fig. 6). Consequently, two constellations or suites of these interactions can be predicted.

Prediction (A). When the external and/or internal phosphorus loading is high, this increases the phosphorus concentration directly and suppresses indirect effects (2) to (5) which decrease the phosphorus concentration. Therefore indirect effects (i) to (iii), which cause an increase in the phosphorus concentration, become dominant. As a result, the in-lake phosphorus concentration remains high (unchanged or increased) and no reduction of the annual average biomass of phytoplankton can be expected.

Prediction (B). When the external and internal phosphorus loading is low, this contributes to a low phosphorus concentration and enhances indirect effects (2) to (5) which also decrease the phosphorus concentration. Therefore indirect effects (1) to (5) become dominant relative to indirect effects (i) to (iii). As a result, the in-lake phosphorus concentration decreases as a consequence of biomanipulation, and the annual average biomass of phytoplankton is also expected to decrease.

The validity of these assumptions about the role of indirect effects of biomanipulation on phosphorus metabolism of the water-body, and their consequences for the phytoplankton biomass, can be checked against the results of two independent long-term biomanipulation experiments, using entire lakes of quite different phosphorus loadings (Benndorf *et al.* 1988;

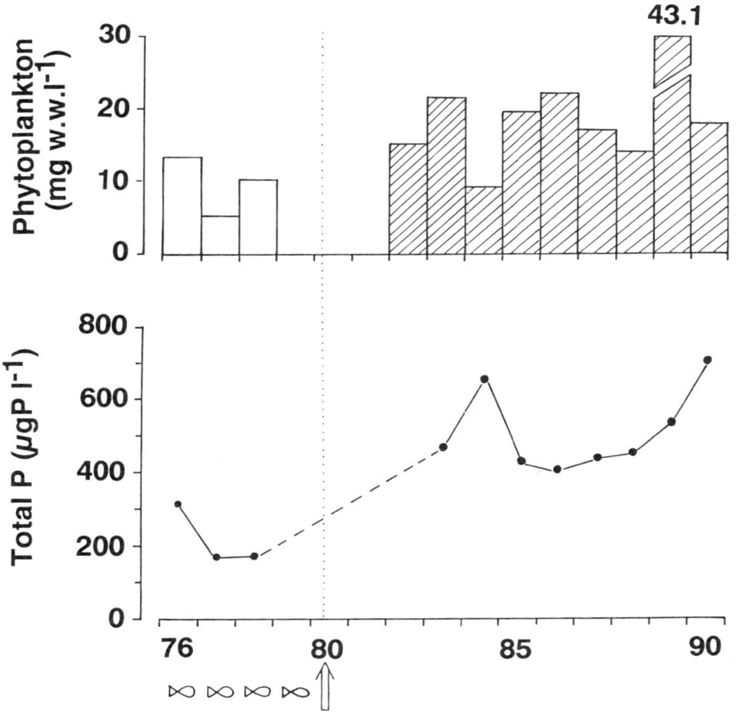

Figure 7. Total phosphorus (µg P l^{-1}) and phytoplankton biomass (mg wet weight per litre of water) before (fish symbols, left, up to 1980) and during (right, after 1980) biomanipulation in the hypereutrophic Bautzen Reservoir (Germany). Volume-weighted vertical averages are shown. (Note: Secchi depth, which is not shown here, was significantly increased due to a change in phytoplankton composition during biomanipulation; see Benndorf *et al.* 1988)

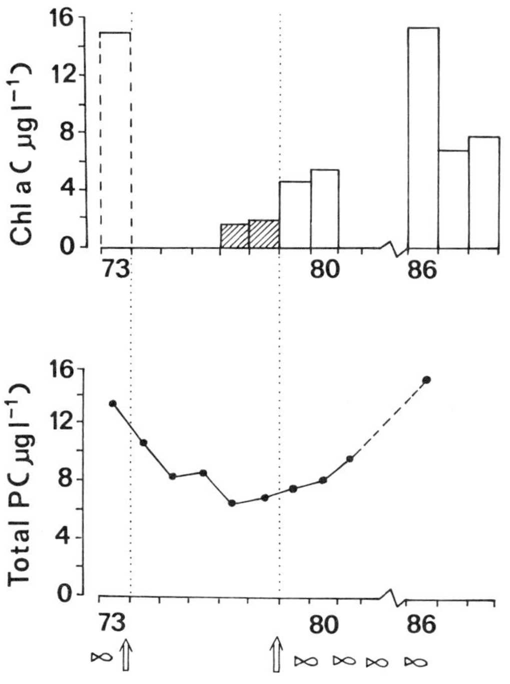

Figure 8. Total phosphorus (µg P l^{-1}) and phytoplankton biomass (as chlorophyll-*a* µg l^{-1}) before and after (fish symbols, up to 1973 and after 1978) as well as during (1974 to 1978) biomanipulation in the oligotrophic Lilla Stockelidsvatten (Sweden). Volume-weighted averages are shown. (From Stenson 1988).

Stenson 1988). A comparison of Figures 7 and 8 reveals complete agreement between predictions (A) and (B) and the respective observations on the hypereutrophic Bautzen Reservoir and the oligotrophic Lilla Stockelidsvatten. Likewise, model simulations for Bautzen Reservoir and the oligotrophic Lake Stechlin also show agreement with predictions (A) and (B), (Benndorf & Miersch 1991).

Phosphorus-loading threshold between predictive constellations (A) and (B)

There is obviously a threshold for phosphorus loadings which separates predictions (A) and (B). The existence of this threshold was first hypothesized by Benndorf (1987) and it has been confirmed recently by Søndergaard *et al.* (1990), Jeppesen *et al.* (1990), Sanni & Waervagen (1990) and Benndorf & Miersch (1991). This threshold is assumed to be relatively variable from lake to lake, rather than having a fixed value. But from the available information cited above, it can be estimated that the threshold should range between 0.5 and 2.0 g total P m^{-2} year^{-1}. If the phosphorus loading is below this threshold, the combination of direct and indirect effects listed under prediction (B) should ensure the long-term success of biomanipulation. On the other hand, if the threshold is exceeded, only a short-term or temporary success can be expected from biomanipulation.

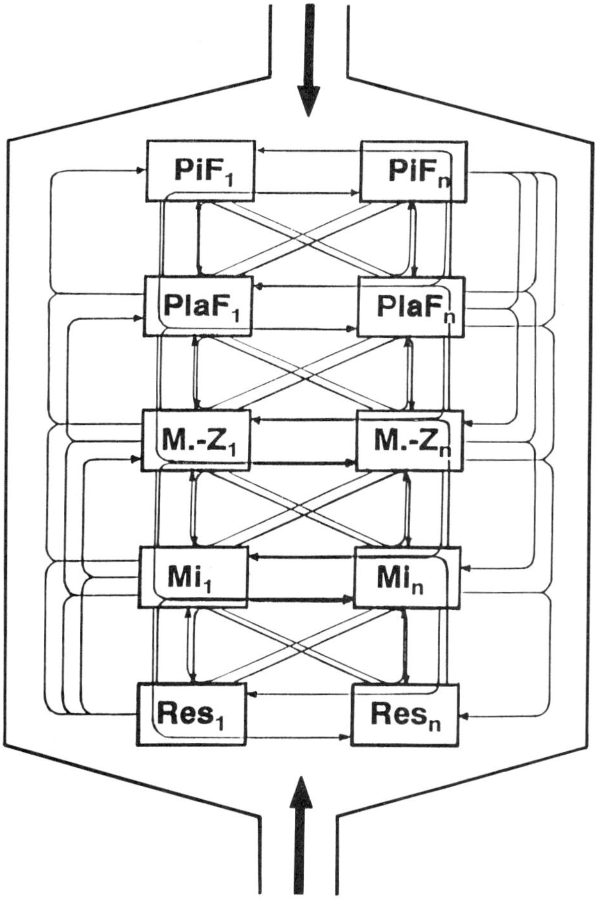

Figure 9. Schematic representation of the food web depicted in Fig. 5, showing all couplings by indirect effects. The high complexity of the system should be controlled simultaneously from the bottom and from the top.

Conclusions

(1). Indirect effects play an essential role in structuring the processes (interactions) of food webs. They can enhance or decrease the stability of biomanipulation measures.
(2). The control of indirect effects is absolutely necessary to stabilize biomanipulation measures, but this is much more difficult than the control of direct effects and needs a deeper understanding of the structuring mechanisms of food webs. For biomanipulation, the problem is how to selectively support the positive indirect effects.
(3). The probability seems not to be high that the lower trophic levels (especially the phytoplankton biomass) can be reduced by controlling the food web exclusively from the top.
(4). The numerous indirect effects can be used optimally (and the positive indirect effects can be selectively enhanced) if strong top–down control (biomanipulation) is combined with maximal limitations at the bottom of the food web, e.g. limitation by phosphorus or light (Fig. 9). This combination is much more efficient than each of the two control strategies alone.

This article is based on research project 30 F 100 701, which has been supported by the Federal Ministry of Research and Technology (BMFT). But only the author is responsible for the article's contents.

References

Benndorf, J. (1979). Kausalanalyse, theoretische Synthese und Simulation des Eutrophierungsprozesses in stehenden und gestauten Gewässern. Dissertation B, Dresden University of Technology, Dresden. 165 pp.

Benndorf, J. (1987). Food web manipulation without nutrient control: a useful strategy in lake restoration ? *Schweizerische Zeitschrift für Hydrologie*, **49**, 237-248.

Benndorf, J. (1988). Biomanipulation. Special Issue of *Limnologica*, **19** (1), 110 pp.

Benndorf, J. (1990). Conditions for effective biomanipulation; conclusions derived from whole-lake experiments in Europe. In *Biomanipulation – Tool for Water Management* (eds R. D. Gulati *et al.*), pp. 187-203. Kluwer Academic Publishers, Dortrecht.

Benndorf, J. & Miersch, U. (1991). Phosphorus loading and efficiency of biomanipulation. *Verhandlungen der Internationalen Vereinigung für Theoretische und Angewandte Limnologie*, **24**, 2482-2488.

Benndorf, J., Schultz, H., Benndorf, A., Unger, R., Penz, E., Kneschke, H., Kossatz, K., Dumke, R., Hornig, U., Kruspe, R. & Reichel, S. (1988). Food web manipulation by enhancement of piscivorous fish stocks: long-term effects in the hypertrophic Bautzen reservoir. *Limnologica*, **19**, 97-110.

Carpenter, R. S. (Editor) (1988). *Complex Interactions in Lake Communities*. Springer Verlag, New York. 283 pp.

Carpenter, S. R., Elser, M. M. & Elser, J. J. (1986). Chlorophyll production, degradation and sedimentation: implications for paleolimnology. *Limnology and Oceanography*, **31**, 112-124.

Crowder, L. B., Drenner, R. W., Kerfoot, W. C., McQueen, D. J., Mills, E. L., Sommer, U., Spencer, C. N. & Vanni, M. J. (1988). Food web interactions in lakes. In *Complex Interactions in Lake Communities* (ed. S. R. Carpenter), pp. 141-160. Springer Verlag, New York.

Dini, M. L., O'Donell, J., Carpenter, S. R., Elser, M. M., Elser, J. J. & Bergquist, A. M. (1987). *Daphnia* size structure, vertical migration, and phosphorus redistribution. *Hydrobiologia*, **150**, 185-191.

Esjmont-Karabin, J. (1984). Phosphorus and nitrogen excretion by lake zooplankton (Rotifers and Crustaceans) in relationship to individual body weights of the animals, ambient temperature and presence or absence of food. *Ekologia Polska*, **32**, 3-42.

Gilbert, J. J. (1985). Competition between rotifers and *Daphnia*. *Ecology*, **66**, 1943-1950.

Gliwicz, Z. M. (1986). Predation and the evolution of vertical migration in zooplankton. *Nature*, **320**, 746-748.

Gulati, R. D., Lammens, E. H. R. R., Meijer, M.-L. & van Donk, E. (Editors) (1990). *Biomanipulation – Tool for Water Management*. Kluwer Academic Publishers, Dortrecht. 628 pp.

Jagtman, E., Hosper, S. H., Meijer, M.-L. & van Donk, E. (1988). The role of fish stock management in eutrophication control in shallow lakes in the Netherlands. *EIFAC Symposium on Management Schemes for Inland Fisheries, Göteborg, Sweden, 31 May - 3 June, 1988.*

Jeppesen, E., Jensen, J. P., Søndergaard, M., Mortensen, E., Sortkjaer, O. & Olrik, K. (1990). Fish manipulation as a lake restoration tool in shallow, eutrophic, temperate lakes. 2: threshold levels, long-term stability and conclusions. In *Biomanipulation - Tool for Water Management* (eds R. D. Gulati *et al.*), pp. 219-227. Kluwer Academic Publishers, Dortrecht.

Kerfoot, W. C. & Sih, A. (Editors) (1987). *Predation: Direct and Indirect Impacts on Aquatic Communities.* University Press of New England, Hanover and London. 386 pp.

Köhler, J., Köhler, A. & Benndorf, J. (1989). Biomanipulation im Experimentalgewässer Gräfenhain: Veränderte direkte und indirekte Effekte nach fünfjähriger Versuchsdauer. *Acta Hydrochim Hydrobiologie*, **17**, 633-646.

Lammens, E. H. R. R., Gulati, R. D., Meijer, M.-L. & van Donk, E. (1990). The first biomanipulation conference: a synthesis. In *Biomanipulation – Tool for Water Management* (eds R. D. Gulati *et al.*), pp. 619-627. Kluwer Academic Publishers, Dortrecht.

Lampert, W. (1987). Vertical migration of freshwater zooplankton: indirect effects of vertebrate predators on algal communities. In *Predation: Direct and Indirect Impacts on Aquatic Communities* (eds W. C. Kerfoot & A. Sih), pp. 291-299. University Press of New England, Hanover and London.

Lampert, W., Fleckner, W., Rai, H. & Taylor, B. E. (1986). Phytoplankton control by grazing zooplankton: a study on the spring clear-water phase. *Limnology and Oceanography*, **31**, 478-490.

McQueen, D., Post, J. R. & Mills, E. L. (1986). Trophic relationship in pelagic ecosystems. *Canadian Journal of Fisheries and Aquatic Sciences*, **43**, 1571-1581.

Miller, T. E. & Kerfoot, W. C. (1987). Redefining indirect effects. In *Predation: Direct and Indirect Impacts on Aquatic Communities* (eds W. C. Kerfoot & A. Sih), pp. 33-37. University Press of New England, Hanover and London.

Mitchell, S. F., Hamilton, D. P., MacGobbon, W. S., Nayar, P. K. B. & Reynolds, R. N. (1988). Interrelations between phytoplankton, submerged macrophytes, black swans (*Cygnus atratus*) and zooplankton in a shallow New Zealand lake. *Internationale Revue der Gesamten Hydrobiologie*, **73**, 145-170.

Olsen, Y. (1988). Fish elimination in Lake Haugatjonn: the contribution of the zooplankton in reducing algal biomass.

Nordic Symposium "Ecological Lake Management", Lahti, Finland, August 29-31, 1988.

Peters, R. H. (1987). Metabolism in *Daphnia. Memorie dell' Istituto Italiano di Idrobiologia Dott. Marco de Marchi*, **45**, 193-243.

Sanni, S. & Waervagen, S. B. (1990). Oligotrophication as a result of planktivorous fish removal with rotenone in the small, eutrophic, Lake Mosvatn, Norway. In *Biomanipulation - Tool for Water Management* (eds R. D. Gulati *et al.*), pp. 263-274. Kluwer Academic Publishers, Dortrecht.

Sommer, U., Gliwicz, Z. M. & Lampert, W. (1985). The PEG model of seasonal succession of plankton in fresh waters. *Archiv für Hydrobiologie*, **106,** 433-471.

Søndergaard, M., Jeppesen, E., Mortensen, E., Dall, E., Kristensen, P. & Sortkjaer, O. (1990). Phytoplankton biomass reduction after planktivorous fish reduction in a shallow eutrophic lake: a combined effect of reduced internal P-loading and increased zooplankton grazing. In *Biomanipulation - Tool for Water Management* (eds R. D. Gulati *et al.*), pp. 229-240. Kluwer Academic Publishers, Dortrecht.

Stenson, J. A. E. (1988). Fauna structure and water quality. *Aqua Fennica*, **18**, 179-184.

Uehlinger, U. & Bloesch, J. (1987). The influence of crustacean zooplankton on the size structure of algal biomass and suspended and settling seston (Biomanipulation in Limnocorrals II). *Internationale Revue der Gesamten Hydrobiologie*, **72**, 473-486.

Werner, E. E., Gilliam, J. F., Hall, D. J. & Mittelbach, G. G. (1983). An experimental test of the effects of predation risk on habitat use in fish. *Ecology*, **64**, 1540-1548.

Wright, D. I. & Shapiro, J. (1984). Nutrient reduction by biomanipulation: an unexpected phenomenon, and its possible cause. *Verhandlungen der Internationalen Vereinigung für Theoretische und Angewandte Limnologie*, **22,** 518-524.

The significance of sedimentation and sediments to phytoplankton growth in drinking-water reservoirs

DIETRICH UHLMANN[1] AND HEIDEMARIE HORN[2]

[1] *Laboratory of Hydrobiology, University of Technology, Mommsenstrasse 13, DO 8027 Dresden, Germany*

[2] *Working Group on Limnology, Saxon Academy of Sciences, Neunzehnhain, P.O. Box 30, DO 9343 Lengefeld, Germany*

In the mesotrophic–eutrophic Saidenbach Reservoir in Saxony, the nanoplankton and cyanobacteria have increased at the expense of diatom dominance, due to a doubling of the external phosphorus load in the last 15 years. However, the phosphorus sedimentation flux is still very high (up to 80% of the input), corresponding to more than 2 g m^2 d^{-1} in terms of dry weight. There is a strong correlation between the abundance of diatoms in the euphotic zone and their sedimentation flux (with a delay of about 2 weeks). Only about 25% of the deposited material could be clearly attributed to plankton biomass; the remainder resulted from flocculation and precipitation processes or directly from the inflow of clay minerals. The ash content of the deposited material was high (73%). Thus the sedimentation flux can be considered to operate as an internal water-treatment/oligotrophication process within the lake.

The neighbouring Neunzehnhain Reservoir still has a very clear water with a transparency up to 18 m depth. Though the sediment was not much lower than Saidenbach sediment in total phosphorus and total numbers of bacteria, sulphide was always absent and the ratio of Fe^{2+} to Fe^{3+} was very low in the upper (0 – 5 cm) layer. Thus the external and internal phosphorus loads do not attain the critical level necessary to induce a "phosphorus – phytoplankton" feedback loop.

Introduction

Saidenbach Reservoir is situated in the Erzgebirge Hills, Saxony, and has been in operation since 1933. It is one of the principal sources of drinking-water for the municipality of Chemnitz. Its full water level is at 439 m, the catchment has an area of 76 km^2 and is mainly used for agriculture. The area of the reservoir is 1.46 km^2 and its volume is 22.4 million m^3, with a mean depth of 15 m and maximum depth 45 m. The catchment of the neighbouring oligotrophic Neunzehnhain Reservoir has an area of 12.8 km^2 and is covered by forest (mainly spruce, *Picea*). Its area is 0.29 km^2 and the volume is 2.5 million m^3, with a mean depth of 10.3 m and maximum depth 33 m.

Saidenbach Reservoir is longitudinally separated, with a pre-reservoir and underwater pre-reservoirs arranged in series in the inflow area. These not only serve as traps for suspended solids but also for phosphorus (Figs 1 – 3). Although design criteria for pre-reservoirs, concerning their capacity to immobilize phosphorus, have been described by Uhlmann & Benndorf (1980) and Benndorf & Pütz (1987 a,b), the quantitative aspects of the processes governing the retention, immobilization and sedimentation of phosphorus in process reservoirs arranged in series are still largely unknown. Phosphorus retention may range between 20% (Stimpfig *pers. comm.*) and more than 50% if different pre-reservoirs are compared. These also have to cope with suspended solids, the load of which is a reflection of the intensity of erosion processes in the catchment (Fig. 4).

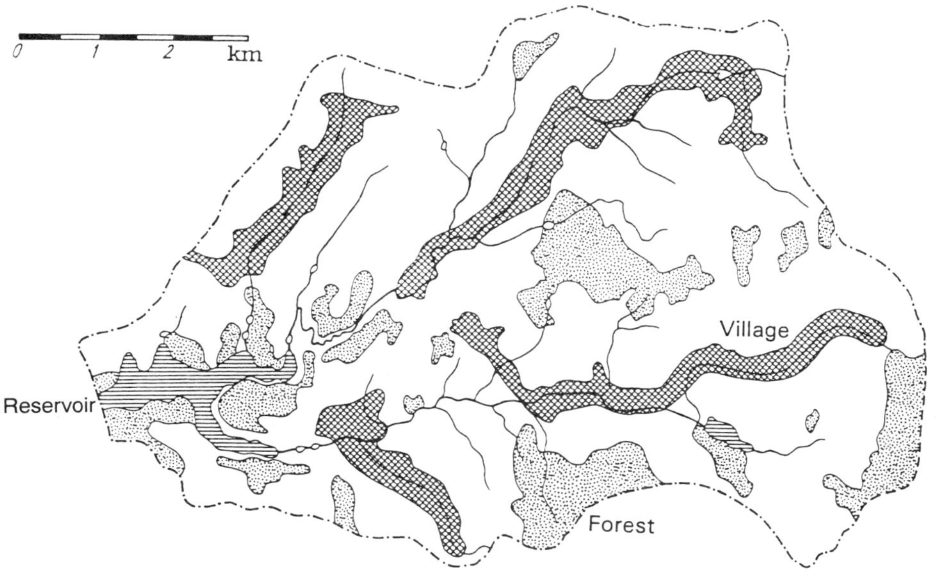

Figure 1. Saidenbach Reservoir (horizontal shading) and its catchment area, showing forest (stippled) and villages (cross-hatched).

In the following account the processes governing the sedimentation of phytoplankton biomass and phosphorus, as factors which decrease the residence time of P-compounds in the euphotic layer, will be dealt with in some detail. The chemical analyses of water and sediments referred to here were performed as described in "Ausgewählte Methoden" (1986); the sediment trap was developed by Horn & Horn (1990).

Phytoplankton growth and sedimentation

One of the principal processes which govern the sedimentation of phosphorus in mesotrophic to eutrophic drinking-water reservoirs like Saidenbach is the incorporation of phosphate by diatoms. These have a high sinking velocity and thus decrease the residence time of phosphorus in the euphotic zone. Investigations on Saidenbach also dealt with the question: to what extent are the seasonal changes in the abundance of phytoplankton followed by corresponding changes in the sedimentation flux, in terms of phytoplankton biomass and phosphorus, or suspended matter? As is evident from Figure 5, there was a surprisingly close correlation between the "biovolume" of the phytoplankton, and its sedimentation flux, with a phase shift of about 2 weeks between the two maxima. On the other hand, the sedimentation flux of suspended solids in general was not clearly correlated with the phytoplankton sedimentation. This is in accordance with the low proportion (about 25%) of the deposited material which clearly can be attributed to the phytoplankton biomass. This is also comparable to the low proportion (less than 26%) in the deposited material recorded by Hamilton-Taylor *et al.* (1984) from Lake Windermere in England. In Saidenbach the phosphorus load has doubled in the last 15 years. As a result, dissolved silica was depleted at the end of the diatom growth period. This was the trigger for a substantially increased abundance of cyanobacteria like *Aphanizomenon* and green algae, with a high proportion of nanoplanktonic forms. This resulted in an oxygen minimum in the metalimnion and at the surfaces of the corresponding sediment layers, and in a longer residence time for phosphorus in the euphotic zone. The minimum Secchi depth, which in the past has been 4.9 m, is now 3.8 m.

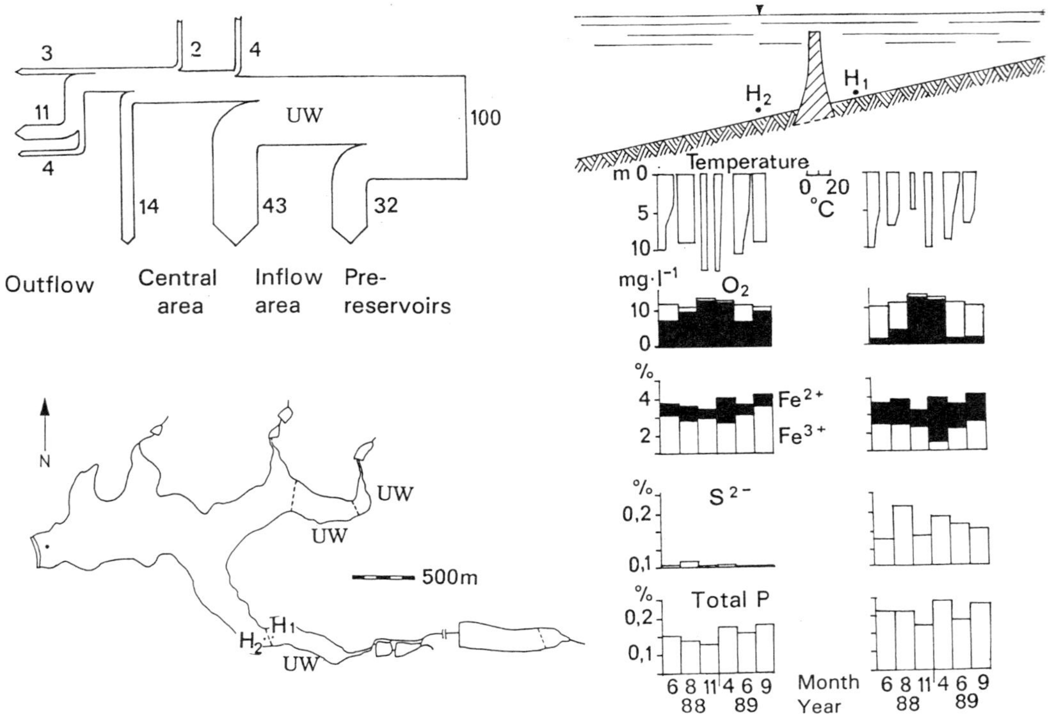

Figure 2. Some general features of Saidenbach Reservoir. Bottom left: plan view showing positions of one large and six small above-ground pre-reservoirs, and three under-water pre-reservoirs (UW). Above left: approximate proportions (numbers, %) of total phosphorus sedimentation in the longitudinal axis of the reservoir (after Hofmann 1972 and Petersohn 1987), although the phosphorus flux in the outflow (shown as 3+11+4=18%) may attain about 25% according to more recent calculations. Right: some properties of the water-body and sediments above (H1) and below (H2) the dam of the Haselbach underwater pre-reservoir. Variations in water level are evident from the temperature (°C)/depth (m) diagrams. Oxygen concentrations (mg l^{-1}) refer to the water layer 1m above the bottom. The white portion of the columns denotes the oxygen deficit below saturation.

Phosphorus sedimentation

The sedimentation flux of particulate matter (including phosphorus) seems to be the most important loss process with regard to phytoplankton growth in Saidenbach. In lakes and reservoirs with a short residence time for water, there is a comparatively low proportion of phytoplankton biomass in the total sedimentary flux. This has been observed in lakes in Austria and Switzerland, and also in Lake Windermere. As the allochthonous suspended solids (see Fig. 4) normally have a high concentration of iron or aluminium in catchments with igneous rock, they may operate as adsorbents, flocculants and precipitants, and thus play a significant role in the nutrient balance. The sedimentation flux in Saidenbach was, on average, about 2.33 g m^2 d^{-1} (as dry weight) and this may be considered to be a significant internal oligotrophication/water-treatment process within the lake. The deposition of the particulate matter, and thus phosphorus sedimentation, attains a particularly high level in the inflow area of the lake.

In most of the lakes mentioned above, the phosphorus sedimentation flux attains a high level (Bloesch & Sturm 1986; Gächter *et al.* 1989; Kleiner & Stabel 1989). This likewise has been observed in reservoirs (Porcalova 1989).

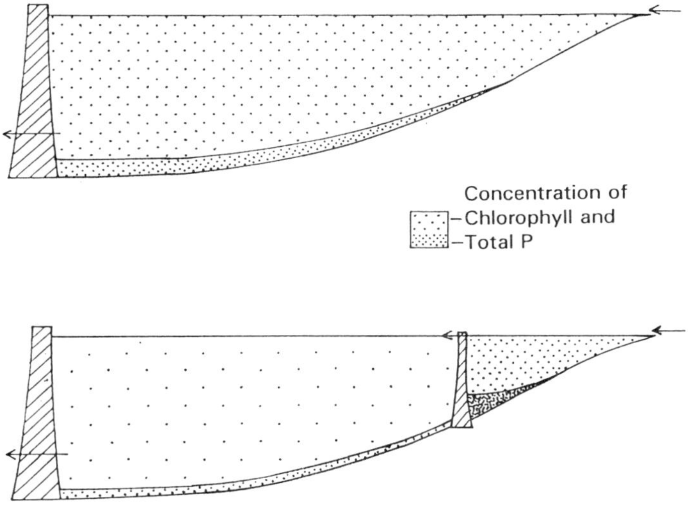

Figure 3. Diagrammatic representation of the effect of pre-reservoirs, showing an increase in phytoplankton growth and sedimentation, and a decreased loading of the adjacent main reservoir. The relevant processes of phosphorus and nitrogen elimination are: incorporation of N and P into biomass; absorption/binding of P to Fe, Ca and Al; formation of $MgNH_4PO_4$; nitrification/denitrification.

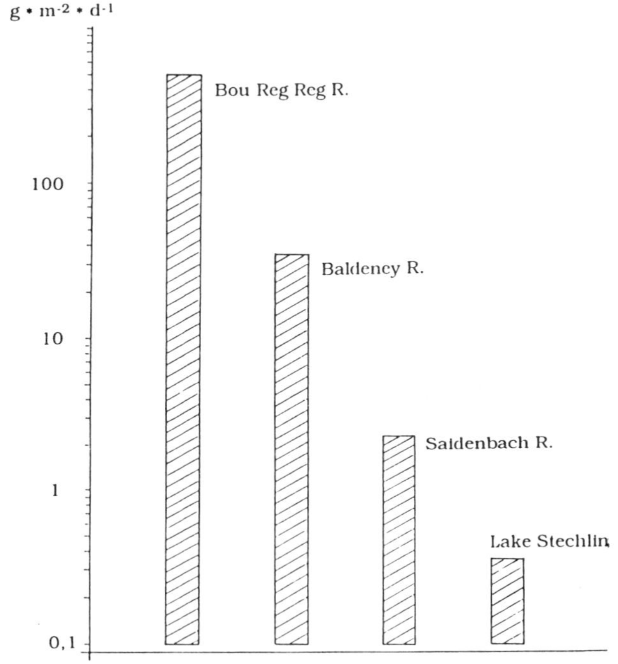

Figure 4. Comparison of the deposition rates of suspended solids (g dry wt m^{-2} d^{-1}) in three reservoirs and one lake with different intensities of erosion in their catchments. Bou Regreg is a large drinking-water reservoir in Moroc (Abouzaid *et al.* 1984). Baldeney Reservoir is an impoundment of the River Ruhr in a densly populated and highly industrialized area (Imhoff 1990), and Lake Stechlin is a deep oligotrophic lake of glacial origin with a small catchment area (data from Proft, *pers comm.*).

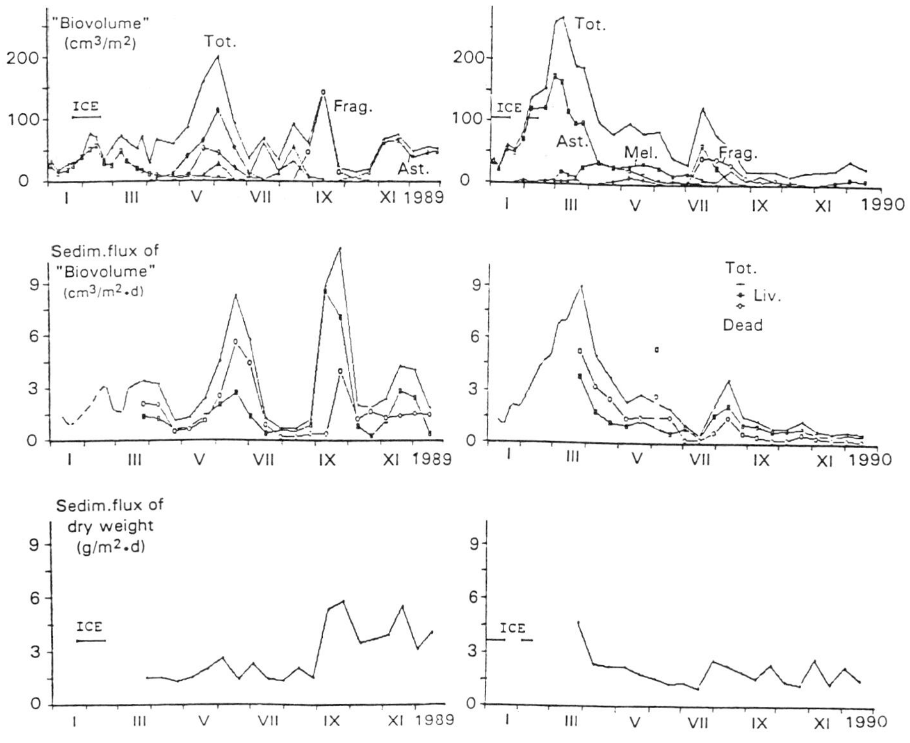

Figure 5. Seasonal variations of biomass (cm^3 m^{-2}), sedimentation flux (cm^3 m^{-2} d^{-1}) and total sedimentation flux (g dry wt m^{-2} d^{-1}) in Saidenbach Reservoir. —•—, total phytoplankton; □, *Asterionella*; ❍, *Fragilaria*; ■, *Melosira*.

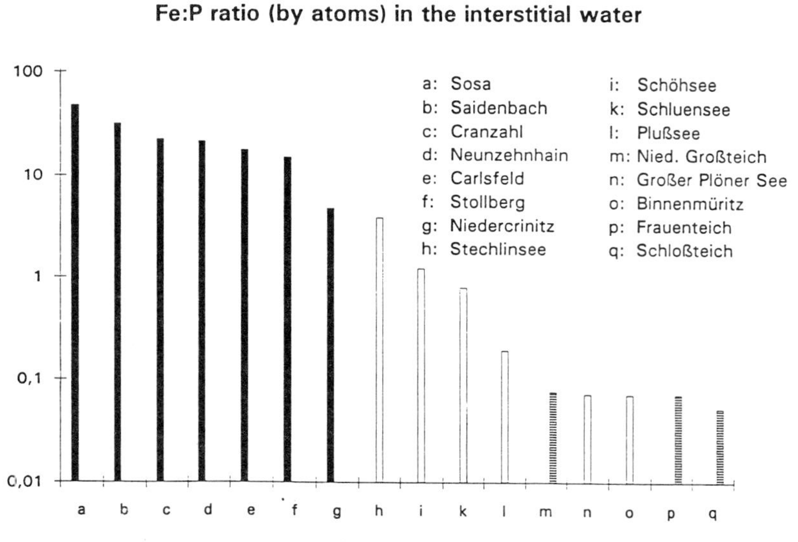

Figure 6. The atomic ratio of iron to phosphorus in the interstitial water of bottom sediments in different types of water-bodies: ■, reservoirs; □, lakes; ▤, carp ponds. The numerical data were drawn from Rühle (1971) and Guderitz (unpublished).

Sediments of reservoirs normally have a significantly higher iron content than sediments of lakes with a long water-residence time. In most of the drinking-water reservoirs presented in Figure 6, the atomic Fe:P ratio is essentially higher than 20:1, whereas it is below 1:10 in hypertrophic carp ponds and some eutrophic lakes. Furthermore, a substantial part of the iron is bound to sulphide in these conditions and thus is not available for the immobilization of phosphorus. In Saidenbach the (inorganic) sulphur concentration of the sediment was 0.38%. This is an indication of a comparatively high sulphide content, but as the iron concentration averages about 4% its major part is still available to bind phosphate.

Both Saidenbach and Neunzehnhain reservoirs have very high nitrate concentrations in the hypolimnion. The concentration of nitrate-nitrogen was nearly always higher than 4.5 mg l^{-1}. This means that the equivalent nitrate oxygen concentration was higher than that of molecular oxygen. Laboratory experiments with sediment cores showed that this nitrate assists in the immobilization of phosphate (Hupfer & Uhlmann 1991). Its velocity of diffusion from the water into the sediment is very high. As it is normally available in excess, there is only a low risk that an overshoot in phosphorus release occurs due to nitrate depletion. The immobilization of phosphate by the addition of nitrate was reversed by the presence of antibiotics. This indicates that microbial processes play an important role in the immobilization of phosphate, in addition to the chemical binding and adsorption which was described fifty years ago (Mortimer 1941). The existence of the oxidised barrier layer of the sediment – water interface, with its high content of trivalent iron, thus also depends upon the flux of nitrate oxygen.

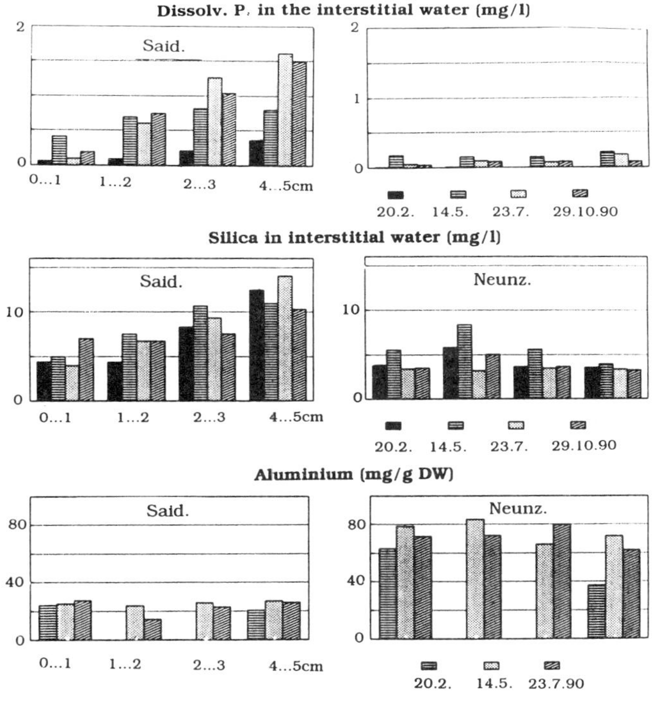

Figure 7. Comparison of dissolved reactive phosphorus (mg l^{-1}) and silica (mg l^{-1}) in the interstitial water of surface sediments and aluminium (mg g^{-1} dry wt) in the sediments of Saidenbach and Neunzehnhain reservoirs. The adjoining columns denote sampling dates in 1990 (20 February, 14 May, 23 July, 29 October) and each set represents samples taken at 0–1, 1–2, 2–3 and 4–5 cm depth in the sediments.

Figure 7 illustrates some differences between the Saidenbach and Neunzehnhain reservoirs. The concentration of dissolved phosphate in Neunzehnhain is obviously below the critical level which would be necessary to initiate a self-maintaining cycle of phosphorus release, phytoplankton growth, and sedimentation of phosphorus, together with reducing power (biomass). The vertical gradient of silica concentration in the interstitial water of Saidenbach corresponds to its growth-limiting role with regard to *Fragilaria crotonensis*.

The aluminium content of Neunzehnhain sediment was surprisingly high (Fig. 7) and obviously the result of acid atmospheric deposition. At times, the concentration of aluminium hydroxide flocs in the water-body was so high that these could be observed by the naked eye. In contrast the dissolution of aluminium from the soil is much less significant in the Saidenbach catchment, where the addition of lime for agriculture has been practised over many decades.

The impact of acid depositions originating from combustion processes is indicated by sulphate concentrations in the tributaries, which are much greater than would be expected from the natural geochemical conditions. This anthropogenic increase in sulphate was observed 35 years ago, but the effects only now become clearly evident. The significance of sulphate reduction is apparent from Figure 2, where the sampling points H1 and H2 refer to locations above and below the dam of the Haselbach underwater pre-reservoir. A longitudinal profile is presented on the right-hand part of the text-figure. The variations in water level are evident from the temperature profiles. At the bottom of the pre-reservoir there was a lower oxygen concentration and higher iron and sulphide concentrations, but nevertheless the sediment has a high phosphorus content, indicating that the pre-reservoir acts as a phosphorus trap.

Table 1. *Aspects of the phosphorus budget for Saidenbach Reservoir (downstream from the pre-reservoirs) in 1989 and 1990.*

A. Annual mass flow (kg year^{-1}), load per unit area (g m^{-2} year^{-1}) and mean eplilimnion concentrations (mg m^{-3}) of total phosphorus and dissolved reactive phosphorus.

Year	Total P Mass flow	Load	Concn	Dissolved reactive P Mass flow	Load	Concn
1989	2725.52	2.161	24.4	1027.6	0.815	5.54
1990	1535.95	1.197	20.5	528.6	0.412	3.81

B. Rates of sedimentation (cm day^{-1}) of total phosphorus (TP), dissolved reactive phosphorus (DRP) and particulate phosphorus (Part. P). calculated from the Vollenweider model (after Benndorf 1979) and measured with a sediment trap.

Year	Calculated DRP	TP	Measured Part. P	TP
1989	32.4	16.3	101	48
1990	24.5	10.8	79	45

C. Average total phosphorus load (g m^2 year^{-1}) to the reservoir (1) and above the sediment trap at 39 m depth (2), sedimentation flux (3) and (4) the flux expressed as a percentage of the load at 39 m (i.e. [(3)/(2)] × 100).

Year	(1)	(2)	(3)	(4)
1989	2.161	5.842	3.470	59.4
1990	1.197	3.246	2.929	90.2

D. Annual mass balance of total phosphorus, showing: (1) initial concentration (kg per total reservoir volume), (2) mass flow (kg year^{-1}), (3) loss in the outflow (kg year^{-1}), (4) theoretical initial concentration for the following year (kg per total reservoir volume), (5) elimination of TP from the water (kg per total reservoir volume) and (6) percentage eliminated (i.e. [(5)/(2)] × 100).

Year	(1)	(2)	(3)	(4)	(5)	(6)
1989	706.7	2726	854	2578	2081	76.4
1990	496.7	1536	460	1572	1182	77.0
1991	389.8	–	–	–	–	–

In Table 1, data on phosphorus sedimentation in Saidenbach are presented against a background of phosphorus mass-flow and concentration. The velocity of sedimentation measured by a sediment trap was still much higher than that expected from former calculations (12.1 cm d^{-1} for dissolved reactive phosphorus, Benndorf 1979). Data for the sinking velocity of total and particulate phosphorus have been compiled in Figure 8; the seasonal variations were lower than expected. The mass balance calculation (Table 1D) is for two succeeding years with very different hydrological conditions. Nevertheless the total phosphorus sedimentation fluxes were similar and comparable to data for sub-alpine lakes. In Saidenbach, phosphorus elimination was 76% in 1989 and 77% in 1990, corresponding to the levels measured in former years and confirming the significant role of phosphorus sedimentation in this material's balance sheet. This effect is amplified by the hypolimnetic outflow, which is a sink for dissolved reactive phosphorus. A very high phosphorus retention capacity has also been recorded for several lakes elsewhere, such as Lake Biwa (Tezuka 1986).

The sinking velocity of phosphorus in Saidenbach (Fig. 8) is equivalent to 280 – 370 m per year. As P-carriers, mineral particles with a high sorption/binding capacity are still much more important than phytoplankton biomass. If the numerical value of the phosphorus sedimentation flux is divided by the sediment flux in terms of dry weight, we obtain 0.3% P. This is equal to the concentration actually measured in the upper layers of the bottom sediment of Saidenbach (Fig. 9).

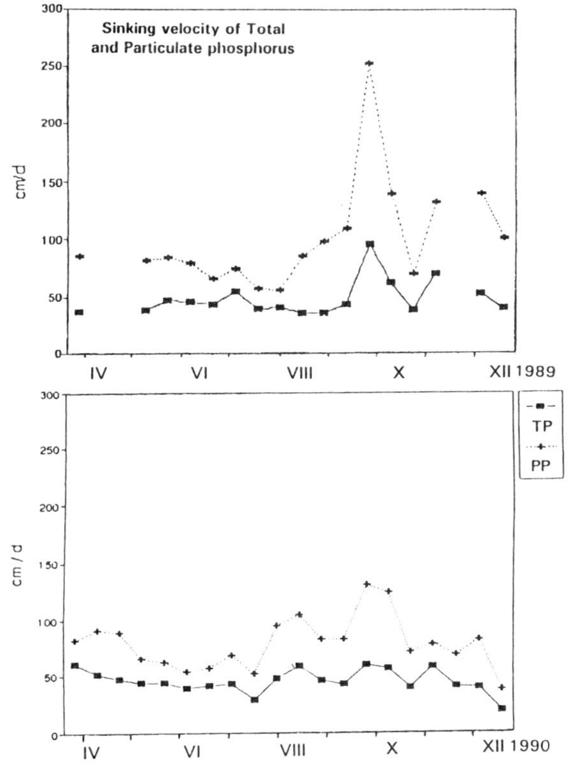

Figure 8. Seasonal variations of phosphorus sedimentation (cm d^{-1}) in Saidenbach Reservoir, in 1989 and 1990. ■, total phosphorus; +, particulate phosphorus.

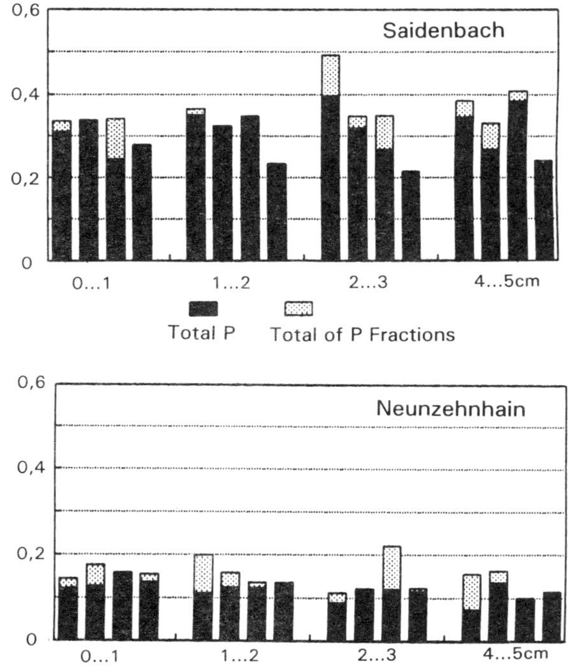

Figure 9. Total phosphorus (percent dry wt) of the surface sediments of Saidenbach and Neunzehnhain reservoirs. The four adjoining colums denote sampling dates in 1990 (as in Fig. 7) and each set represents samples taken at 0–1, 1–2, 2–3 and 4-5 cm depth in the sediments. (Fractionation after Uhlmann *et al.* 1990)

Sediment composition

The significant role of the bottom sediment as a store of phosphorus may be illustrated by a comparison with the concentrations in the interstitial water and in the hypolimnion water (Fig. 10). The concentration gradient between sediment and hypolimnetic water covers more than five orders of magnitude. This indicates that the mobilization of only a very small proportion of the sediment phosphorus would be equivalent to a very strong P-load surge. Though all three levels of phosphorus are lower in the oligotrophic Neunzehnhain than they are in Saidenbach (Fig. 10), the concentration of phosphorus in the sediment of Neunzehnhain is also still so high that a very slight mobilization of phosphorus would be an abundant source of eutrophication. Such a mobilization, however, does not occur in this type of reservoir, and the dissolved reactive phosphorus concentration was always very low in the water of Neunzehnhain. Figure 10 indicates that the phosphorus concentration in the sediment of Saidenbach was higher than in Neunzehnhain by a factor of 2. On the other hand, the phytoplankton biomass was, on average, higher by about two orders of magnitude.

Surprisingly, there was no significant difference between the numbers of bacteria per ml of dry weight or of fresh sediment in the two reservoirs. Likewise the activity of esterases (fluorescein diacetate) was not significantly different (Fig. 11). This underlines the finding of Hupfer (1990) that the results obtained with fluorescein diacetate are closely related to microbial biomass. However, they can not be used as a measure of potential *in situ* activity. The latter is better indicated by the loss in breaking strength of cellulose fibres, the activity of

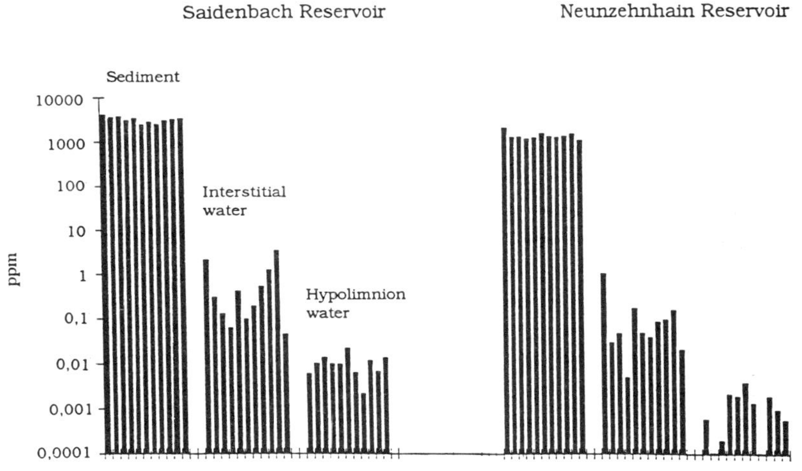

Figure 10. The gradient of phosphorus concentrations (parts per million) between water and sediment in the mesotrophic–eutrophic Saidenback Reservoir and the oligotrophic Neunzehnhain Reservoir. The columns show corresponding results for eleven samples of sediment (0–5 cm), interstitial water and hypolimnion water, respectively.

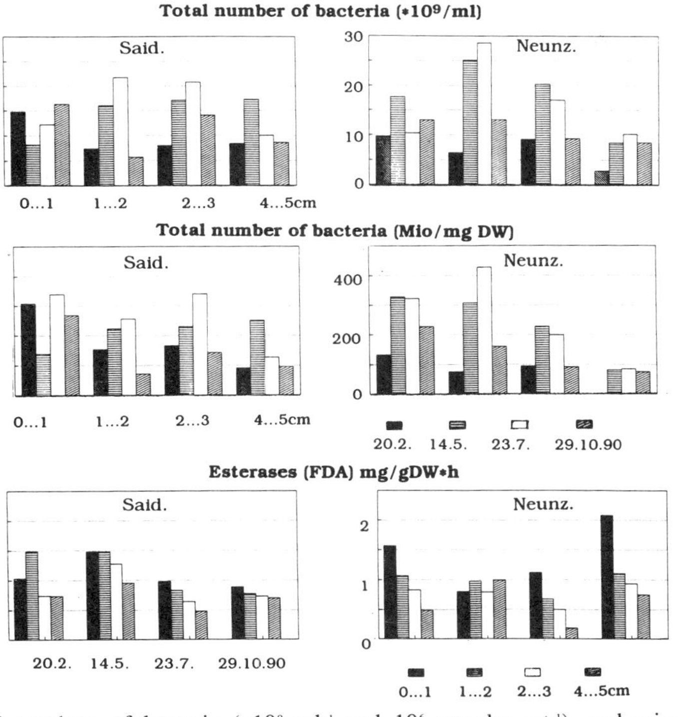

Figure 11. Total numbers of bacteria (x10^9 ml^{-1} and 10^6 mg dry wt^{-1}), and microbial cleavage of fluoresceine diacetate (mg g^{-1} dry wt h^{-1}) in samples from the surface sediments of Saidenbach and Neunzehnhain reservoirs. The four adjoining columns denote sampling dates in 1990, (as in Fig. 7) and each set represents samples taken at 0–1, 1–2, 2–3 and 4–5 cm depth in the sediments.

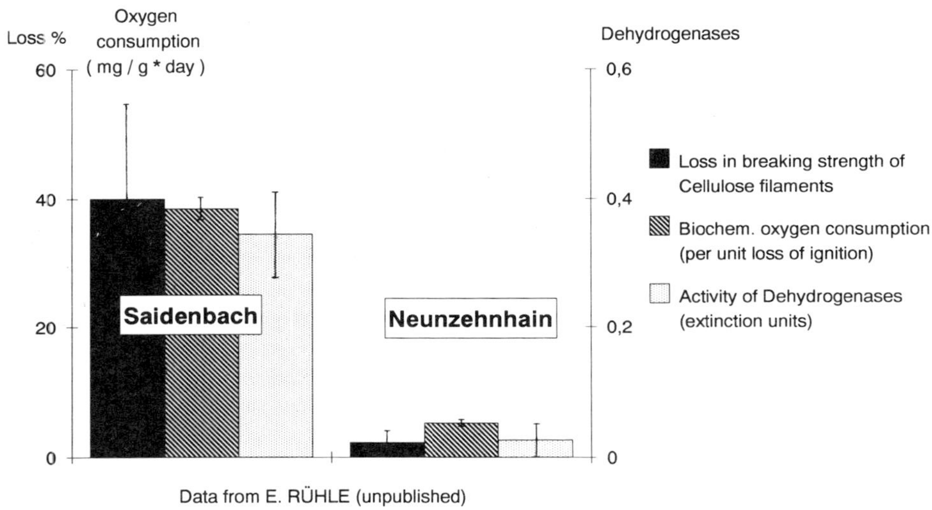

Figure 12. A comparison of some criteria for *in vitro* microbial activity of the surface sediments of Saidenbach and Neunzehnhain reservoirs (after E. Rühle, unpublished), showing: ■, the loss (%) in breaking strength of cellulose filaments; ▨, biochemical oxygen demand (per unit loss of ignition, mg g^{-1} d^{-1}); □, activity of dehydrogenases (extinction units)

dehydrogenases, and the biochemical oxygen demand of the sediment (Fig. 12). Obviously the majority of bacteria in Neunzehnhain sediment are in a dormant form; their activity seems to be both substrate- and transport-limited. These bacteria probably play a significant role in the immmobilization of phosphorus. This is indicated by the comparatively high proportion of NaOH-soluble non-reactive phosphorus in the sediments of both reservoirs (see also Hupfer & Uhlmann 1991). As in activated-sludge samples from sewage treatment plants with biological phosphorus elimination, this fraction may contain microbial polyphosphates (see Uhlmann *et al.* 1990), and it is supposed that the situation is similar in seston particles and in bottom sediments with similarly very high concentrations of bacteria. (Unfortunately, direct detection of polyphosphates in the sediment samples by means of NMR was impossible because of interference caused by iron). In Neunzehnhain, there is obviously no stimulus for the bacteria to mobilize their phosphorus reserves due to the lack of suitable substrates.

Conclusions

A longitudinal separation of a reservoir into cells arranged in series may significantly contribute to the immobilization of phosphorus in the first cells. There are two types of P-carriers: (1) biomass, either living or dead, or in an altered form as faecal matter or detritus, and (2) mineral particles with a high binding capacity for phosphorus. The influx of iron compounds is relevant to the inactivation of both phosphate and sulphide. Due to the high flushing rate, the Fe:P ratio is normally overstochiometric. Other important chemicals which are essential to the lake's internal water-treatment process are aluminium and nitrate. In catchment areas with limestone dominance, phosphorus sedimentation is governed by adsorption or chemical binding to calcium carbonate. Acid atmospheric depositions rich in sulphur influence the phosphorus balance of the well-buffered mesotrophic to eutrophic Saidenbach and the oligotrophic Neunzehnhain reservoirs in different ways (Fig. 13). Whereas

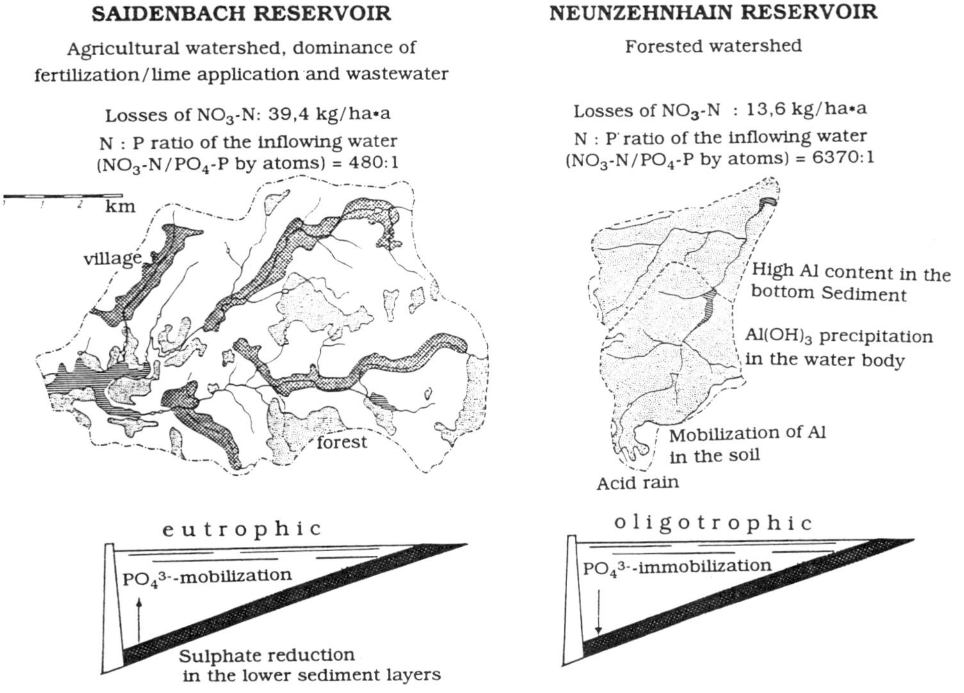

Figure 13. Summary of the effects of atmospheric deposition and land-use on two drinking-water reservoirs of different trophic character.

the natural sulphate concentration of surface waters in the central Erzgebirge hills should be very low, due to the character of the rocks, it is now available in excess as a component which promotes the mobilization of phosphate in sediments. On the other hand, sulphide formation is negligible in Neunzehnhain sediment due to the lack of "reducing power"(H-donors). On the contrary, the acid rainwater promotes the mobilization of aluminium in the poorly-buffered soils, and this results in the formation of aluminium hydroxide in the reservoir. This is equivalent to a high potential of phosphate inactivation. The phosphorus load to the reservoir is far below the level which would be necessary to trigger self-maintaining cycles of phytoplankton growth and phosphate regeneration.

The authors express their gratitude to the Ministry of Research and Technology, Bonn, for funding the research project (BMFT 30 F 100701). They are grateful to Dr W. Horn, Dr L. Paul and M. Hupfer for helpful suggestions and to C. Appelt, P. Ackermann, K. Köhler and M. Weishaupt for technical assistance.

References

Abouzaid, H., Landner, L. & Falkowski, T. (1984). Eutrophication of the Sidi Mohammed Ben Abdellah Reservoir in Morocco. *Water Quality Bulletin*, **9**, 37-41, 71.

Ausgewählte Methoden der Wasseruntersuchung (1986), Bd. 1, *Chemische, Physikalisch-chemische und Physikalische Methoden*. Fischer Verlag, Jena.

Benndorf, J. (1979). A contribution to the phosphorus loading concept. *Internationale Revue der Gesamten Hydrobiologie*, **64**, 177-188.

Benndorf, J. & Pütz, K. (1987a and b). Control of eutrophication of lakes and reservoirs by means of pre-dams. I. Mode of operation and calculation of the nutrient elimination capacity. II. Validation of the phosphate removal model and size optimization. *Water Research*, **21**, 829-838, 839-842.

Bloesch, J. & Sturm, M. (1986). Settling flux and sinking velocities of particulate phosphorus (PP) and particulate organic carbon (POC) in Lake Zug, Switzerland. In *Sediments and Water Interactions* (ed. P. G. Sly), pp. 481-496. Springer, Heidelberg and New York.

Gächter, R., Mares, A., Grieder, E., Zwyssig, A. & Höherer, P. (1989). Auswirkungen der Belüftung und Sauerstoffbegasung auf den P-Haushalt des Sempachersees. *Wasser, Energie, Luft*, **81**, 481-496.

Hamilton-Taylor, J. Willis, M. & Reynolds, C. S. (1984). Depositional fluxes of metals and phytoplankton in Windermere as measured by sediment traps. *Limnology and Oceanography*, **29**, 695-710.

Hofmann, R. (1972). Untersuchungen zum Phosphathaushalt der Talsperre Saidenbach. Dissertation Fakultät Bau-, Wasser-und Forstwesen TU Dresden.

Horn, W. & Horn, H. (1990). A simple and reliable method for the installation of sediment traps in lakes. *Internationale Revue der Gesamten Hydrobiologie*, **75**, 269-270.

Hupfer, M. (1990). Enzymatische Aktivitäten in Gewässersedimenten. (Unpublished Research Report, Institute of Hydrobiology, University of Dresden).

Hupfer, M. & Uhlmann, D. (1991). Microbially mediated phosphorus exchange across the mud-water interface. *Verhandlungen der Internationalen Vereinigung für Theoretische und Angewandte Limnologie*, **24**, 2999-3003.

Kleiner, J. & Stabel, H. H. (1989). Phosphorus transport to the bottom of Lake Constance. *Aquatic Science*, **51**, 181-191.

Mortimer, C. H. (1941, 1942). The exchange of dissolved substances between water and mud in lakes, I and II. *Journal of Ecology*, **29**, 280-329; 30, 147-201.

Petersohn, D. (1987). Das Phytoplankton im Längsschnitt der Talsperre Saidenbach in Abhängigkeit von biotischen und abiotischen Faktoren. Dissertation Fakultät Bau-, Wasser-und Forstwesen TU Dresden.

Porcalova, P. (1989). Sedimentation in the Rimo Reservoir. *Archiv für Hydrobiologie Beihefte Ergebnisse der Limnologie*, **33**, 349-353.

Rühle, E. (1971). Sedimentuntersuchungen an Trinkwassertalsperren des Erzgebirges. Dissertation Fakultät Bau-, Wasser-und Forstwesen TU Dresden.

Tezuka, Y. (1986). Does the seston of Lake Biwa release dissolved inorganic nitrogen and phosphorus during aerobic decomposition ? Its implication for eutrophication. *Ecological Research*, **1**, 293-302.

Uhlmann, D. & Benndorf, J. (1980). The use of primary reservoirs to control eutrophication caused by nutrient inflows from non-point sources. In *Land-use Impacts on Lake and Reservoir Ecosystems, Proceedings MAB 5 Workshop May 26 - June 2, 1978, Warsaw, Poland* (eds N. Duncan & J. Rzoska), pp. 152-195. Facultas Verlag, Vienna.

Uhlmann, D., Röske, I., Hupfer, M. & Ohms, G. (1990). A simple method to distinguish between polyphosphate and other phosphate fractions of activated sludge. *Water Research*, **24**, 1355-1360.

Calcite covering of sediments as a possible way of curbing blue-green algae

HELMUT KLAPPER
GKSS Institute for Water Research Magdeburg, D-O-3010 Magdeburg, Gouvernementsberg 1, Germany

Natural calcite precipitation in lakes is a well-known control mechanism of eutrophication. In hard-water lakes, calcite deposits on the flat bottoms of shallow lakes and near the shores of deeper lakes resulted from biogenic decalcification during the millenia after the last glacial period.

The objective of a new restoration technology is to intensify the natural process of precipitation by utilizing the different qualities of calcareous mud layers. In a pilot experiment in Lake Rudower See, East Germany, phosphorus-poor deeper layers of the sediments were flushed out and spread over the phosphorus-rich uppermost sediments, to promote the co-precipitation of calcite with phosphorus from the water-column.

Introduction: phosphorus elimination as a tool against blue-green algae

Several different methods have been used for curbing blooms of blue-green algae by drastically reducing the phosphorus concentration in the water column. From a series of field experiments on nutrient precipitation in lakes, using ferric or aluminium salts, the most promising so far is to seal the sediments with phosphorus-binding cations. To hinder the development of bloom-forming nitrogen-fixing species of Cyanophycea, the ratio of N:P should be kept higher than 10 by low phosphorus concentrations. These conditions were achieved in a 100-ha lake near Magdeburg after application of 400 tonnes of aluminium sulphate in liquid form (5 grams Al^{3+} per m^3). Prior to the precipitation of phosphorus, *Aphanizomenon flos-aquae*, *A. gracile* and *Anabaena lemmermannii* formed water-blooms which polluted the bathing areas of the lake. Following the addition of aluminium sulphate, algal blooms have been absent for 5 years and only some colonies of *Microcystis* and *Limnothrix* in low concentrations were found at the end of the recreation seasons. Secchi-depth transparency of the water in summertime increased from year to year. Also, macrophytes were re-established in the lake and filter-feeding cladocerans are now able to eat and control the more digestible phytoplankton species. An overdose of aluminium (twice the amount needed to precipitate the phosphorus), formed a thin hydroxide layer on the sediment surface and prevented the re-solution of phosphorus into the overlying layer of water during anaerobic conditions (Roenicke 1991).

Calcite precipitation: a natural control mechanism

Despite its usefulness, from both ecological and economic standpoints the application of ferric or aluminium salts is a relatively difficult and expensive tool for improving water quality. Therefore we have tried to find a more natural, easier and cheaper technology for reducing the phosphorus concentrations in lake water. Hard-water lakes demonstrate such a natural process, in the form of calcite precipitation or biogenic decalcification. The main process of interest is a co-precipitation of phosphorus containing biological and colloidal matter.

Because of calcite precipitation in early spring, hard-water lakes may be more oligotrophic than soft waters containing similar concentrations of phosphorus. The aim of the new restoration technology is to intensify this natural process by utilizing the different qualities of calcareous sediment layers to provide an artificially enhanced internal precipitation of phosphorus (Strong & Eadie 1978; Benndorf *et al.* 1983; Stabel 1986; Koschel 1989; Kleiner 1988, 1990)

Calcite flushing in a polymictic lake

Investigations were first made on Lake Rudower See in East Germany. This 170-ha shallow lake lies in a tectonic trench. The deeper parts of the lake basin have been filled up by lake lime, forming a flat surface at 5–6 m depth. Samples were taken at different points on the lake, both on the sediment surface, with an Ekman grab sampler, and in 2–3 metres of sediment with a chamber collector and a boring bar. Here the sediments have lower concentrations of phosphorus and organic matter but higher concentrations of cations with phosphorus-binding properties (see Fig. 1 and Table 1).

Table 1. *Sediment quality (means of eight sampling stations, shown in Figure 1) in Lake Rudower See*

Depth	PO_4^{3-} (mg g dry wt^{-1})	Fe (mg g dry wt^{-1})	Ca^{2+} (mg g dry wt^{-1})	Ignition loss (%)
Surface	4.2	7.28	175	28.6
2–3 metres	2.25	8.69	218	16.2
Difference (%)	-46.4	+19.4	+24.6	-43.4

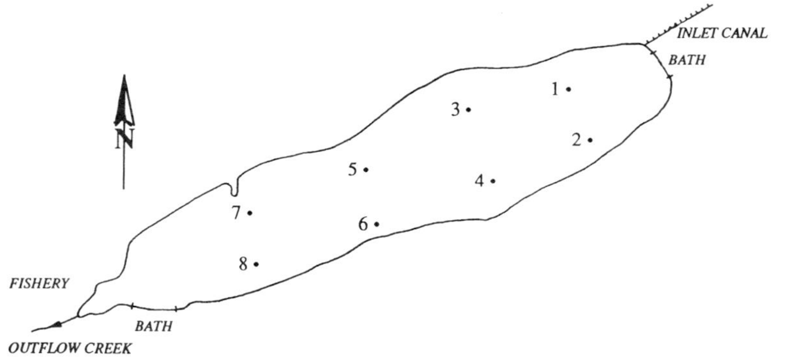

Figure 1. Eight sampling points for sediments in Lake Rudower See; BATH shows the bathing areas at each end of the lake.

The main constituent of all the samples is calcite ($CaCO_3$). The lake lime in deeper layers is very compact and cannot be pumped out directly. Therefore it was flushed out with the help of rinsing jet equipment and then the water–calcite suspension was pumped and spread over the phosphorus-rich youngest sediments (see Fig. 2).

The generation of some unevenness on the lake bottom, in the form of holes or trenches, is a wanted side-effect. These holes act as sediment traps, in which the newly-forming sludge from sedimented plankton collects and becomes locally concentrated, so adverse influences on water quality are localized and the re-solution of phosphorus is diminished (see Fig. 3).

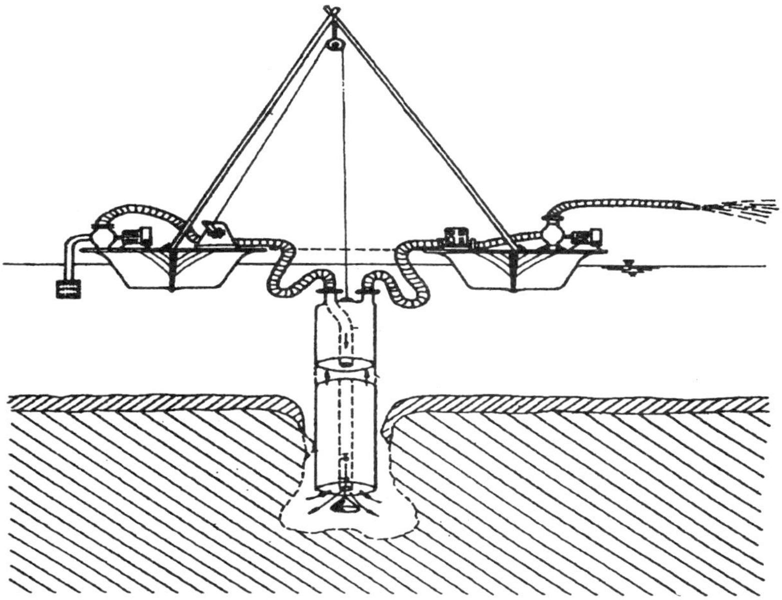

Figure 2. Drill-flushing method for covering phosphorus-rich recent sediments with a layer of calcite from deeper sediments.

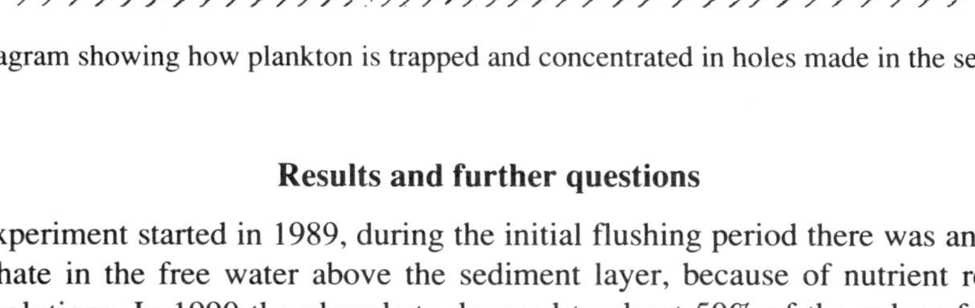

Figure 3. Diagram showing how plankton is trapped and concentrated in holes made in the sediments.

Results and further questions

In a pilot experiment started in 1989, during the initial flushing period there was an increase of orthophosphate in the free water above the sediment layer, because of nutrient release from interstitial solutions. In 1990 the phosphate dropped to about 50% of the values for 1988, the year prior to flushing with calcite, and in 1991 orthophosphate concentrations reached values near zero (Fig. 4).

So far the expected effects on blue-green algae have not materialized, although quantitative measurements of biomass in the lake are lacking for the period prior to flushing with calcite. *Microcystis* still forms water blooms, so the quality goal is not yet reached.

A series of theoretically possible mechanisms besides that of phosphorus limitation need to be investigated in the near future. (1), Is it possible to counteract a rise in pH caused by higher carbonate hardness and buffer capacity, and thus overcome the natural advantage of

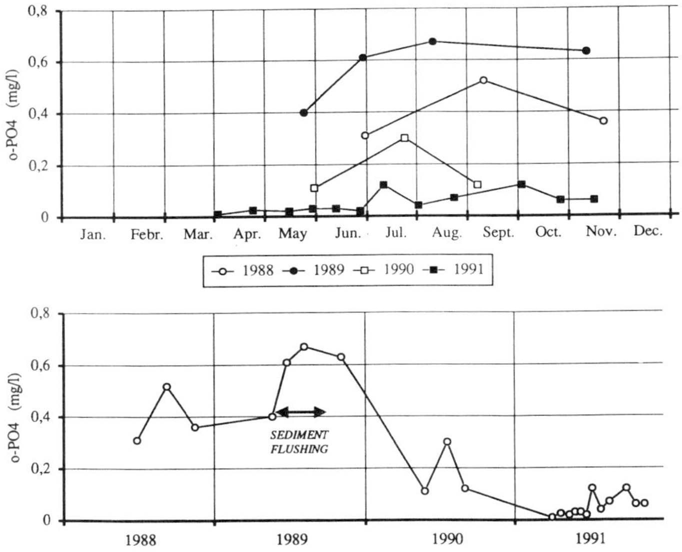

Figure 4. Orthophosphate concentrations (mg l^{-1}) in the free water of Lake Rudower See, shown chronologically (below) for 1988-1991, and compared between years (above) for ○, 1988, ●, 1989, □, 1990 and ■, 1991.

blue-green algae which can photosynthesise at high pH values? (2), Is there any evidence that carbon dioxide is increased by oxidation of methane when this is rinsed into the free water when flushing the calcite? (3), Can blue-green algae be curbed by covering or burying their bottom-resting cells with lime? (4), After calcite flushing, will the desired silicious algae compete more successfully with blue-green algae because of higher concentrations of liberated silicon? (5), Can other stabilization mechanisms be used for combatting blue-green algae, e.g. higher stocks of suitable herbivorous fishes and/or increased growth of macrophytes? (6), What can be done to prevent the dramatic retrogression of reed beds? (7), Is it in fact possible to curb blue-green algae in a polymictic lake that is no more than 6 metres deep and therefore has a high availabilty of phosphorus from the lake bottom?

Of course, observations on the effects of this first experiment will be continued. From other restoration experiments it is known that major responses occur mostly after a time lag of several years, when a new balance level is rather quickly achieved.

Preparation for the next application on a deep dimictic lake

In the meantime a second experiment is being prepared for the 50-m deep dimictic Lake Arendsee, for which good limnological data are available. Calcite deposition has been found in the littoral region of this lake. The aim is to flush this calcite from the littoral and spread it over the deepest parts of the lake, where there are soft, phosphorus-rich sediments.

The biological goal is to combat the frequent occurrence of *Oscillatoria rubescens*. A technical problem is how to transport the flushed calcite over distances of more than 1 kilometre to the required places. A large floating plastic tube, towed by boat, should be suitable for this purpose.

This report is the first on a new approach to restore hard-water lakes. If it is successful, some 100 similar lakes (in Germany) may be restored by this technology.

References

Benndorf, J., Proft, G. & Recknagel, F. (1983). Calcite precipitation as a natural control mechanism of eutrophication. *Archiv für Hydrobiologie*, **98**, 380-408.

Kleiner, J. (1988). Coprecipitation of phosphate with calcite in lake water: A laboratory experiment modelling phosphorus removal with calcite in Lake Constance. *Water Research*, **22**, 1259-1265.

Kleiner, J. (1990). Calcite precipitation-regulating mechanism in hardwater lakes. *Verhandlungen der Internationalen Vereinigung für Theoretische und Angewandte Limnologie*, **22**, 638-643.

Koschel, R. (1989). Pelagic calcite precipitation and trophic state of hardwater lakes. *Archiv für Hydrobiologie, Beiheft Ergebnisse der Limnologie*, **33**, 713-722.

Roenicke, H. (1991). Langzeiteinfluss einer Fällmittelapplikation auf die Nährstoffdynamik und Phytoplanktonsukzession in einem Magdeburger Baggersee. *Limnologentagung der SIL und DGL, Mondsee/Österreich. Tagungsband, 205-209.*

Stabel, H.-H. (1986). Calcite precipitation in Lake Constance. *Limnology and Oceanography*, **31**, 1081-1093.

Strong, A. E. & Eadie, B. J. (1978). Satellite observation of calcium carbonate precipitation in the Great lakes. *Limnology and Oceanography*, **23**, 877-887.

Development of trophic state and phytoplankton in an oligo–mesotrophic drinking-water reservoir (Kleine Kinzig) in the northern Black Forest, Germany

EBERHARD HOEHN

Limnology Bureau Heohn (LBH), Glümerstr, 2a, D/w-7800 Freiburg, Germany

Since 1989, intensive studies have been made on a relatively new (1983–84) oligotrophic reservoir and its pre-reservoir in the Black Forest. This paper briefly reports on the hydrochemistry, especially annual variations in phosphorus loadings, and the seasonal development of phytoplankton in 1989 and 1990.

Introduction

The drinking-water reservoir Kleine Kinzig has only been in existence since 1983–84, and it is the only reservoir of its kind in Baden-Würtemberg. The geological formation of its catchment area, which is situated in the northern Black Forest, consists of Bunter sandstone. A rare form of unbuffered water, low in ions and nutrients, is found here; Kleine Kinzig is one of the few oligotrophic lakes in Baden-Württemberg. The rainfall amounts to 1,600 – 2,000 mm per annum. The main reservoir has a maximum depth of 60 m and a storage volume of $13x10^6$ m^3.

An intensive limnological investigation of the reservoir has been carried out since 1989. Key problems of these investigations are acidification and algal nutrients, and their effects on the biocoenoses (phytoplankton (E. Hoehn), zooplankton (S. Schmidt-Halewicz) and fish (U. Dost)). On the largest inlet (Kleine Kinzig) there is a pre-reservoir (Z_{max} = 10 m, volume 125,000 m^3) which also serves as a trap to reduce the input of turbidity and nutrients into the main reservoir. This pre-reservoir was investigated in the same way as the main reservoir. The following presents the first results of the development of the phytoplankton biomass together with the observed phosphorus concentrations. The effectiveness of the pre-reservoir as a nutrient trap is hereby assessed.

Methods

In 1989 and 1990, the phytoplankton biomass and spectrum of species (at 0 to 20 m in the main reservoir and surface to bottom in the pre-reservoir), chlorophyll, oxygen, conductivity and under-water light were examined at intervals of 7–10 or 14 days, in each case at the deepest point near the dam. Chemical analysis (ionic balance, phosphorus fractions, silica, DOC and UV-absorption) was done monthly for the pelagic habitat and every fortnight for the three most important inlets. Primary production (O_2 method) was determined monthly. The quantity of water in the inlets and outlets was registered continuously, as was global radiation. The bulk precipitation in the hydrological year 1990 was investigated hydrochemically (ionic balance, phosphorus fractions and DOC).

Figure 1. Hydrochemistry of the Kleine Kinzig pre-reservoir and its inlet during 1989 and 1990.
(A), Inlet flow (upper, litres s^{-1}) and water retention-times (lower, days) for the epilimnion (—), the entire pre-reservoir (. . . . .) and the "theoretical" retention-time (70% o-PO_4 elimination) (——•——•——).
(B), Inlet phosphorus concentrations (upper, µg P l^{-1}) and phosphorus loading (lower, g P d^{-1}) for total phosphorus (—), orthophosphate-P (. . . . .), total dissolved phosphorus (——•——•——) and particulate phosphorus (hatched).

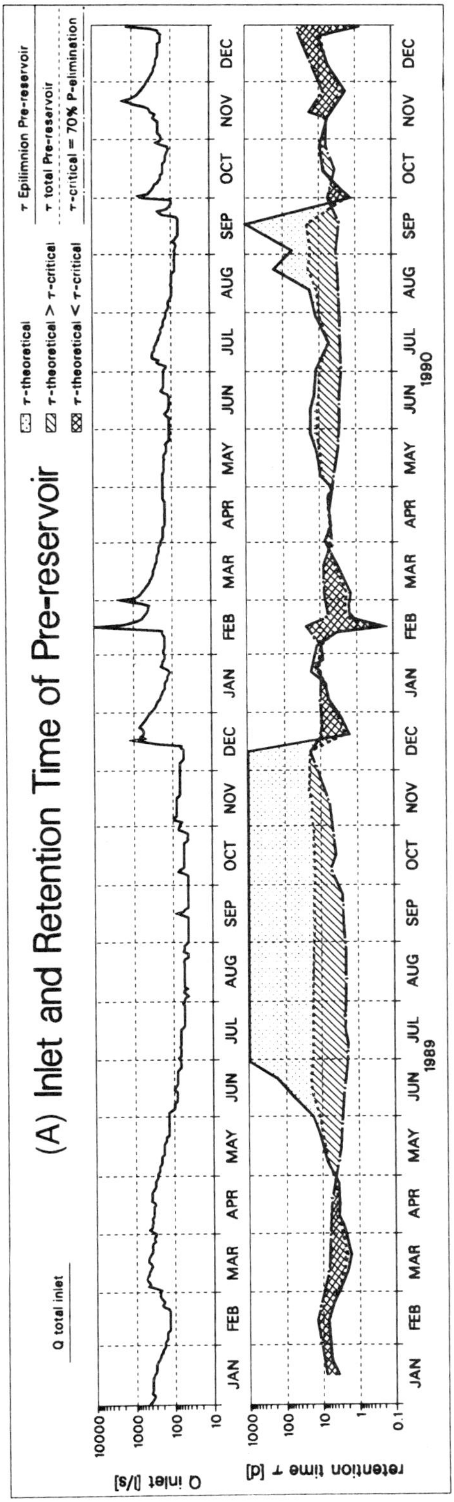
(A) Inlet and Retention Time of Pre-reservoir
Q total inlet
τ-theoretical
τ-theoretical > τ-critical
τ-theoretical < τ-critical
τ Epilimnion Pre-reservoir
τ total Pre-reservoir
τ-critical = 70% P-elimination
Q inlet [l/s]
retention time τ [d]
10000
1000
100
10
0.1
JAN FEB MAR APR MAY JUN JUL AUG SEP OCT NOV DEC
1989
1990

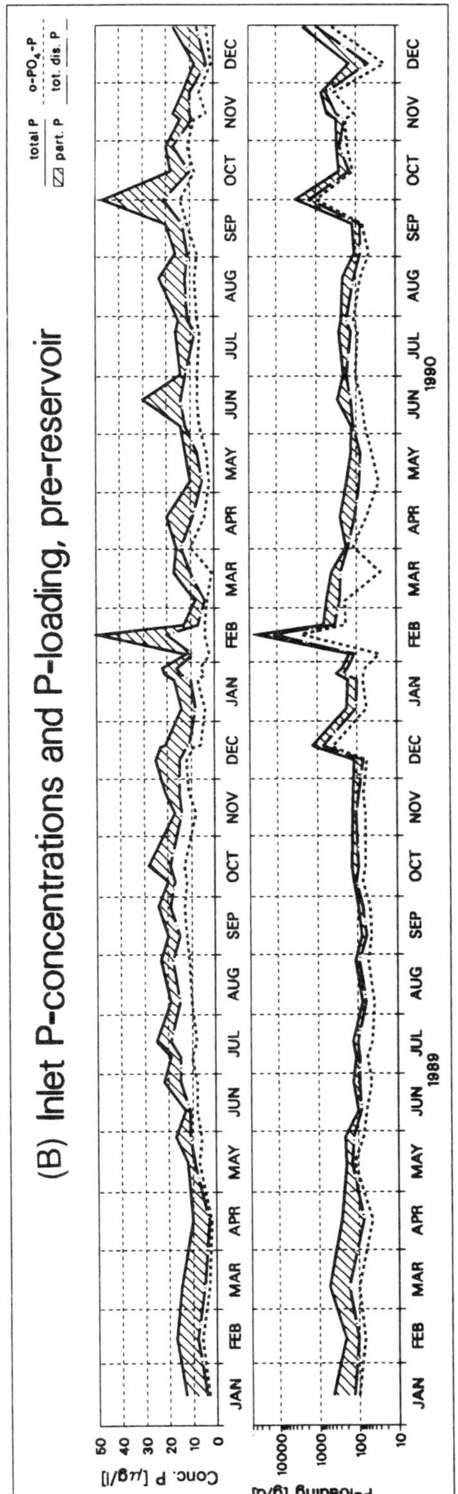
(B) Inlet P-concentrations and P-loading, pre-reservoir
total P
o-PO4-P
part. P
tot. dis. P
Conc. P [µg/l]
P-loading [g/d]
50
40
30
20
10
0
10000
1000
100
10
JAN FEB MAR APR MAY JUN JUL AUG SEP OCT NOV DEC
1989
1990

Results

Hydrochemistry

The discharge from the inlets has an essential influence on the hydrochemistry of both the pre-reservoir and the main reservoir. It regulates the water retention-time in the pre-reservoir and therefore has a significant control over biological activity (algal biomass and productivity). Owing to the physical relief of this Black Forest region, floods drain away quickly, as occurred, for example, in February 1990 (Fig. 1A). The theoretical water retention-time τ was calculated for the epilimnion and for the total pre-reservoir. Differing from Benndorf *et al.* (1975), the epilimnion was assumed to be the upper 5-metre layer of the pre-reservoir, because 1% of the incident radiation ordinarily could be measured at a depth greater than 5 m. In the epilimnic layer, 0–5 m, the mean underwater radiation was 29.5% of the incident radiation at the surface. It was supposed that the epilimnic layer discharges only via the overflow sill. Hence the water retention-time (dotted area, Fig. 1A) must range between the values for the total pre-reservoir (dashed line, Fig. 1A) and the values for the epilimnion (solid line, Fig. 1A).

The curve, τ–crtical = 70% PO_4–elimination in Figure 1A, shows the theoretical retention-time during which, according to Benndorf *et al.* (1975) and Fachbereichstandard der DDR (1983), just 70% of the orthophosphate was eliminated by algal activity. This so-calculated retention-time is also defined as the critical retention-time, because at lower values of τ-critical the loss by dilution is greater than can be compensated by an increase of the phytoplankton (Benndorf *et al.* 1975).

At times of inlet discharge higher than 110 litres s^{-1} the theoretical retention-time is less than the critical retention-time (cross-hatched area, Fig. 1A). Thus the increase of algae is only possible when the inlet discharge is lower than 100 litres s^{-1} (simple hatched area, Fig. 1A).

Due to a leaking weir at the outlet construction of the pre-reservoir, a volume of approximately 66 litres s^{-1} is drained off by hypolimnic waters. If in dry summers the discharge of the entire outlet falls beneath this quantity then no more water can drain off via the overflow sill. In this way the thermal and hydrological conditions in the epilimnion of the pre-reservoir undergo a drastic change, as the water retention-time is extremely long (July-November 1989). At this time, high phosphorus-loadings occur in the pre-reservoir as the orthophosphate concentrations reach their highest levels in the inlet, exceeding 12 $\mu g\ l^{-1}$, especially in 1990 (Fig. 1B). Concurrently there was a large increase in phytoplankton biomass, especially in the summer of 1989 (Fig. 2).

From the data available it is not possible to establish any constant trend for the loss rate of orthophosphate. On average about 56% was removed. However, normally there is no loss of total phosphates; in fact on average there is an accumulation of 6%. As a rule the pre-reservoir is only effective as a nutrient trap when the inlets are discharging very little water.

When discharge rates are low, as in the summer of 1989, algal growth can be limited by the consumption of available carbon, due to a high rate of primary production and dense populations of phytoplankton. As a result, pH values rise above 10, whereby aluminium becomes increasingly soluble; such conditions constitute a toxic environment for fish!

Turbid materials washed in with floods are also an important factor, because they carry adsorbed phosphates into the reservoir. The concentrations and loadings of particulate phosphates during flooding, particularly in February 1990 (see hatched area in Fig. 1B), illustrate this. Directly after flooding in February 1990 the proportion of these particulate phosphates increased three-fold in the main reservoir.

The volume-weighted concentrations of phosphorus fractions in the open (bulk) precipitation (Table 1) indicate that atmospheric fertilization may be postulated for the main reservoir, as orthophosphate is not detectable in the epilimnion in the summer.

Table 1. *Some volume-weighted mean concentrations and amounts of nitrogen and phosphorus deposited in open precipitation at Kleine Kinzig in the hydrological year, 1990.*

Mean pH 4.93 (12 μ equiv. $H^+ l^{-1}$, 195.8 mol $H^+ ha^{-1}$), conductivity 41.9 μS cm^{-1}.

Total amount of rainfall: 1631.9 litres m^2

Fractions	Concentrations (μg l^{-1})	Amounts (g ha^{-1})
NH_4–N	387	6321
NO_3–N	319	5207
Total N	706	11528
PO_4–P	4	61
Dissolved P	7	120
Particulate P	11	174
Total P	18	294

Phytoplankton

During 1990, adsorptively-bound phosphates in the main reservoir provided extra nutrients for algae and eutrophication increased, especially after the winter floods of 1990. Thus the diatom *Asterionella formosa* appeared for the first time in the spring of 1990 and simultaneously the silica concentrations decreased (also see Sommer 1989). Throughout the early summer, the cells of *Asterionella formosa* sank to deeper water levels, and the population diminished greatly when the "clearwater phase" was reached at the end of June 1990. This phase developed more rapidly in 1990 than in 1989 and it lasted for a shorter period of time; in 1989 algal biomass remained relatively high through July and August (Fig. 2). According to the PEG model (Sommer *et al.* 1986), this indicates development towards the mesotrophic state. Whether or not these factors, together with the appearance of *Asterionella formosa*, can be attributed to the discharge (flooding) alone and are therefore reversible, will have to be ascertained by further investigations.

The biomass of Dinophyceae (main species *Gymnodium uberrimum*) increased significantly in the second half of 1990 (Fig. 3). On the other hand the chlorococcal species *Dictyosphaerium pulchellum*, which had been relatively abundant in the main reservoir in 1989, almost disappeared in 1990, and there was also a distinct decrease in Cryptophyceae (*Cryptomonas erosa*, *C. pyreniodifera* and *Rhodomonas minuta*). The appearance of *Synura* sp. in autumn was observed again in 1990. In both 1989 and 1990 the development of the total biomass could be represented as a double-peaked curve, with maxima in spring and autumn (Fig. 3). As expected in the pre-reservoir, times with growth of phytoplankton coincide with times when the theoretical retention time exceeded the critical retention time (Fig. 2).

In 1989, the summer maximum in the pre-reservoir was distinctly characterized by Phytomonadina (*Chlamydomonas insolita*, *Chl. pseudopertusa*) and the diatom *Fragilaria capacina* (= *Synedra rumpens*). A large proportion of the summer biomass also consisted of the tetrasporal (chlorococcal) species *Pseudosphaerocystis lacustris* (Fig. 2). Dinophyceae were only detectable in the pre-reservoir towards the close of the summer maximum (*Peridinium inconspicuum*). This group of algae apparently has a longer generation time, which was rendered possible during the dry summer of 1989 due to the longer water retention-time in the pre-reservoir. The decline of the summer maximum is accompanied by an increase in the appearance of the meso-zooplankton (Schmidt-Halewicz, *pers. comm.*). Its development is also

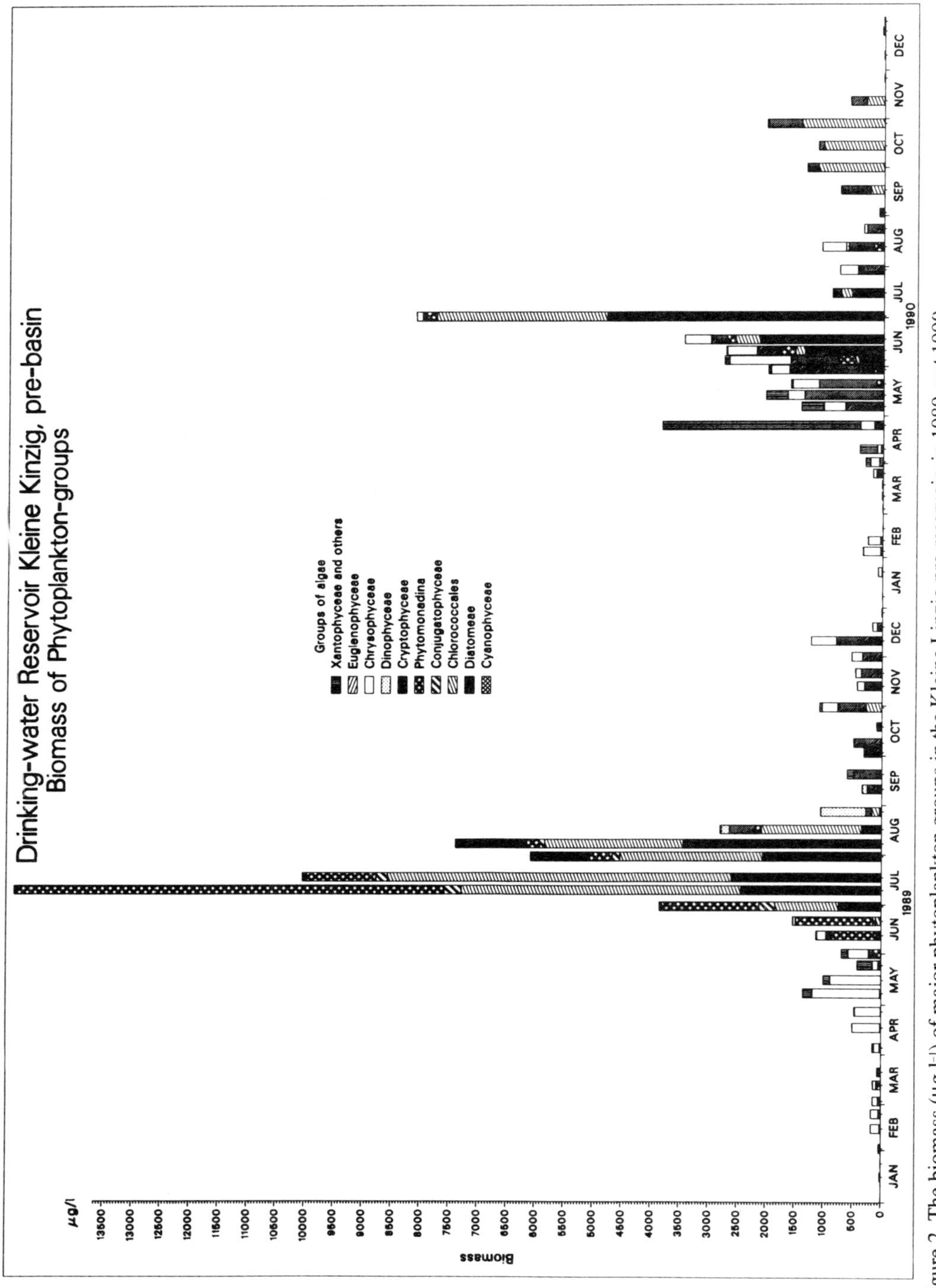

Figure 2. The biomass (µg l⁻¹) of major phytoplankton groups in the Kleine Linzig pre-reservoir, in 1989 and 1990.

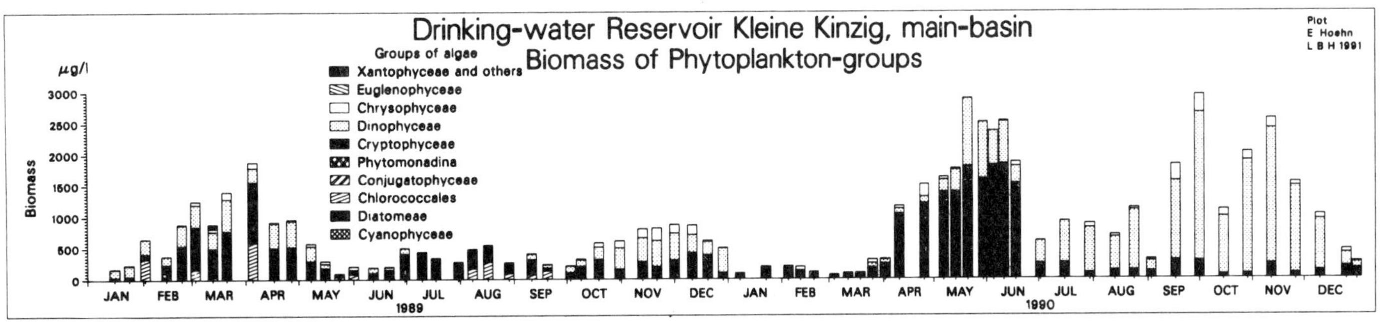

Figure 3. The biomass (µg l^{-1}) of major phytoplankton groups in the Kleine Kinzig main reservoir, in 1989 and 1990.

dependent on a long water retention-time. The decrease of the summer biomass in 1989 and 1990, despite low water discharge rates, was due to the grazing pressure of the zooplankton.

The pattern of the development of the algal biomass in the pre-reservoir shifted from a single-peaked curve in summer 1989 towards double-peaked maxima in early summer and the autumn of 1990. *Asterionella formosa* appeared thereby in large amounts in the pre-reservoir, constituting up to 60% of the total biomass (Fig. 2). There was also an increase in cryptomonads. The Phytomonadina were not able to develop into the large biomass that occurred in 1989. However an unknown species of Chrysophyceae or Xanthophyceae (*Nephrochloris* sp.?) attained a large biomass in the spring of 1990. The autumn maximum in the pre-reservoir was formed in particular by the chlorococcal species *Ankyra lanceolata*, present in only small numbers in the main reservoir in the previous year.

This research project was financed in particular by the Association for Water Supply Kleine Kinzig (WKK). Thanks to the following for doing some of the chemical analyses: Regional Institution for Protection of the Environment (LfU), Baden-Württemberg, (The Institute of Lake Research, Langenargen and The Institute for Water and Refuse Management, Karlsruhe). Special thanks also to: Dr E. Rott, Institute of Botany, University of Innsbruck, Dr R. Kümmerlin, LfU, Institute of Lake Research, Langenargen, and Vinzenz Maurer, Institute of Zoology, University of Berlin.

References

Benndorf, J., Pütz, K., Krinitz, H. & Henke, W. (1975). Die Funktion der Vorsperren zum Schutz der Talsperren vor Eutrophierung. *Wasserwitschaft-Wassertechnik*, **25**, H1, 19-25.

Fachbereichstandard der DDR (1983). Nutzung und Schutz der Gewässer – Stehende Binnengewässer – Nährstoffelimination in Vorsperren. *TGL* **27885/05**.

Sommer, U. (1989). The role of competition for resources in phytoplankton succession. In *Plankton Ecology* (ed. U. Sommer). Springer, Berlin.

Sommer, U., Gliwicz, Z. M., Lampert, W. & Duncan, A. (1986). The PEG-model of seasonal succession of planktonic events in fresh waters. *Archiv für Hydrobiologie*, **106**, 433-471.

Changes of water quality and sediment phosphorus of a small productive lake following decreased phosphorus loading

S. I. HEANEY[1], J. E. CORRY AND J. P. LISHMAN

Freshwater Biological Association and Institute of Freshwater Ecology, Far Sawrey, Ambleside, Cumbria, LA22 0LP, England.

Esthwaite Water is the most productive or eutrophic lake in the English Lake District. Since 1945 its water quality has been determined from weekly or biweekly measurements of temperature, oxygen, plant nutrients and phytoplankton abundance. The lake receives phosphorus from its largely lowland-pasture catchment, sewage effluent from the villages of Hawkshead and Near Sawrey, and from a cage-culture fish farm. From 1986 phosphorus has been removed from the sewage effluent of Hawkshead which was considered to contribute between 47% and 67% of the total phosphorus loading to the lake. At the commencement of phosphorus removal regular measurements of phosphorus in the superficial 0–4 cm layer of lake sediment were made from cores collected at random sites. Since 1986 the mean annual concentration of alkali-extractable sediment phosphorus has decreased by 23%. This change is not significant at the 5% level but nearly so. There has been no marked change in water quality over this period. Summer dominance of blue-green algae which arose in the early 1980s after decline of the previous summer forms, *Ceratium* spp., has been maintained. Improvement in water quality is unlikely to be achieved at the present phosphorus loading.

Introduction

There is general agreement that the degree of phosphorus loading to many lakes determines the level of algal production or eutrophication. The relationship between phosphorus loading and algal biomass was assessed by Vollenwider (1968), and the causal effect of phosphorus on algal production was illustrated in the whole-lake experimental studies of Schindler (1974) and in the large-scale mesocosms by Lund (e.g. Lack & Lund 1974). These and other studies have resulted in remedial measures being taken to reduce the external phosphorus inputs to many lakes suffering the severe effects of eutrophication.

Numerous case studies now exist of the reduction of anthropogenic inputs of phosphorus to lakes and reservoirs (e.g. Sas 1989). It is apparent that success in reducing the biomass of algal populations is variable and rarely immediate. A complicating factor is that phosphorus stored in the lake sediment under conditions of anoxia (e.g. Mortimer 1941, 1942; Heaney, Smyly & Talling 1986) and at high pH (e.g. Mackereth, 1953; Drake & Heaney 1987) may be released to the overlying water as an internal load. The release of such phosphorus to the water column at critical times of phosphorus depletion may give rise to algal production in addition to that expected for external loading alone.

The present paper describes the effect of the reduction of sewage-borne phosphorus to a small English lake. Esthwaite Water is generally agreed to be the most productive or eutrophic

[1] *Present address: Aquatic Sciences Research Division, Department of Agriculture, Newforge Lane, Belfast, BT9 5PX, Northern Ireland.*

lake in the English Lake District (Pearsall 1921; Jones 1972; Gorham *et al.* 1974). The lake is a grade 1 Site of Special Scientific Interest and together with some fenland to its north, including Priest Pot, has been included in the list of Wetlands of International Importance under the Ramsar convention. Despite artificial enrichment by sewage and fish-farm effluents, the lake is one of the best examples in England and Wales of a mesotrophic system. At its northern end is a well-developed and studied hydrosere. The lake contains a nationally rare plant species (*Najas flexilis* (Willd.) Rotsk, & W. L. E. Schmidt) and other species of restricted distribution. The lake has been the subject of much study (e.g. Heaney, Smyly & Talling 1986, and references therein). More recently, the phosphorus content of its sediments and their potential for remobilization (Drake & Heaney 1987) and the long-term physical, chemical and biological changes (Heaney *et al.* 1988; Talling & Heaney 1988) have been investigated.

Following a report to the Nature Conservancy Council (Talling & Heaney 1983) which concluded that the sewage treatment plant at Hawkshead was the major point source of phosphorus input to the lake, a decision was taken by the North West Water Authority (NWW) to prevent most of the sewage-borne phosphorus from entering the lake by undertaking chemical precipitation of this nutrient at the sewage works. At the same time it was agreed that the phosphorus removal would be accompanied by a programme of research to determine the effect of decreasing external phosphorus load to the lake on water quality and algal growth. This research had three objectives. The first was to provide a better estimation of the amounts of total phosphorus and soluble reactive phosphorus entering and leaving the lake. This work was undertaken by NWW (Agar *et al.* 1988) and showed that, prior to phosphorus removal, the contribution of phosphate-phosphorus in the effluent from Hawkshead sewage treatment plant was only 47% of the total phosphate-phosphorus loading to the lake, compared to the value of 67% estimated by Talling & Heaney (1983). It was apparent that the earlier study had underestimated the non-sewage contribution from the catchment.

It is the second two objectives that are the subject of the present paper. One was to determine from regular sampling of the lake any changes of water quality before and after phosphorus removal. The other was to determine any changes in the potential internal sediment load of phosphorus, by measuring the sodium hydroxide extractable phosphorus content of the sediment of the lake for a number of years before and after phosphorus removal at the sewage treatment works.

Methods

Water quality

Regular sampling was undertaken at a buoy near the deepest position of the lake (A in Fig. 1). Vertical profiles of water temperature (± 0.1°C) and oxygen (resolution 6 μmol l^{-1}) were measured *in situ* using a combined thermistor and oxygen electrode.

Secchi disc transparency was obtained with a white disc of 30 cm diameter. Water samples for chemical analyses and algal counts were collected using a weighted algal tube (Lund 1949) from the 0–5-m layer. A sub-sample was immediately fixed in the field with Lugol's iodine for subsequent algal counts using the Utermöhl inverted microscope technique (Lund *et al.* 1958).

Total phosphorus, soluble reactive phosphorus, ammonia-nitrogen and dissolved silicon were analysed as described in Mackereth *et al.* (1989) with the NH_4-N method adapted for use with a discrete analyser, and ascorbic acid as the reductant for dissolved silicon. Nitrate-nitrogen was analysed as described by Hilton & Rigg (1983).

Chlorophyll-*a* was determined after separation of algae on glass-fibre filters (Whatman GF/C), extraction with boiling methanol, and estimation as the chlorophyll optical equivalent (Talling 1974).

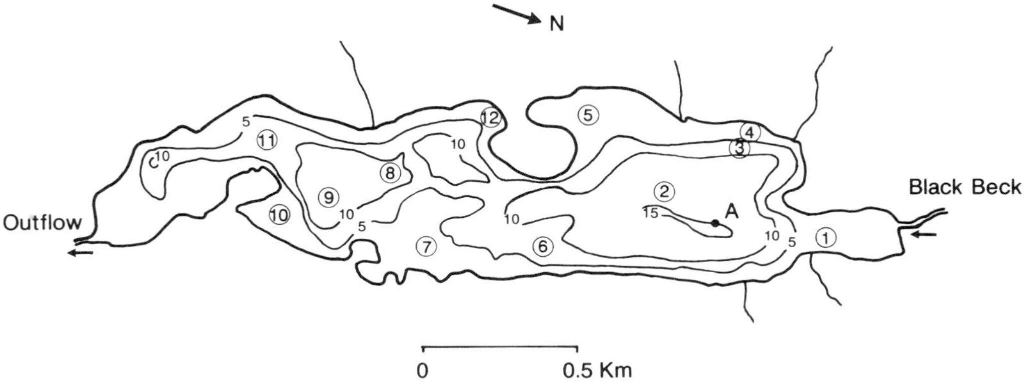

Figure 1. Map of Esthwaite Water showing bathymetric contours (in m), approximate sampling positions for obtaining cores (1–12), and sampling buoy for water quality data (A).

Sediment analysis

Sediment cores were collected according to a randomly stratified sampling design near the positions indicated in Figure 1. Twelve cores were taken where possible for each sample which included three deep (> 5 m water depth) and three shallow (< 5 m water depth) sites in each of the two main basins of the lake. For a period only eleven cores were obtained due to the loss of a corer of imperial dimensions which could not be replaced. Cores were taken using a Jenkin sampler, and the upper 0–4 cm sediment layer was obtained using a hydraulically-operated extruder (Jones & Ohnstad 1982).

The total phosphorus content of the sediment was determined on a sub-sample of 0.1 g dried material from the upper 0–4 cm layer. Each sub-sample was digested to dryness in a teflon beaker after the addition of 2 x 1 ml Aristar nitric acid. Further digests were made in the following order: 2 x 2 ml Analar perchloric acid (70%), 2 x 2 ml Analar hydrofluoric acid (48%) and 1 x 2 ml Analar perchloric acid (70%). The residue was taken up in 50 ml of 0.1 m hydrochloric acid. Suitable aliquots were neutralized and phosphorus measured using the lower sensitivity method for phosphate in Mackereth *et al.* (1989).

Inorganic soil phosphate or alkali-extractable phosphorus was determined on further sub-samples from the 0–4 cm layer after the method of Petersen & Corey (1966). Approximately 2 g wet sediment was accurately weighed into 10 ml plastic tubes. At the same time a separate sub-sample was taken for dry weight determination and dried at 60°C for 1-2 days. 8 ml of 0.125 M NaOH was added to each tube containing the weighed 2 g of wet sediment, stoppered and mixed gently until homogenised. The tubes were fitted onto racks in a Denby RM200 rotator and rotated at *c.* 20 r.p.m. for 17 hours to extract the phosphate. It was determined by experimentation that no phosphate was obtained after further extractions. The suspension was then centrifuged for 15 min at 2400 r.p.m. and the supernatant transferred to another plastic tube. 40 μl of concentrated H_2SO_4 was added to the suspension and swirled gently to flocculate the organic matter, and then centrifuged again for 15 min. at 2400 r.p.m. A suitable aliquote was then taken for analysis of phosphate as described above.

Results

Changes in sediment phosphorus content

Analysis of total phosphorus is very time consuming and two such measurements were only made on samples collected shortly after phosphorus removal from the sewage effluent began during April 1986, and again during early 1990. The sample for 1986 was collected on 11 June but, as the core from site 1 was not available for analysis, a value for a core obtained on 9 July from this site is used. Table 1 gives the results for both total phosphorus and alkali-extractable phosphorus (AEP) at individual sites. The mean values and standard deviations for total phosphorus and AEP, respectively, for the 1986 sample are 4152 ± 1811 and 2040 ± 1383, and for the 1990 samples they are 3791 ± 1821 and 1500 ± 1334 μg P per g sediment dry weight. These show decreases in the mean total phosphorus and AEP of 9 and 26% respectively, although the actual concentration decreases are similar at 361 and 507 μg P per g sediment dry weight.

Table 1. *Concentrations of total phosphorus (TP) and alkali-extractable phosphorus (AEP) in the top 0–4 cm layer of sediments at 12 sites in Esthwaite Water (see Fig. 1) in 1986 and 1990*

Samples were taken on 9 July 1986 (site 1) and 11 June 1986 (sites 2–12), and all sites were sampled again on 7 March 1990. Sites with an asterisk are where the sediments lie below deeper water (5–15m). Values for TP and AEP are in μg P per g dry sediment; AEP is also given as a percentage of TP at each site.

Sampling site	1986			1990		
	TP	AEP	AEP/TP, (%)	TP	AEP	AEP/TP, (%)
1	2735	1153	42	2966	857	29
2*	5278	3108	59	4719	2771	59
3*	3385	1522	45	3672	1193	32
4	3462	1430	41	3678	1025	28
5	2487	943	38	1647	475	29
6*	4114	1707	41	3518	1087	29
7	3464	1299	38	2427	585	24
8*	5853	4018	67	5916	2990	51
9*	9001	5372	60	8180	4684	57
10	3243	1241	38	2659	684	26
11	3663	1321	36	4113	1618	39
12	3143	1360	43	1999	428	21

Table 2. *Mean concentrations ± standard deviation of total phosphorus (TP) and alkali-extractable phosphorus (AEP) in sediments of Esthwaite Water, in 1986 and 1990*

Results are given for whole lake sediment (12 sites combined), sediment from the north and south basins of the lake (6 sites each), shallow sediment (6 sites lying under 0–5 m depth of water), and deep sediment (6 sites lying under 5–15 m depth of water); N is the number of sediment cores Values for TP and AEP are in μg P per g dry sediment; AEP is also given as a percentage of TP at each site.

Year	Site	N	TP	AEP	AEP/TP (%)
1986	Whole lake	12	4152 ± 1819	2040 ± 1383	49
1990		12	3791 ± 1821	1533 ± 1299	40
1986	Shallow	6	3089 ± 339	1238 ± 173	40
1990		6	2563 ± 720	675 ± 230	26
1986	Deep	6	5216 ± 2084	2843 ± 1625	55
1990		6	5020 ± 1776	2390 ± 1378	48
1986	North basin	6	3577 ± 1013	1643 ± 767	46
1990		6	3367 ± 1016	1235 ± 793	37
1986	South Basin	6	4782 ± 2323	2437 ± 1802	52
1990		6	4216 ± 2415	1832 ± 1695	43

Table 2 indicates the changes in AEP from the two major basins and different depth strata in the lake relative to those for total phosphorus. For the whole lake between 1986 and 1990, AEP as a percentage of total phosphorus declined by 11% from 46 to 35%. The largest change occurred in the 0–5-m depth stratum where there was a decrease of 14% compared to only 6% for the 5–15-m stratum. There was little difference between the north and south basins.

Although total phosphorus was analysed on only two occasions, EAP was measured on samples collected monthly for the first year and thereafter at three-monthly intervals. As there were very large differences in the concentration of sediment phosphorus throughout the lake, indicating a contagious distribution, arithmetic mean values of samples and subsets are given in Figures 2-4 with 95% confidence limits calculated using Taylor's power law transformation (Elliott, 1977). On the few occasions when the analysis for a site was not obtained, a value was assigned by taking the mean of the time-adjacent values for the same site. These interpolated values have been used in all statistical analysis. Comparison of time-series using the analysed values only, and with interpolated values included, show similar overall trends (Fig. 2). The incorporation of interpolated values, however, gives less erratic mean values for samples, especially where analyses for site 9 were missing. This deep-water site always contained at least twice the amount of sediment phosphate found at any other site, and its absence from a sample-collection always decreased the sample mean.

The changes in mean phosphate values for all samples against time are given in Figure 2. The 95% confidence limits are very wide for each date, indicating the large differences in phosphate concentration of the superficial sediment throughout the lake. There was a general decline in the mean phosphate concentration between 1986, when phosphate removal from the sewage effluent began, and 1989. The mean annual phosphate content for 1986–1989 (Table 3) indicates a year-by-year gradual decline in phosphate concentration and a 23% reduction in the phosphate content of the sediment over this time interval.

Table 3. *Annual mean concentrations of alkali-extractable phosphorus (AEP) in sediments of Esthwaite Water, in 1986 and 1990*

Means for AEP, with 95% confidence limits and variances, are given for sediment cores taken at 12 sampling sites on various occasions in 1986 to 1989; N is the number of sub-samples (sediment cores).
Values are in µg P per g dry weight of sediment; the last column shows the cumulative decrease in AEP as a percentage of the mean for 1986.

Year	N	Mean	95% CL	Variance	Decrease (%)
1986	120	2081	1784 – 2387	269501	—
1987	72	1927	1541 – 2313	2685468	7
1988	48	1723	1332 – 2114	1800830	17
1989	49	1611	1220 – 2002	1782732	23

To determine if there was a statistically significant decrease in the phosphate content of the sediment between 1986 and 1989 the mean annual concentrations for both years were compared using the "normal deviate (d) test". For both years the number of sub-samples were sufficiently great that the *central-limit theorum* was considered applicable. Thus it was assumed that the probability distribution of the sample means around the sediment mean was normal, with the overall mean equal to the sediment mean (Elliott 1977). The null hypothesis was that both samples came from the same sediment and therefore their means were not significantly different. The calculated value for d was 1.91, indicating that the sample means were not significantly different, although nearly so, at the 5% level. The variances were not significantly different at the 5% level, with the variance ratio, F, = 1.51.

Differences in sediment phosphate between shallow-water (0–5 m) and deep-water (5–15 m) sites is shown in Figure 4. This shows that shallow-water sites contain much less phosphate than deep-water sites.

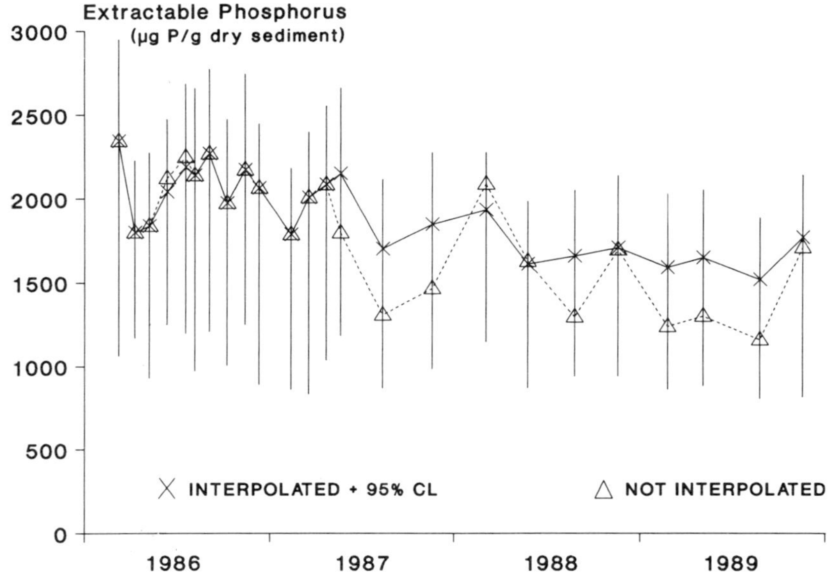

Figure 2. Time-series of the mean concentration of alkali-extractable phosphorus in the 0–4 cm layer of sediment from samples of cores from Esthwaite Water. Values with 95% CL are for samples including interpolated values (x); values for samples not including interpolated values (Δ) are given for comparison (further explanation in the text).

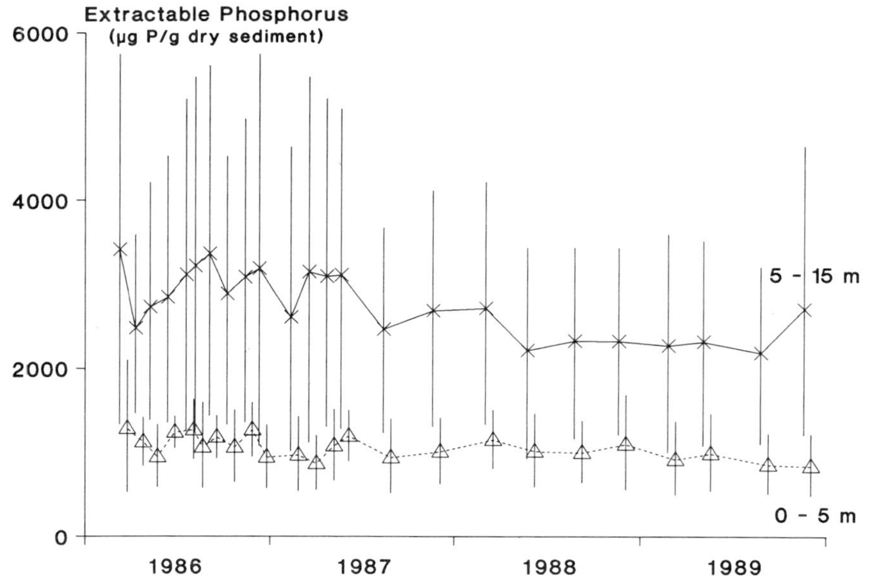

Figure 3. Time-series of the mean concentration and 95% CL of alkali-extractable phosphorus in the upper (0–4 cm) sediment layer of cores from Esthwaite Water, separated into sub-samples underlying shallow (Δ, 0–5 m) and deep (x, 5–15 m) water. Note that values for shallow water samples are offset to the right.

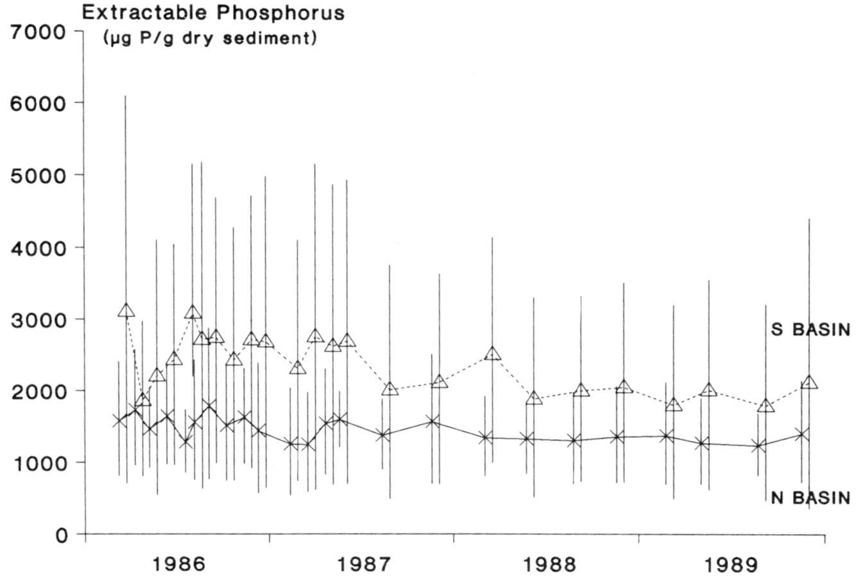

Figure 4. Time-series of the mean concentration and 95% CL of alkali-extractable phosphorus in the upper (0–4 cm) sediment layer of cores from Esthwaite Water, separated into sub-samples from its north (x) and south (Δ) basins. Note that values for the south basin are offset to the right.

Esthwaite Water is divided into two main basins separated by the peninsula of Strickland Ees, with a third and smaller basin at its southern end (Fig. 1). In this study cores were only taken from the north and south basins. The mean annual values of sediment phosphate in the south basin were consistently higher than those in the north basin (Figure 4). Using all the data for 1986–1989 the difference between the sample means from the two basins was highly significant ($d = -5.19$; $P < 0.01$) and the variances were significantly different at the 0.2% level ($F = 8.09$). The consistently very high values of sediment phosphorus at site 9 contributed strongly to the between-basin differences. If values for site 9 are excluded, the sample means for the two basins are not significantly different ($d = -1.75$; $P > 0.05$), although the variances were ($F = 4.25$; $P < 0.002$).

Changes in lake water quality

(a) Phosphorus. Seasonal changes in the concentration of total phosphorus in the 0–5-m water layer at the sampling buoy (A, Fig. 1) are shown in Figure 5. There is little pronounced seasonality, although there tends to be a minimum at the end of April associated with the decline and sedimentation of the spring diatom population, and an increase during late summer when blue-green algal blooms occur. Winter concentrations are variable and to some degree influenced by weather-dependent resuspension of sediment.

The mean annual concentrations of total phosphorus (Table 4) surprisingly show that the value for 1986 was lower than for subsequent years. The mean annual value for 1988 was significantly higher, at the 5% level, than for 1986 (Table 5).

As with total phosphorus, the mean annual concentrations of soluble reactive phosphorus show that the value for 1988 was significantly greater than that for 1986 (Table 4). Seasonal changes in the concentration of soluble reactive phosphorus during 1986–1989 (Fig. 6) show the expected winter maxima and summer minima. The latter, although generally close to the limit of detection of the method (0.6 µg SRP l^{-1}), show a number of short-lived increases associated with the decline and mineralisation of the algal populations (c.f. Fig. 7).

Table 4. *Annual mean concentrations of total phosphorus (TP), soluble reactive phosphorus (SRP) and chlorophyll-*a *in lakewater (0–5 m depth) in Esthwaite Water, in 1986 to 1989*

Means (µg per litre of lakewater) and 95% confidence limits for TP, AEP and Chl*a* are given for 42–44 samples taken each year at the sampling buoy (see Fig. 1).

Values are also given for Secchi-disc transparency (m depth of water) for 40–44 samples in each year.

	1986		1987		1988		1989	
TP	26.7	(24.2–29.2)	29.7	(26.5–31.9)	30.1	(28.4–31.8)	29.3	(27.7–30.9)
SRP	2.5	(1.4–3.6)	3.7	(2.4–5.0)	4.6	(3.0–6.2)	3.9	(2.5–5.3)
Chl*a*	22.0	(16.2–27.8)	24.4	(18.8–30.0)	26.7	(20.6–32.8)	25.3	(20.4–30.2)
Secchi	2.7	(2.3–3.1)	2.5	(2.1–2.9)	1.9	(1.6–2.3)	2.3	(2.0–2.6)

Table 5. *Results of statistical comparisons between annual means shown in Table 4 for total phosphorus (TP), soluble reactive phosphorus (SRP), chlorophyll–*a *and Secchi-disc transparency in Esthwaite Water, in the years 1986 to 1989*

d is the normal deviate (*d* test) and *F* is the variance ratio.

Asterisks denote significance at the 95% level (*), 98% level (**) and 99.8% level (***); all other values are not significant at the 95% level.

	1987		1988		1989	
	d	*F*	*d*	*F*	*d*	*F*
TP	– 1.82	1.34	– 2.25*	2.13**	– 1.74	2.30***
SRP	– 1.44	1.30	– 2.18*	2.03**	– 1.59	1.51
Chl*a*	– 0.60	1.06	– 1.14	1.09	– 0.88	1.42
Secchi	0.92	1.37	3.61***	1.91**	2.00**	2.31***

*(b) Light attenuation and chlorophyll-*a. The depth at which a white Secchi disc of 30 cm diameter disappears from view when lowered into the water gives a useful measure of transparency. For Esthwaite Water there is usually an inverse correlation between Secchi depth and chlorophyll-*a* concentration, as phytoplankton biomass is a major agent of light attenuation (Talling & Heaney 1983). Changes, with time, of Secchi transparency and the concentration of chlorophyll-*a* are shown in Figure 7. There is considerable interannual variation, with the usual "clear water phase" present in May– June 1986 and 1987, after the sedimentation of spring diatom growth, but poorly developed in 1988 and 1989 due to the early growth of blue-green algal populations.

The mean values for Secchi transparency and chlorophyll-*a* concentration are given in Table 4. As with total phosphorus and soluble reactive phosphorus, the mean annual value of chlorophyll-*a* for 1988 was the highest of the four years, although not significantly higher than in 1986 (Table 5).

(c) Major algal species. There was no marked change of algal quality during the period of decreased sewage input. The major algal genera producing the chlorophyll-*a* maxima are shown in Figure 7. The spring maxima consisted of *Asterionella formosa* Hass. and *Melosira italica* subsp. *subarctica* O. Müller (= *Aulocoseria subarctica* (O. Müller) Haworth). The overall biomass of the spring diatom growth is generally determined in this lake by silicon availability. Summer communities were dominated by the blue-green algae *Aphanizomenon flos-aquae* f. *gracile* Ralphs ex Born. et Flah., *Anabeaena flos-aquae* Bréb. ex Born. et Flah.,

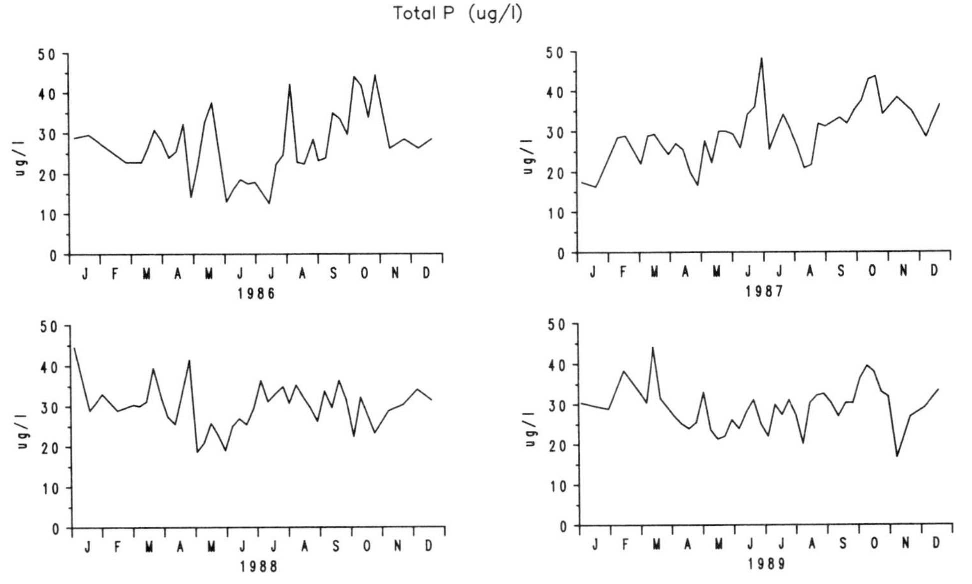

Figure 5. Time-series for concentrations of total phosphorus in the upper (0–5 m) water layer of Esthwaite Water during the years 1986–1989.

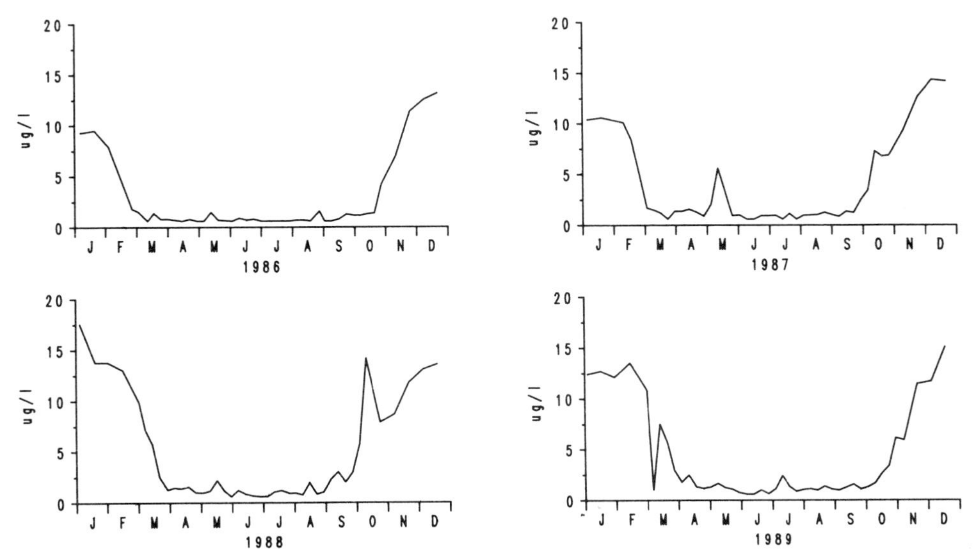

Figure 6. Time-series for concentrations of soluble reactive phosphorus in the upper (0–5 m) water layer of Esthwaite Water during the years 1986–1989.

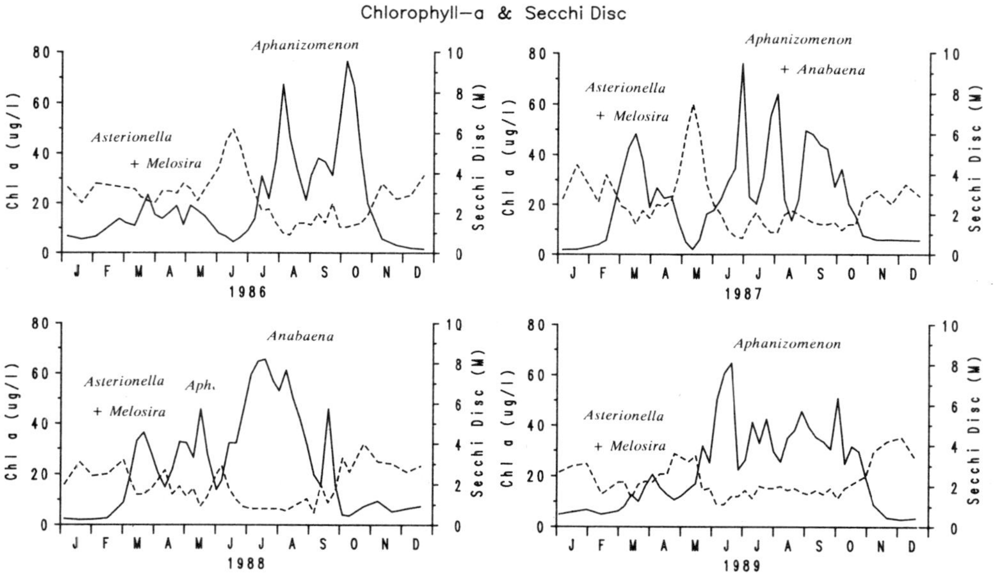

Figure 7. Time-series for concentrations of chlorophyll-*a* in the 0–5 m layer (—) and for Secchi disc transparency (---) in Esthwaite Water during the years 1986–1989. The dominant algal genera during chlorophyll-*a* maxima are indicated.

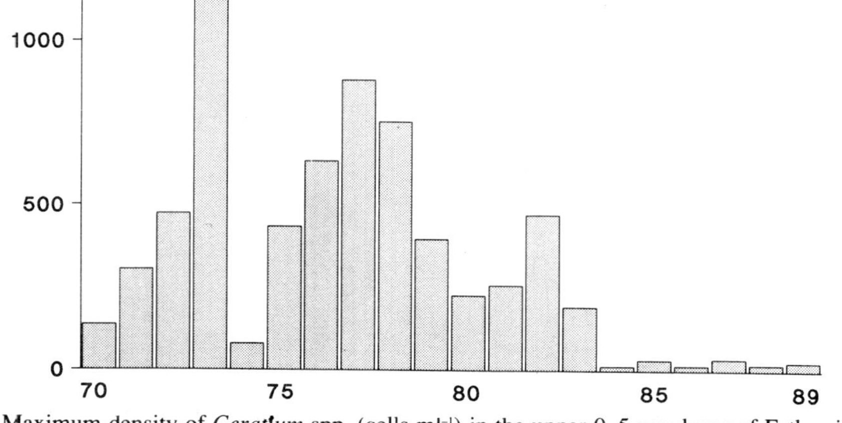

Figure 8. Maximum density of *Ceratium* spp. (cells ml^{-1}) in the upper 0–5 mm layer of Esthwaite Water from 1970 to 1989. The great decrease in maximum annual density between 1983 and 1989 is the result of parasitism caused by the biflagellate fungus *Aphanomycopsis cryptica*.

and *A. solitaria* Kleb. There was considerable interannual variability in the timing of maxima of certain species of blue-green algae, particularly *Aphanizomenon flos-aquae*. This was caused in part by an unusual grazing episode during summer 1987, by the protozoan ciliate *Nassula*, together with the relatively long hydraulic residence time of the lake, enabling the persistence of slow-growing and poorly-grazed winter populations (Canter, Heaney & Lund 1990). The dinoflagellates *Ceratium* spp. which dominated the summer populations during the 1970s and early 1980s produced relatively small populations (Fig. 8) because of repeated parrasitism by the biflagellate fungus *Aphanomycopsis cryptica* Canter (Canter & Heaney 1984; Heaney *et al.* 1988).

Discussion

These results complement the study of Agar *et al.* (1988) which determined the external inputs of soluble reactive phosphorus and total phosphorus to Esthwaite Water between 1985/1986 and 1987/1988, including estimates of the discharge from the cage-culture fish farm into the south basin. This showed that over this period the total phosphorus input had decreased by 1240 kg year^{-1}. The total phosphorus contribution from the fish farm was estimated at *c.* 620 kg year^{-1} of which about 60% is likely to be in the soluble phosphorus fraction and directly available for algal growth (Foy & Rosell 1991a). Agar *et al.* (1988) also concluded that, following phosphorus removal at the sewage treatment works, there was a decrease in the estimated annual net uptake of phosphorus by the lake, from 1830 kg for 1985/1986 to 330 kg for 1987/1988.

The results of the present study and that of Agar *et al.* (1988) thus indicate that the reduction in external phosphorus loading has been compensated for by an increase in internal loading from the sediments as the lake approaches a new equilibrium state. It is hardly surprising that until this is reached there have been no appreciable changes in the mean annual concentrations of lakewater total and soluble reactive phosphorus, and chlorophyll-*a*. The new equilibrium state, when it is reached, will depend upon the overall external loading derived from diffuse catchment inputs, sewage-borne phosphorus, now markedly reduced, and inputs from the fish farm. Phosphorus contribution from the latter is currently being determined and is dependent upon the level of fish production and the phosphorus content of the fish food (Foy & Rosell 1991b).

From the sediment core analyses, the mean lake sediment phosphate decreased by 358 µg P per g dry sediment between 1986 and 1988. Using a value of 90% for the water content of the upper 0–4 cm sediment layer, and a conservative estimate of 1 km^2 for the area of lake sediment, it can be calculated that about 1430 ($= 1 \times 10^6 \times 10^4 \times 10^{-1} \times 358 \times 10^{-9}$) kg of phosphorus will have been given up in *two* years by the lake sediment. Acknowledging the possible errors in such an estimation, it can be compared with a similar reduction in the total external *annual* input of 1240 kg phosphorus estimated for almost the same period by Agar *et al.* (1988).

The annual external loading of total phosphorus and soluble reactive phosphorus for Esthwaite Water was estimated by Agar *et al.* (1988) to be 1.65 and 0.85 g P m^{-2} respectively. These values, following the reduction of sewage-borne phosphorus, would maintain the lake's classification as eutrophic on the basis of the Vollenweider's earliest graphical relationship of phosphorus loading, phytoplankton abundance (as chlorophyll-*a*), and mean lake depth (Vollenweider 1968). However, on the basis of the greater range of variables considered by Vollenweider & Kerekes (1984), the lake would be classified as mesotrophic with regards to nitrogen and phosphorus levels but eutrophic with regard to chlorophyll-*a* concentration and water clarity.

The greatest change in algal quality within the lake took place during 1983 prior to the reduction of phosphorus loading. Before this date the summer phytoplankton was dominated by populations of *Ceratium* spp. (Heaney *et al.* 1988). Thereafter populations of blue-green algae or cyanobacteria have formed prolonged growths throughout summer and autumn and also occasionally during late spring following a high level of inoculum carry-over from the previous year. As a result there has been a general increase in the annual duration and abundance of phytoplankton growth as shown by chlorophyll-*a* concentration (Talling 1992). This change of quantity and quality of phytoplankton has resulted from a progressive enrichment of the lake (Heaney *et al.* 1988; Talling & Heaney 1988) but more significantly because of the decline of populations of the dinoflagellates *Ceratium* spp. as a result of persistent parasitism by the biflagellate fungus *A. cryptica*. Diseases of phytoplankton and microbial grazing are well documented but the demise of a regular dominant phytoplankter by such means is rarely recorded. The failure of populations of *Ceratium* spp. may in part be due to their relatively slow growth rate, making them unable to recover from fungal attack before nutrient and light resources are used by competing populations of cyanobacteria.

The impact of phosphorus removal for changes of water quality has yet to be fully realized. Indeed it may be reasonably questioned if it ever will with the present phosphorus loading, although the removal of the sewage-borne phosphorus is nevertheless a step in the right direction. Even with continuous year-round removal of phosphorus by the sewage treatment works, at present not undertaken during the winter months, it is likely to be some years yet before an improvement in water quality, i.e. a reduction of summer algal biomass and an increase in water transparency, can be expected. The magnitude of the phosphorus input to the lake from the fish farm may be critical in determining the ultimate success or failure in improving water quality in this lake.

We wish to thank North West Water and the Nature Conservancy Council for financial support, the Institute of Freshwater Ecology of the Natural Environment Research Council for facilities, Drs J. F. Talling, FRS, J. W. G. Lund, FRS and Hilda Canter-Lund for their interest and support and Mrs Jean McCauley for typing the manuscript.

References

Agar, G., Banks, C., Critchley, R. F., Devitt, S., Featherstone, T. A., Mycock, E., Phillips, R. G. & Pinder, J. R. (1988). Nutrient enrichment of Esthwaite Water and the Hawkshead Sewage Works. North West Water, Planning and Engineering Department Report PR88/10. 25 pp.

Canter, H. M. & Heaney, S. I. (1984). Observations on zoosporic fungi of *Ceratium* spp. in lakes of the English Lake District; importance for phytoplankton population dynamics. *New Phytologist*, **97**, 601-612.

Canter, H. M., Heaney, S. I. & Lund, J. W. G. (1990). The ecological significance of grazing on planktonic populations of cyanobacteria by the ciliate *Nassula*. *New Phytologist*, **114**, 247-263.

Drake, J. C. & Heaney, S. I. (1987). Occurrence of phosphorus and its potential remobilization in the littoral sediments of a productive lake. *Freshwater Biology*, **17**, 513-523.

Elliott, J. M. (1977). Some methods for the statistical analysis of samples of benthic invertebrates. 2nd edition. *Scientific Publications of the Freshwater Biological Association*, **25**, 156 pp.

Foy, R. H. & Rosell, R. (1991a). Loadings of nitrogen and phosphorus from a Northern Ireland fish farm. *Aquaculture*, **96**, 17-30.

Foy, R. H. & Rosell, R. (1991b). Fractionation of phosphorus and nitrogen loadings from a Northern Ireland fish farm. *Aquaculture*, **96**, 31-42.

Gorham, E., Lund, J. W. G., Senger, J. E. & Dean, W. E. Jr. (1974). Some relationships between algal standing crop, water chemistry, and sediment chemistry in the English Lakes. *Limnology and Oceanography*, **19**, 601-617.

Heaney, S. I., Lund, J. W. G., Canter, H. M. & Gray, K. (1988). Population dynamics of *Ceratium* spp. in three English lakes, 1954–1985. *Hydrobiologia*, **161**, 133-148.

Heaney, S. I., Smyly, W. J. P. & Talling, J. F. (1986). Interactions of physical, chemical and biological processes in depth and time within a productive English lake during summer stratification. *Internationale Revue der Gesamten Hydrobiologie*, **17**, 441-494.

Hilton, J. & Rigg, E. (1983). Determination of nitrate in lake water by the adaptation of the hydrazine-copper reduction free method for use on a discrete analyser: performance statistics and an instrument-induced difference from segmented flow conditions. *Analyst*, **108**, 1026-1028.

Jones, J. G. (1972). Studies on freshwater microorganisms: phosphatase activity in lakes of differing degrees of eutrophication. *Journal of Ecology*, **60**, 777-791.

Jones, J. W. G. & Ohnstadt, F. R. (1982). The Jenkin surface-mud sampler user manual. *Occasional Publications of the Freshwater Biological Association*, **15**, 45 pp.

Lack, J. J. & Lund, J. W. G. (1974). Observations and experiments on the phytoplankton of Blelham Tarn, English Lake District. The Experimental Tubes. *Freshwater Biology*, **4**, 399-415.

Lund, J. W. G. (1949). Studies on *Asterionella* 1. The origin and nature of cells producing seasonal maxima. *Journal of Ecology*, **37**, 389-419.

Lund, J. W. G., Kipling, C. & Le Cren, E. D. (1958). The inverted microscope technique of estimating algal numbers and the statistical basis of estimation by counting. *Hydrobiologia*, **11**, 143-170.

Mackereth, F. J. H. (1953). Phosphorus utilization by *Asterionella formosa* Hass. *Journal of Experimental Botany*, **4**, 296-313.

Mackereth, F. J. M., Heron, J. & Talling, J. F. (1989). Water analysis: some revised methods for limnologists (second impression). *Scientific Publications of the Freshwater Biological Association*. **36**, 124 pp.

Mortimer, C. H. (1941-2). The exchange of dissolved substances between mud and water in lakes; I and II. *Journal of Ecology*, **29**, 280-329; III and IV. **30**, 147-201.

Pearsall, W. H. (1921). The development of vegetation in the English Lakes, considered in relation to the general evolution of glacial lakes and rock basins. *Proceedings of the Royal Society, London*, B, **92**, 259-284.

Petersen, G. W. & Corey, R. B. (1966). A modified Cheng and Jackson procedure for routine fractionation of inorganic soil phosphates. *Proceedings of the Soil Science Society of America*, **30**, 563-565.

Sas, H. (1989). *Lake Restoration by Reduction of Nutrient Loading: Expectations, Experiences, Extrapolation–Coordination*. Academia Verlag Richarz GMBH.

Schindler, D. W. (1974). Eutrophication and recovery in experimental lakes: Implications for lake management. *Science, New York*, **184**, 897-899.

Talling, J. F. (1974). Photosynthetic pigments. General outline of spectrophotometric methods: specific procedures. In *A Manual on Measuring Primary Production in Aquatic Environments*, 2nd Ed. (ed. R. A. Vollenweider), pp. 22-26. Oxford. Blackwell Scientific Publications.

Talling, J. F. (1992). Comparative seasonal changes, and inter-annual variability and stability, in a 26-year record of total phytoplankton biomass in four English lake basins. *Hydrobiologia* (in press).

Talling, J. F. & Heaney, S. I. (1983). Evaluation of historical data on the nutrient status of Esthwaite Water, Cumbria. Commissioned Report from the Freshwater Biological Association to the Nature Conservancy Council. 44 pp.

Talling, J. F. & Heaney, S. I. (1988). Long-term changes in some English (Cumbrian) lakes subjected to increased nutrient inputs. In *Algae and the Aquatic Environment* (ed. F. E. Round), pp. 1-29. Biopress, Bristol.

Vollenweider, R. A. (1968). *Scientific Fundamentals of the Eutrophication of Lakes and Flowing Waters, with Particular Reference to Nitrogen and Phosphorus as Factors in Eutrophication*. O.E.C.D. Tech. Report. DAS/CSI/68.27, 15 pp. 34 figs.

Vollenweider, R. A. & Kerekes, J. (1980). The loading concept as basis for controlling eutrophication philosophy and preliminary results of the OECD programme on eutrophication. *Progress in Water Technology*, **12**, 5-38.

Eutrophication in the River Meuse

J.-P. DESCY
Unit of Freshwater Ecology, Department of Biology, Facultés Universitaires Notre-Dame de la Paix, Rue de Bruxelles 61, B-5000 Namur, Belgium

The severe problems caused by large phytoplankton populations in the River Meuse date back to the beginning of the 1980s. However, no clear relationship can be established between an increase of algal growth and dissolved nutrient concentrations, at least in the Belgian part of the river. Most probably, plankton algae start developing in France, utilizing large inputs of phosphorus from some of the tributaries: this point will be investigated further, as well as the effect of a reduction in the releases of phosphorus.

A mathematical model helps to understand the main factors which control algal growth: underwater light, temperature, discharge and grazing by zooplankton. The last is a major loss process in summer and, as shown by recent observations, may trigger a seasonal succession leading to dominance by large phytoplankton taxa.

With regard to water quality, eutrophication is a major problem in drinking-water treatment (filter clogging, etc.) and large numbers of decomposing algae may adversely affect the oxygen budget of the river. On the other hand, algal photosynthesis is the most important oxygen source at periods of low discharge, and reduced algal production may result in dramatic oxygen decreases in heavily polluted stretches of the river.

Introduction

The eutrophication problems in large rivers, leading to the development of huge amounts of phytoplankton, have received less attention than in lakes. The same is true for the ecology of river phytoplankton. Some of the first studies on production and composition of suspended algae communities in rivers were done 20 years ago in Britain with, for instance, the study of the Rivers Thames and Kennet (Lack 1971; Lack & Berrie 1974). Another example is the complete analysis of composition and evolution in time, carried out by Szemes (1964) for the River Danube in Hungary. Since then, most of the attention paid to river phytoplankton – or potamoplankton – has been linked to the eutrophication problem and, from the beginning of the 1980s, numerous publications have dealt with increases of algal biomass, as well as the changes in composition of communities in some large rivers (see Descy 1987 for a brief review). Thanks to many studies, based on field measurements, experimental cultures and modelling at various scales of space and time, the dynamics of potamoplankton are currently well understood, at least with regard to the major mechanisms that control the wax and wane of the algal community as a whole.

In comparison with lakes, large rivers – where suspended algae can develop dense and diverse communities – can be viewed as water-bodies with the following features. (1), Very short residence times which vary with the flow rate, i.e. the river discharge in a given cross-section. (2), Turbulent mixing throughout the year: the water column is typically not stratified, except at very low discharges. (3), There is an important nutrient supply from the watershed and the sediments, so that nutrient depletion is seldom observed.

Furthermore, the high concentration of suspended matter reduces the vertical penetration of light. So, according to Reynolds (1984a, b, 1988), the river phytoplankton should be dominated by "w" or "R" species, able to grow in a strongly disturbed and light-limited environment: it should therefore be similar to the phytoplankton of eutrophic (or mesotrophic) lakes during the major mixing events (spring and autumn circulation), with high availability of nutrients and a steep gradient of light related to the high ratio of mean depth : euphotic zone ($Z_m : Z_{eu}$ ratio). The theory is consistent with observations: the phytoplankton of large rivers is dominated by diatoms (*Asterionella*, *Stephanodiscus* spp.) during spring and autumn, while green algae (mostly Chlorococcales) and cryptomonads may develop best in summmer conditions. Eventually, at very low discharges, a typical succession leading to dominance by buoyant Cyanobacteria (*Microcystis aeruginosa* Kütz., *Oscillatoria agardhii* Gom.) may be observed.

The river discharge is a major integrating factor related to important processes affecting algal communities (rate of turbulent mixing, nutrient supply from the watershed and the sediment, light regime in the water column via the amount of suspended matter). Furthermore, discharge controls phytoplankton growth through a "dilution" process, as stated by Descy *et al.* (1987): in a given river stretch of length x, the "dilution rate" depends on the reciprocal of the cross-section area ($1/A$) and the linear lateral inflow (dQ/dx). Suspended algae can build up significant populations only when their net growth rate (taking into account the loss factors) exceeds the dilution rate. This control of algal development by discharge favours more fast-growing and disturbance-tolerant algae (i.e. the "colonizers" or "pioneers"), at least in fluctuating conditions.

In this paper, particular attention is given to the River Meuse, which flows through three European countries, with regard to different aspects of planktonic eutrophication.

The River Meuse

The Meuse (Fig. 1) rises in the east of France (in the entity called the "Rhin–Meuse" basin) and flows through Belgium and the Netherlands, where it forms a common estuary with the River Rhine, the Dutch Delta, which flows into the North Sea. The total length of the river is 885 km and its catchment area is 36,000 km^2, 40% of which lies in Belgium. The discharge, measured at Ampsin–Neuville (Belgium), is characterized by large variations between winter (typically 200–800 $m^3\ s^{-1}$ and up to 2000 $m^3\ s^{-1}$ in severe floods) and summer (often less than 30 $m^3\ s^{-1}$). The low-flow period can extend from July to late November, and the most important management and water-quality problems occur at that time. The Meuse is a source of drinking water for an estimated five million people in Belgium (mainly for Brussels and Antwerpen) and the Netherlands; in Belgium, the main water-treatment plants are located in Tailfer (Fig. 1) (direct intake from the Meuse) and near Antwerpen, where the water is taken from the Albert Canal. Like many large rivers, the Meuse is also regulated for navigation. In particular, the Belgian section has been and still is submitted to heavy and ecologically harmful management practices, which have eliminated all the bankside vegetation, once well-developed along its entire length and still present at some places 20 years ago.

The land-use on the catchment is mainly devoted to agriculture, forestry and livestock farming; some industrial areas also exist. They are mostly located in Belgium, on the Sambre and around Liège. Further details can be found in Descy & Empain (1984) and Descy & Mouvet (1984).

History of eutrophication in the River Meuse

The first signs of important algal blooms in the Meuse date back to the beginning of the 1980s, when blooms were also reported in the French Meuse and the Albert Canal. In France, the first "crises planctoniques" were detected by an abnormal brown colour of the water, beginning in a stretch at about 150 km from the source (Léglize & Salleron 1988). Since then, the phenomenon has been monitored by measurements of chlorophyll-*a* from May to September: the data show a rise of the biomass maxima year after year, at least up to 1986 (300 mg chl*a* m^{-3}). The blooms were mostly due to "small *Stephanodiscus*", i.e. centric diatoms of the genus

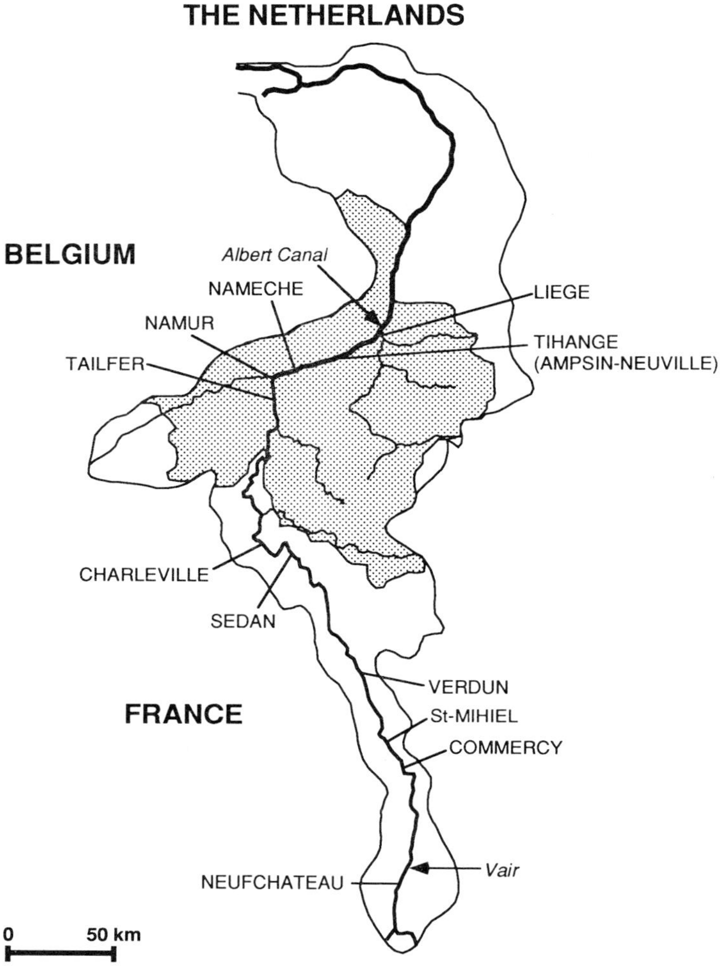

Figure 1. Map of the River Meuse basin, with some important localities cited in the text; the dotted area is the Belgian part of the catchment (about 40% of the total area).

Stephanodiscus (*S. hantzschii* and fo. *tenuis*, *S. parvus*, *S. invisitatus*, etc.). The origin of the blooms was found to be the River Vair, a heavily eutrophicated tributary coming from Vittel, where detergents and phosphoric acid were being used in a factory for cleaning glass bottles to contain, of all things, drinking water!

In Belgium, the alarm was sounded by the AWW, the company which produces drinking water for Antwerp. They found that taste and odour of water from the Albert Canal were getting worse. Thanks to the monitoring of the Meuse, organized by the AWW and dating back to the 1970s, the evolution of algal blooms in time can easily be shown (Fig. 2). Again, there was a steady increase of the algal biomass, particularly from 1980 onwards, even though sporadic maxima were recorded in the 1970s. Meanwhile, various problems due to excess of microscopic algae in the water had arisen in the treatment plants, mainly clogging of the filters at the intakes. Along with these problems, monitoring of chlorophyll-*a* has been done on a daily basis at the CIBE treatment plant at Tailfer (Fig. 3) and the relationship between algal biomass and the frequency of filter-cleaning processes has been clearly established (Fransolet,

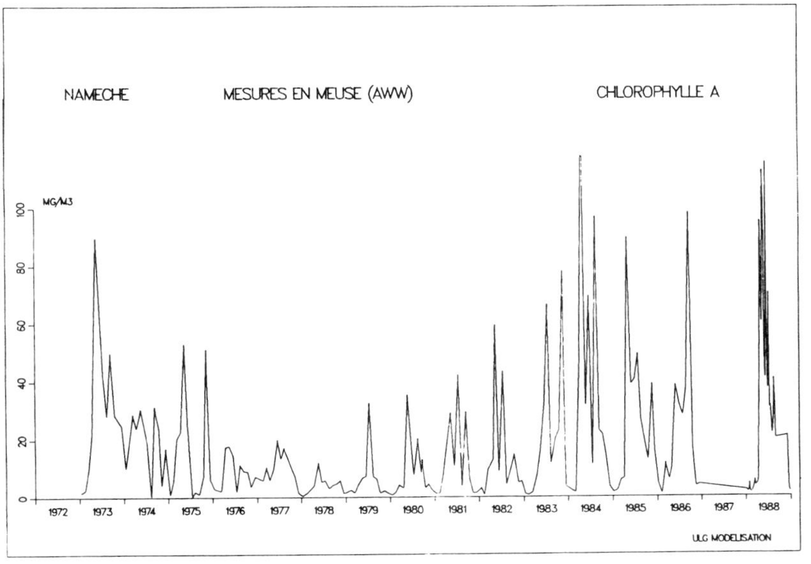

Figure 2. Evolution of phytoplankton biomass (chlorophyll-*a*, mg m^{-3}) in the River Meuse at Namêche (Belgium), from 1973 to 1988. The data (monthly or weekly measurements) are from the Antwerpse Waterwerken (AWW, Antwerp, Belgium) and the graph was supplied by the "Centre Environnement", University of Liége (Belgium).

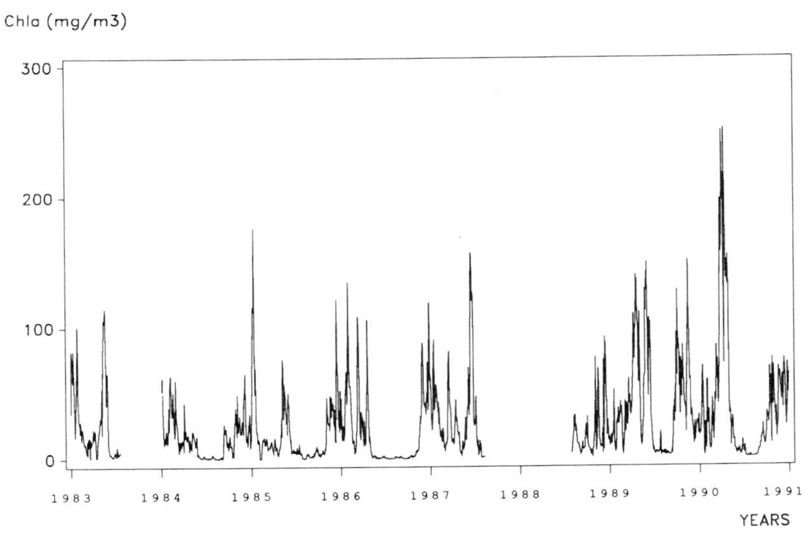

Figure 3. Evolution of phytoplankton biomass (chlorophyll-*a*, mg m^{-3}) in the River Meuse at Tailfer (Beglium), from January 1983 to December 1991. The data (daily measurements) are from the CIBE (Brussels Water Supply Company, Belgium).

pers. comm.). The drinking-water supply was particularly impeded during 1990, when the highest chlorophyll maxima of the past seven years took place.

At this point several questions remain unanswered: what are the factors favouring eutrophication in the Meuse, are they similar in other large rivers, and will eutrophication develop further?

Development of river eutrophication

In large European rivers, planktonic eutrophication has been developing over the past 20 years or so: examples from the rivers Danube, Rhine, Loire and many others can be found in the literature. As in lakes, the factors responsible for excess growth of algae at some periods are clearly nutrients, provided that conditions of light and temperature enable algae to grow. Also as in lakes the main, if not the only responsible nutrient, is phosphorus. Limitation by nitrogen, coming in large amounts from drainage of the watershed, is highly unlikely. By contrast, when a large bloom of diatoms develops, the silica demand may deplete this element so that it may become limiting for the siliceous algae. These conditions may, together with other factors, trigger a community change towards dominance by green algae, when silica depletion lasts for more than a few days (Si is normally supplied in large amounts by the watershed).

If phosphorus is the limiting factor, why did severe eutrophication appear so late? As a matter of fact, the concentration of soluble reactive phosphorus in a river like the Meuse in the Belgian section, has been at saturating levels (> 0.1 mg l^{-1}) for a long time. Various answers have been proposed. For example, there may have been a decrease in toxicity of some micropollutants (e.g. Pb, Cd) which were released in large amounts in the 1970s. However, this holds true for stretches of the river receiving industrial pollutants, but not elsewhere. Climate changes could possibly play a role, but then, why did plankton blooms not occur in 1976? Why has there been a steady increase since 1980?

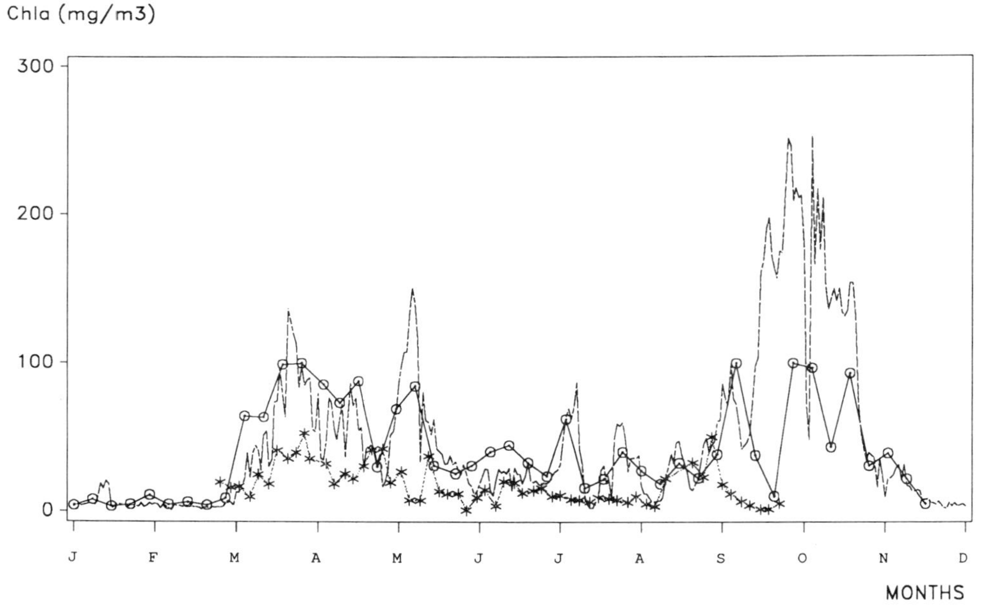

Figure 4. Comparison of the evolution of chlorophyll-*a* (mg m^{-3}) in 1990 at three sites on the River Meuse: ∗, St-Mihiel (France); —, Tailfer (Belgium); ●, Namêche (Belgium). The data for St-Mihiel (Measurements twice a week) were supplied by the "Agence de Bassin Rhin–Meuse", Rozerieulles, France.

For the Meuse, no satisfactory explanation can be found when only the Belgian section of the river is considered. But when events in the French section of the Meuse are taken into account, it becomes clear that the phytoplankton development begins somewhere in France and the algae can grow further downstream, as long as they can maintain a significant net production without undergoing important losses. This is supported by evidence from chlorophyll measurements at distant sites (Fig. 4), as well as by mathematical model simulations (Fig. 5). These models demonstrate that, in summer conditions, without grazing by the zooplankton, the phytoplankton can reach a peak biomass (equivalent to *c.* 150 mg chl*a* m^{-3}) *entirely within* the French territory. What happens downstream is a decrease, partly due to dilution by tributaries (Fig. 5A; see details of the "scenario" in the figure legend). Under other seasonal conditions (Fig. 5B) the (lower) biomass maximum will be achieved further downstream.

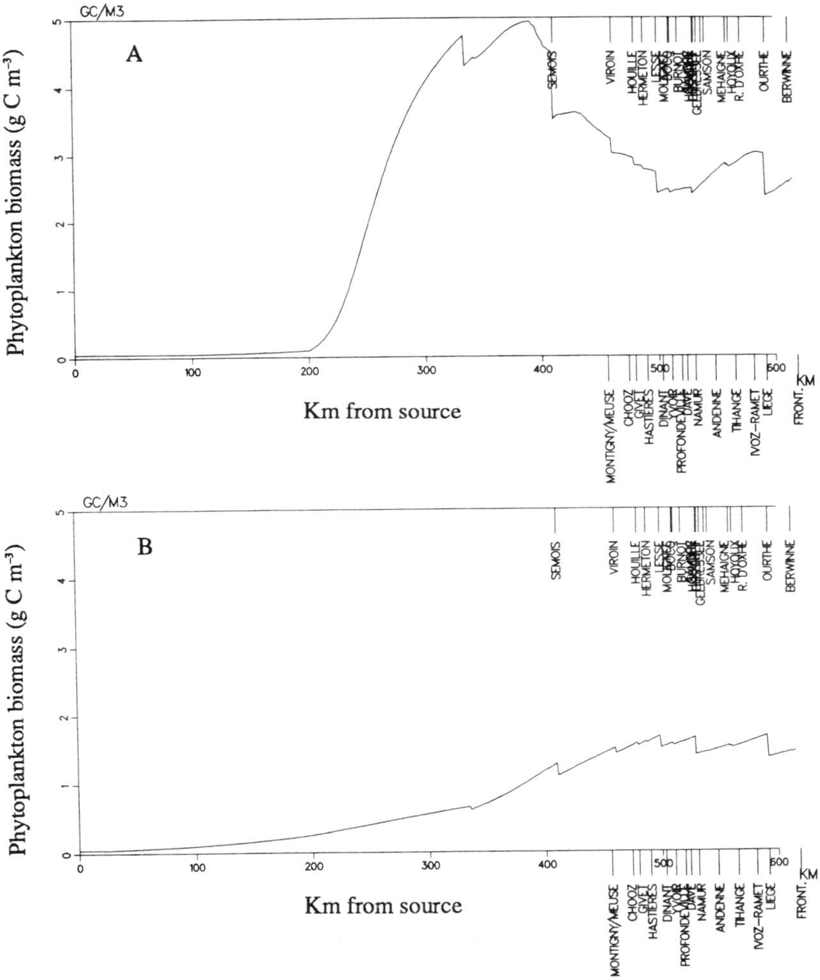

Figure 5. Simulation of the phytoplankton biomass (g C m^{-3}) from the source of the River Meuse in France to the Belgian–Dutch border, according to the PEGASE model (Centre Environnement, University of Liège, Belgium). A, typical summer conditions (20°C, insolation 2200 joules cm^{-2} d^{-1}, discharge 50 m^3 s^{-1}). B, typical spring conditions (15°C, insolation 1800 joules cm^{-2} d^{-1}, discharge 200 m^3 s^{-1}). The main tributaries are indicated on the upper horizontal axis and some localities are shown on the lower horizontal axis.

Even if these simulations can be improved (e.g. by a better physical description of the French Meuse) they show very well how the process is dynamic in the spatial dimension. Hence, any attempt to control eutrophication should begin by taking appropriate measures in France, based on reducing the inputs of phosphorus and taking into account phosphorus uptake, storage and utilization by the algal cells flowing downstream, while their environment (light, temperature, depth of the water column, discharge, presence of zooplankton) is changing along the way.

Reductions of phosphorus are already under way in France, since 1986, for instance by treating some industrial releases (removal of phosphorus in the water-purification plants) or by changing the processing of industrial products (Léglize & Salleron 1988). However, the industrial releases of phosphorus represent only 10 to 34% of the total inputs in the French Meuse; some effort must still be made to reduce inputs in domestic sewage. Nevertheless the outcome is not at all clear, as there is no evidence that reductions of phosphorus will actually make it limiting in the rivers. Furthermore, an international collaboration is still to be established. For instance, very few if any actions have been taken to remove phosphorus inputs to the Meuse in Belgium.

One question remains unanswered: why did eutrophication not develop sooner? The most likely hypothesis is that classic sewage treatment has been designed to eliminate carbon and reduce suspended matter, while releasing huge amounts of readily available phosphorus into the recipient water-bodies. Less allochthonous organic matter and improved light conditions favour primary production: this could explain why river eutrophication is so widespread and receives more attention.

The (harmful) effects of eutrophication in rivers

The effects of eutrophication in rivers are well-known: at the shortest time scales (hours, days), biological activities, which are roughly proportional to the biomass of the organisms, induce variations of dissolved oxygen and pH. Due to reaeration and mixing of the water column, these effects are less dramatic than in lakes. Nevertheless, in some heavily eutrophicated rivers, oxygen depletion may occur every night and eliminates all animal life except a few very tolerant species. Other effects are indirect, like toxicity of ammonia, and changes in the composition of communities (including Cyanobacteria blooms). The decay of dead autochthonous organic matter is a typical harmful consequence which occurs at longer time scales: in the case of planktonic eutrophication, the delayed organic pollution is transported some distance from the point where the growth of algae began. These effects occur either seasonally, when the decrease of sunlight, water transparency and temperature do not permit the maintenance of algal cells, or are due to another mortality factor. A good example may be provided by simulation of the ecological impact of the nuclear power plant of Tihange, upstream of Lièige. In order to determine the rules to be applied to the thermal release of this plant, the "Meuse ecological model" was implemented with "impact functions" describing inhibition and/or mortality of plankton passing through the circuits, as well as thermal effects in the downstream stretch of the river. The simulations, validated afterwards by diel measurements, showed that the operation of the plant during summer results in a maximal oxygen deficit of 2 mg l^{-1} at a variable distance from the plant, depending on the flow rate (Fig. 6). The deficit is due to oxygen consumption by heterotrophic bacteria which decompose the organic matter from dead phytoplankton and zooplankton. As the mortality rate of a function of the proportion of the river flow passing though the plant, a maximal effect occurs when a large amount of phytoplankton drifts downstream from above the plant intake.

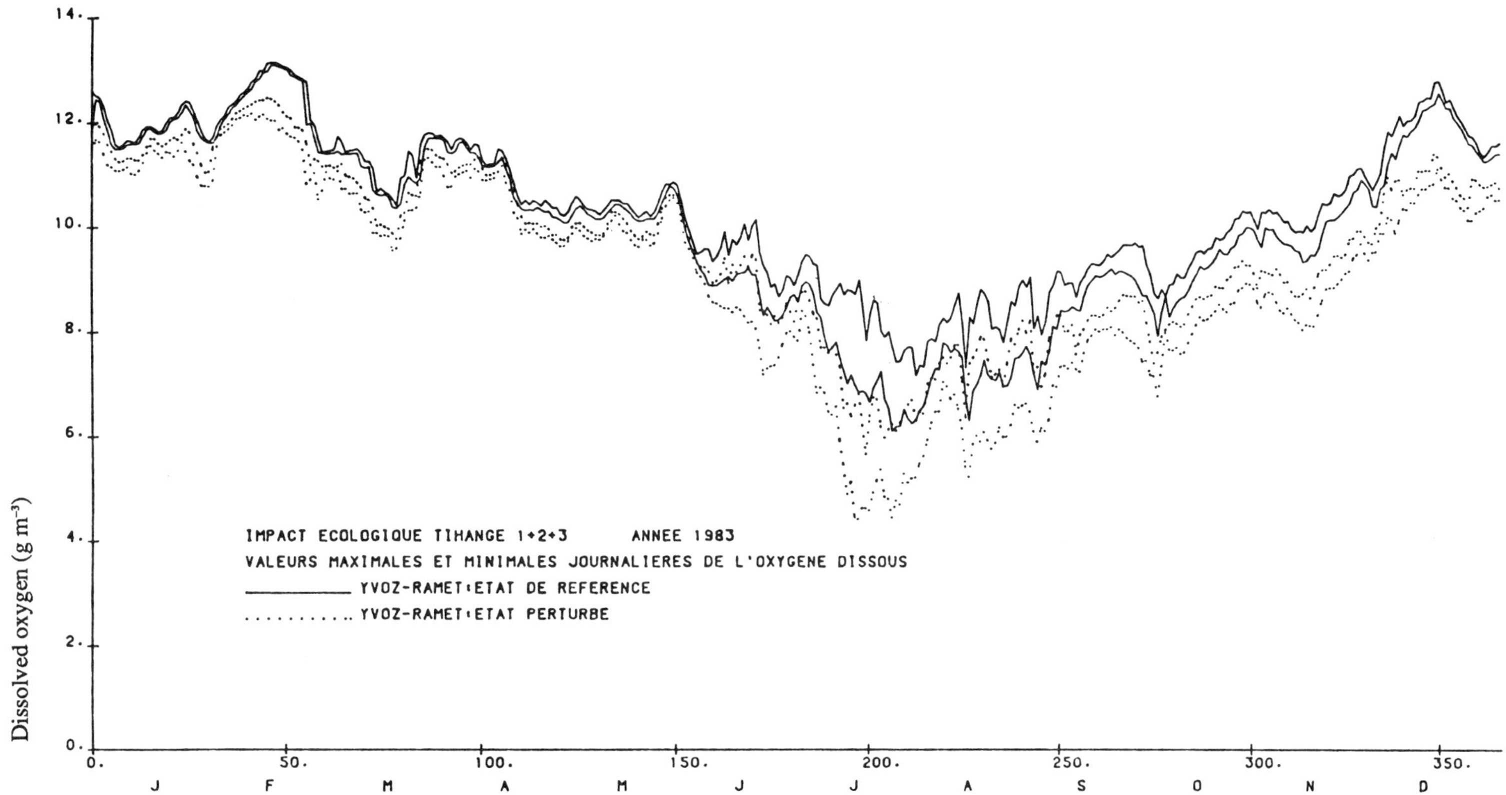

Figure 6. Simulation showing daily maxima and minima for dissolved oxygen concentrations (g m^{-3}) in the River Meuse, during 1983. Simulated values are for upstream (solid lines) and some 15 km downstream (dotted lines) of the Tihange nuclear power plant, The mathematical model used is the "Meuse ecological model", partly described by Descy *et al.* (1987), implemented with impact functions assessing the physiological and lethal effects on the plankton entrained into the plant.

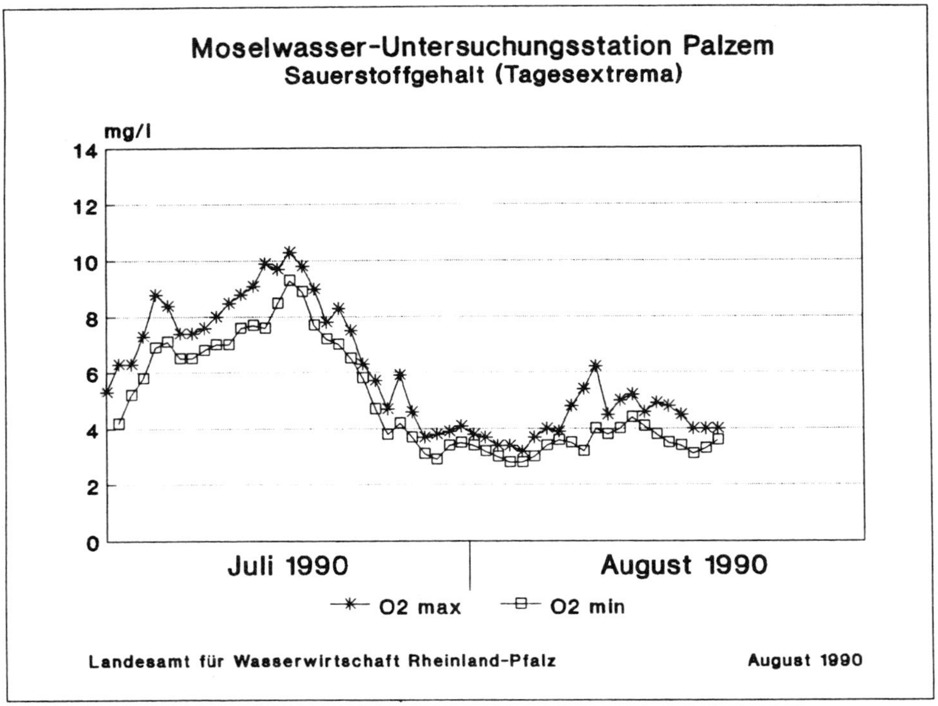

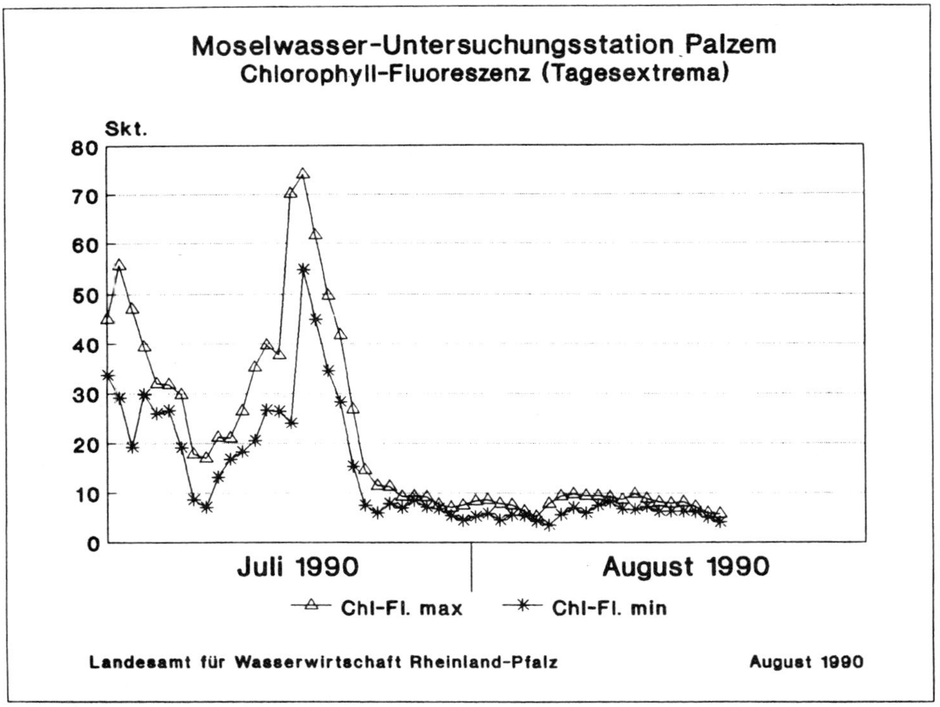

Figure 7. Observations of maxima and minima of dissolved oxygen concentrations (mg l^{-1}, upper diagram) and chlorophyll concentrations (fluorescence measurements, lower diagram) in the River Moselle downstream from the French–German border, in July and August 1990. Data supplied by J.-L. Salleron, "Agence de Bassin Rhin–Meuse", Rozerieulles, France,

Other (useful) eutrophication effects

In large regulated, deep and slow-flowing rivers, photosynthesis is the main process producing oxygen in the water column, while reaeration is negligible in the oxygen budget. The photosynthetic oxygen input is essential for maintaining water quality in stretches of river receiving organic sewage, in order to balance the oxygen demand of decomposer organisms. An example is the Meuse downstream of Liège, which receives a great amount of untreated domestic and industrial sewage. During summer the oxygen level falls to 2 mg l^{-1} (or less!) when the phytoplankton biomass is low (less than 20 mg chl*a* m^{-3}), whereas a concentration exceeding 5 mg l^{-1} is maintained (in daylight) for a biomass greater than 40 mg chl*a* m^{-3}.

Another example showing that a lack of phytoplankton leads to a deficit of the oxygen budget is given by the River Moselle, downstream from the French–German border (Salleron, *pers. comm.*). Following a decrease of the algal biomass, low oxygen concentrations persisted for at least three weeks in August 1990 (Fig. 7). Most likely the phenomenon is due to biotic interactions, i.e. grazing by zooplankton and parasitism by chytrids, which seem to be important in the Moselle (Descy & Willems 1991).

Conclusions and perspectives

A scarcity of algae in large regulated rivers may result in problems caused by poor oxygenation, but an excess of algae is harmful in many ways, resulting in a widespread and uncontrolled decrease of water quality and creating problems for various water-uses by man as well as functional and structural changes in the aquatic ecosystems. Moreover, it must be emphasized that the blooms we see in rivers are only the visible part of the transport of nutrients to the sea, where they will arrive sooner or later. Rivers do not store phosphorus for long periods of time in their sediments, and the phosphorus taken up by the biomass is quickly released into the water. That is why the only realistic and ecologically right way to limit eutrophication is to reduce phosphorus inputs into surface waters. However, in the short term, the measures which can actually be taken to reduce phosphorus inputs to rivers might not limit eutrophication at all. For instance, the 40 to 60% reduction planned by the Rhin–Meuse Water Authority in France will be costly but it might not be effective in reducing algal growth in the Meuse. So the prospects are bleak, unless the policy for controlling eutrophication is one of *selective reduction of effective (in terms of algal growth) sources of phosphorus*, rather than reducing every phosphorus input in the basin. Such a policy puts forward a strategy at the scale of the whole catchment and could be carried out by using tools like PEGASE (abbreviation meaning "planning and management of water treatment"). This is a basin model taking into account the physical characteristics of the rivers, the inputs of carbon, nitrogen and phosphorus from various point and non-point sources, and all the processes which take place in the river, including the functioning of the ecosystem. Whatever the means of the strategy which is finally adopted to overcome eutrophication of large rivers, it should imply an approach at the scale of the whole drainage basin and, in many cases, international collaboration in order to develop common management policies.

References

Descy, J.-P. (1987). Phytoplankton composition and dynamics in the River Meuse (Belgium). *Archiv für Hydrobiologie, Supplement*, **78** (2), (*Algological Studies*, **47**), 225-245.

Descy, J.-P. & Mouvet, C. (1984). Impact of the Tihange nuclear power plant on the periphyton and the phytoplankton of the Meuse River (Belgium). *Hydrobiologie*, **119**, 119-128.

Descy, J.-P. & Empain, A. (1984). 1. Meuse. In *Ecology of European Rivers* (ed. B. A. Whitton), pp. 1-23. Blackwell Scientific Publications, Oxford.

Descy, J.-P., Servais, P., Smitz, J. S., Billen, G. & Everbecq, E. (1987). Phytoplankton biomass and production in the River Meuse (Belgium). *Water Research*, **21**, 1557-1566.

Descy, J.-P. & Willems, C. (1991). Contribution à la connaissance du phytoplancton de la Moselle (France). *Cryptogamie, Algologie*, **12**, 87-100.

Lack, T. J. (1971). Quantitative studies on the phytoplankton of the Rivers Thames and Kennet at Reading. *Freshwater Biology*, **1**, 213-224.

Lack, T. J. & Berrie, A. D. (1974). Phytoplankton production in the Rivers Thames and Kennet at Reading during 1970. In *Light as an Ecological Factor, II.* (eds C. C. Evans, R. Brainbridge & O. Rackham), pp. 43-62. Blackwell Scientific Publications, Oxford.

Léglize, L. & Salleron, J.-L. (1988). L'eutrophisation en rivière. In *Impact de l'Homme sur l'Écosystème Meuse* (eds J.-C. Micha & S. Pilette), pp. 33-38. Presses Universitaires de Namur, Namur.

Reynolds, C. S. (1984a). *The Ecology of Freshwater Phytoplankton*. Cambridge University Press, Cambridge. 384 pp.

Reynolds, C. S. (1984b). Phytoplankton periodicity: the interaction of form, function and environmental variability. *Freshwater Biology*, **14**, 111-142.

Reynolds, C. S. (1988). Functional morphology and the adaptive strategies of freshwater phytoplankton. In *Growth and Reproductive Strategies of Freshwater Phytoplankton* (ed. C. D. Sandgren), pp. 388-433. Cambridge University Press, Cambridge.

Szemes, G. (1967). Das phytoplankton der Donau. In *Limnologie der Donau*, **3** (ed. R. Liepolt), pp. 158-179. E. Schweizerbart'sche Verlag., Stuttgart.

A tool for better management of water quality during periods of algal blooms: the case of the River Oise

T. VANDEVELDE AND V. DA MOLO
Compagnie Générale des Eaux, 52 rue d'Anjou, 75008 Paris, France

Like other rivers in the Paris area, the Oise is subject to important seasonal algal blooms. This eutrophication generates notable problems for the production of drinking-water from a treatment plant on the river at Méry. A mathematical model has been developed to simulate variation in water quality in a pre-treatment storage basin, and another model is currently being adapted to model the River Oise. Integration of the two models should provide a comprehensive tool for predicting variations of phytoplankton and water-quality parameters associated with algal blooms. This will be a decision-aid for optimizing control of the treatment process for providing potable water.

Introduction

Seventy-five percent of water distributed in the Ile-de France region around Paris comes from surface water. The Compagnie Générale des Eaux supplies almost 2 million cubic metres of drinking water per day from three large treatment plants installed on the Seine, Marne and Oise rivers. The treatment plant located on the River Oise at Méry has a capacity of 270,000 m^3 d^{-1}. The treatment plant is fed by water from the river after retention for about 3 days in a 400,000 m^3 storage reservoir.

Like the other rivers in the Paris area, the Oise is subject to important seasonal algal blooms which appear in early April and last until the end of May. Figure 1 shows the evolution of chlorophyll-*a* concentrations at the sampling station of Méry sur Oise. Diatoms dominate at the

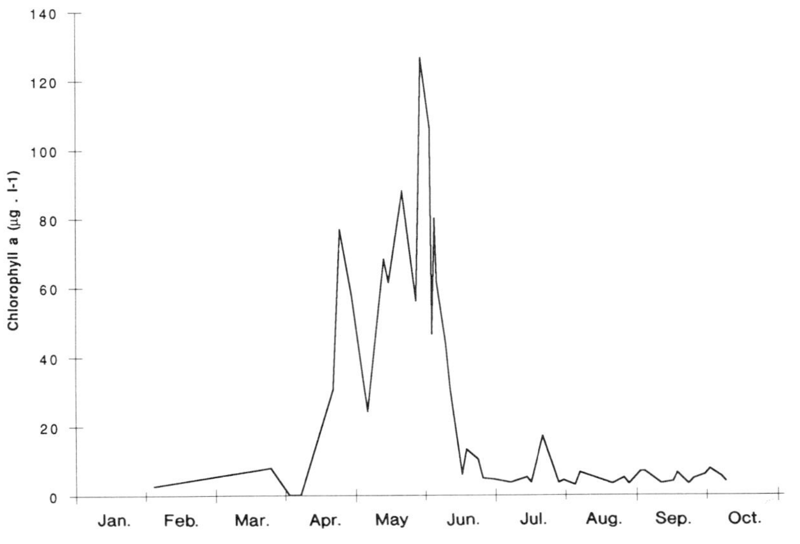

Figure 1. Seasonal evolution of chlorophyll-*a* concentrations in the River Oise at Méry sur Oise.

beginning of the bloom and Chlorophyceae are more important later. A very good and significant correlation between chlorophyll-*a* and phytoplankton cell numbers is found. At other sampling stations on the River Oise we have observed higher concentrations, but these were below 200 $\mu g\ l^{-1}$ chlorophyll-*a* in 1991.

This eutrophication of the river generates notable problems for drinking-water production. Variations in pH, ammonia and oxygen concentration are associated with algal activity and this can appreciably modify the necessary water-treatment parameters. In order to obtain a better understanding of these phenomena, we began to develop a mathematical model to simulate water-quality variation. This work is done in cooperation with Professor Billen and his team at the University of Brussels. The first step of the study, now completed, was to simulate water quality in the storage reservoir at Méry. The second step, currently being developed, is focused on adapting a new model to simulate conditions in the river. Integration of the two models should then give us a comprehensive tool for predicting the variation of phytoplankton and water-quality parameters during the period of algal blooms, providing us with an aid to decision-making for controlling the optimal production of potable water.

The reservoir model

The main objective of the reservoir model was to produce a dynamic description of the most important processes of transformation in the reservoir. The model takes into account sedimentation and phytoplankton dynamics, and also organotrophic bacterial activity and mineralisation of organic matter in the sediment.

The model is composed of four sub-models, enumerated below.

(1). *Sedimentation.* This sub-model relates suspended matter in the reservoir to suspend matter in the river and evaluates net sedimentation.

(2). *Phytoplankton.* This sub-model calculates phytoplankton primary production and chlorophyll content in the reservoir from the chlorophyll concentration in the river, given incident light intensity and temperature. It is linked to sub-model (1), because suspended matter determines the coefficient of light penetration and because phytoplankton also undergoes a process of sedimentation which must be taken into account.

(3). *Organotrophic bacterial activity.* This sub-model evaluates dissolved and particulate organic carbon content, based on concentrations in the river, together with primary phytoplankton production and bacterial decay. It is related to sub-model (2) as phytoplankton production is a source of organic matter.

(4). *Mineralization of organic matter in the accumulating sediments.* This sub-model evaluates the fluxes of nutrients and oxygen through the water–sediment interface, as well as seasonal and long-term variations.

Simulations with the model

The model of the natural processes occurring within the reservoir can be coupled with a simplified hydrodynamical model accounting for the different ways in which the storage reservoir can be operated. Two examples are given below.

In the first example we simulate events where water is drawn out of the reservoir at a constant rate, but is not replaced from the river, so water depth gradually decreases in the reservoir (Fig. 2). In this case, ammonia-nitrogen and phytoplankton increase rapidly. The chlorophyll-*a* concentration reaches approximately 200 $\mu g\ l^{-1}$ after 4 days and there is then a real risk of severe eutrophication.

In the second example we simulate events where the reservoir is shut-off for periods of 5 to 10 days. i.e. there is no inflow or draw-down of water. A strong vertical stratification appears in the reservoir (Fig. 3). The main risk is excessive development of phytoplankton,

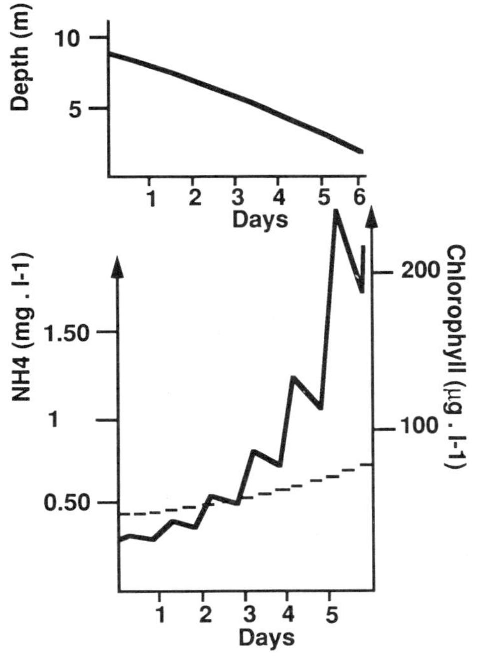

Figure 2. Model simulation of the evolution of water depth (m), ammonia-nitrogen (mg l^{-1}) and chlorophyll-*a* concentrations (µg l^{-1}) in the storage reservoir during a theoretical withdrawal of water at a constant flow rate (50,000 m^3 d^{-1}) in summer-time.

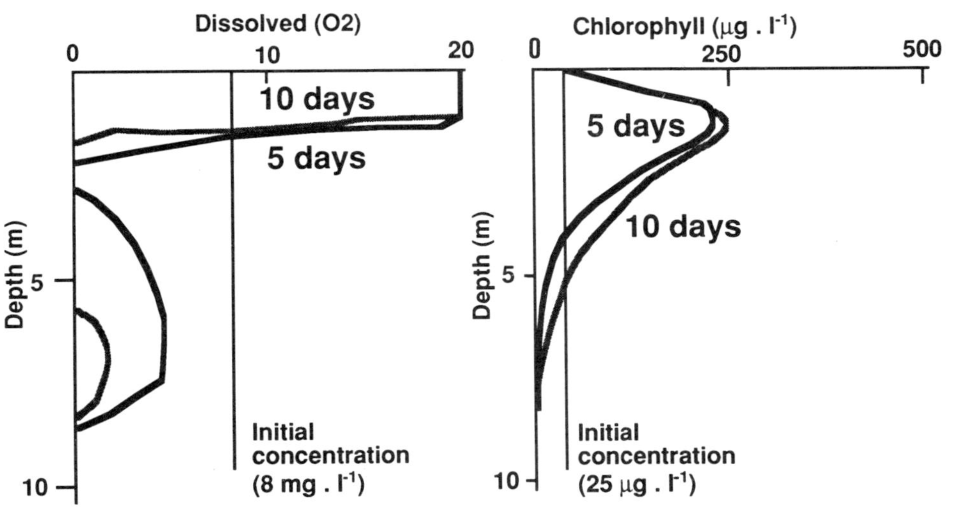

Figure 3. Model simulation of oxygen concentration profiles (mg l^{-1}) and chlorophyll-*a* concentrations (µg l^{-1}) established 5 and 10 days after shutting off the reservoir (zero input and output of water) in summer-time.

accompanied by oversaturation of oxygen in the photic zone (0–4metres depth) and undersaturation in the subphotic zone which may lead to anoxia.

We now turn our attention to the development of a model for the river.

Phytoplankton development in the river

The development of phytoplankton biomass is a result of the competition between biological growth and physical dispersion of cells (Vandevelde *et al* 1987). According to Kierstead & Slobodkin (1953), accumulation of biomass occurs when the net growth rate (including grazing by zooplankton) is higher than the dispersion rate by physical factors.

Light, temperature and nutrients are the factors that control the growth. They are proximal agents controlling cell physiology. Figure 4 gives an example of photosynthetic production at different light intensities in the River Seine during summer. Experiments, based on carbon-14 assimilation, showed the importance of light and allowed us to precisely define the characteristics of production versus photosynthetically active radiation. The variations of production observed between the four sampling dates are mainly explained by a temperature effect. This is confirmed by the good correlation observed between temperature and phytoplankton production in the rivers of the Seine basin, with a marked increase of production when temperature increases (Fig. 5) (for more detail, cf. Hanset *et al.* 1991).

Figure 6 shows an example of the seasonal evolution of zooplankton, with a large increase in numbers of zooplankters at the end of May. This was accompanied by a substantial decrease in phytoplankton, indicating the importance of grazing by zooplankters in determining variation in phytoplankton biomass.

Phytoplankton production is also dependent on the concentrations of nutrients such as phosphorus. In the River Oise, phosphorus is not limiting, but in some of its tributaries the concentrations of phosphorus are very low and could be a limiting factor for phytoplankton growth.

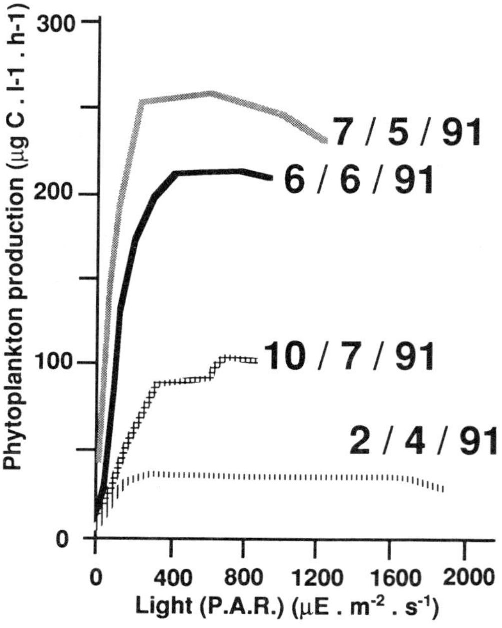

Figure 4. The observed effect of light (photosynthetic active radiation, PAR; $\mu E\ m^{-2}\ s^{-1}$) on phytoplankton biomass (μg carbon $l^{-1}\ h^{-1}$) measured (by ^{14}C assimilation) in the River Seine on four occasions during April to July 1991.

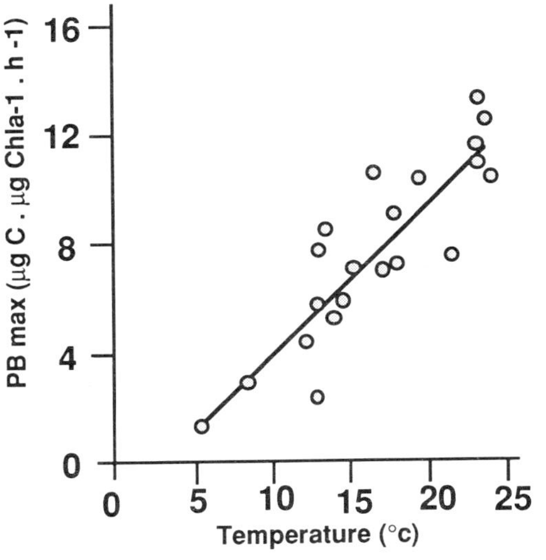

Figure 5. The effect of temperature on phytoplankton production (PB_{max}; µg carbon per µg chlorophyll-*a* h^{-1}) measured in rivers of the Seine basin during 1991.

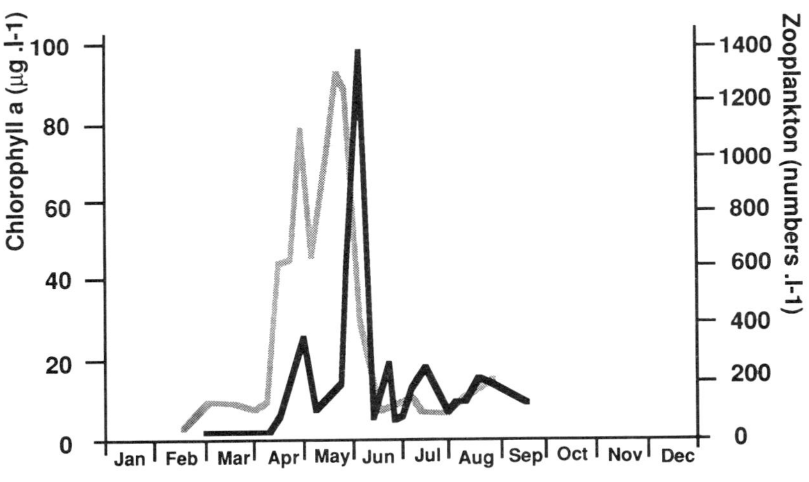

Figure 6. Seasonal evolution of chlorophyll-*a* concentrations (µg l^{-1}) and the impact of grazing by zooplankton (numbers l^{-1}) in the River Seine during 1991.

A biological model for the River Oise

The analysis of the effects of light, temperature and other parameters that influence the production of phytoplankton have been used to produce basic equations for a biological model of the river, shown in Figure 7, where major variables are depicted within circles and major processes are represented in square boxes.

The biological model works in conjunction with a simple hydrological description, based on Strahler's conceptual model (Strahler 1957, 1958) representing a drainage basin by a regular confluence scheme of increasing rank-order tributaries (Fig. 8).

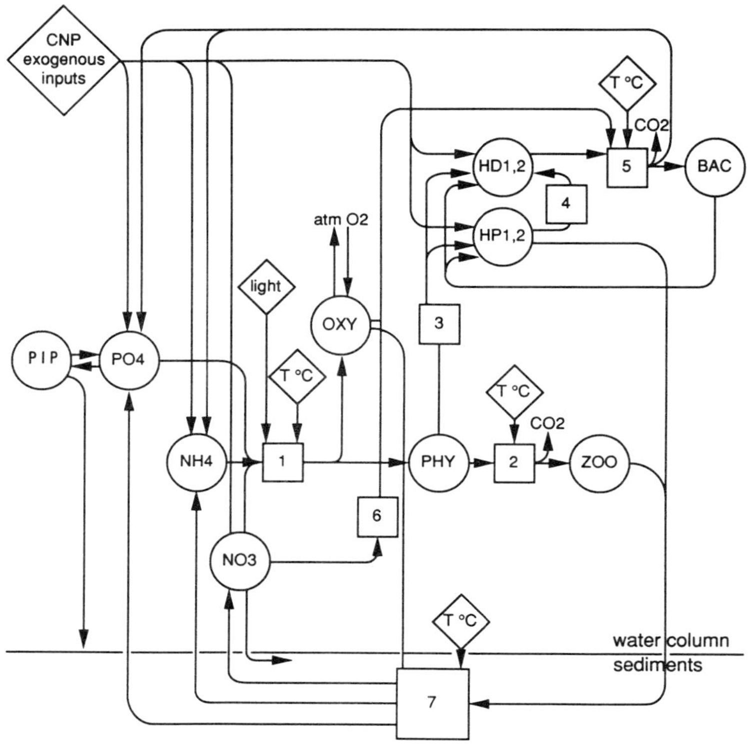

Figure 7. A summary of the biological model for the River Oise.

Circles denote major variables: phytoplankton (PHY), zooplankton (ZOO), heterotrophic bacteria (BAC), dissolved organic matter (HD, 1, 2) and particulate organic matter (HP 1, 2), oxygen (OXY), ammonium (NH_4), nitrate (NO_3), phosphate (PO_4) and inorganic adsorbed phosphorus (PIP).

Numbered squares denote major processes: 1, photosynthesis, prevalence of nutrients, phytoplankton growth; 2, zooplankton grazing, growth, respiration and excretion; 3, algal excretion and lysis; 4, bacterial hydrolysis of particulate organic matter; 5, bacterial hydrolysis and utilisation of dissolved organic matter, bacterial respiration, excretion and growth; 6, denitrification; 7, processes in the sediments which affect the water column (cf. Billen & Servais 1989, 1991).

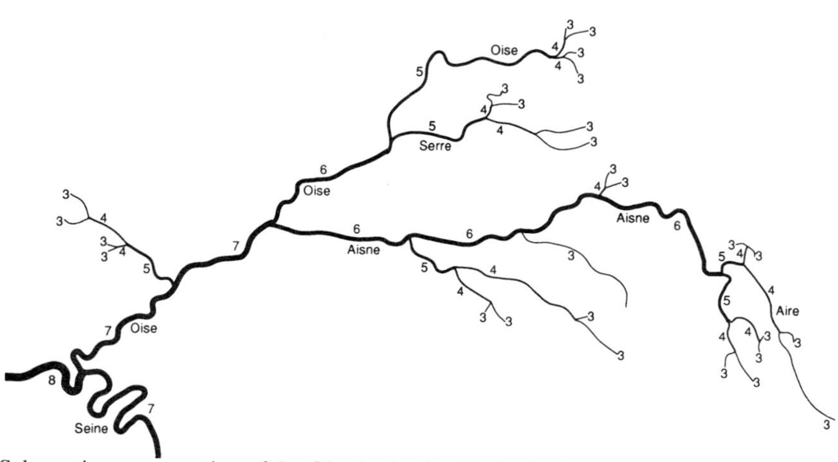

Figure 8. Schematic representation of the Oise basin (16,900 km^2) with hydrological ranking of tributaries (stream orders 1 and 2 not shown).

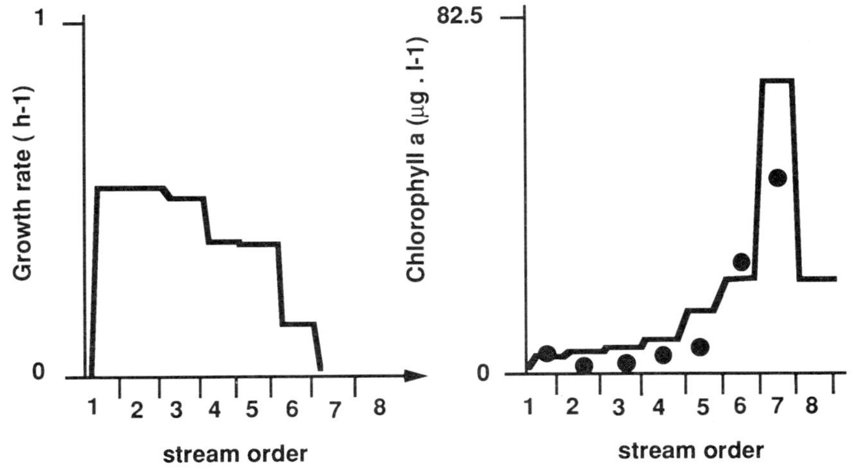

Figure 9. On the right, phytoplankton biomass (μg chlorophyll-*a* l^{-1}) is shown for seven classes of tributaries of the River Seine in April 1991. Measured biomass (points) is plotted against a preliminary simulation (solid line) of biomass using the biological river model shown in Fig. 7. On the left, simulated phytoplankton growth rates are shown against stream rankings.

There is a clear relationship between the rank-order of tributaries and some parameters that control the accumulation of biomass, such as dilution rate, light or hydrodynamics. Field sampling showed that the phytoplankton biomass is low in tributaries with a low hydrological ranking. In fact, biomass development is greatest in rivers where there is a favourable balance between phytoplankton growth and dilution rate. An initial trial simulation with the model gave reasonable agreement with measured phytoplankton biomass (chlorophyll-*a*) in the River Seine; the largest biomass occurred in the river ranked seventh (Fig. 9).

The next step of our work will be to complete the model for the River Oise and link it with the model for the storage reservoir.

References

Billen, G. & Servais, P. (1989). Modélisation des processus de dégradation bactérienne de la matière organique en milieu aquatique. In *Micro-organisms dans les Écosystèmes Océaniques* (eds Bianchi *et al.*), pp. 219-245. Masson.

Billen, G. (1991). Protein degradation in aquatic environments. In *Electroenzymes in Aquatic Environments, No. 7* (ed. R. Chrost), pp. 123-143. Springer Verlag.

Hanset, Ph., Garnier, J., Coste, M. & Billen, G. (1991). Ecophysiologie du phytoplancton dans le reseau hydrographique de la Seine. Rapport PIREN-Seine 1/91/03.

Kierstead, M. & Slobodkin, L. B. (1953). The size of water masses containing plankton blooms. *Journal of Marine Research*, **7**, 141-147.

Strahler, A. N. (1957). Quantitative analysis of watershed geomorphology. *Geophysical Union Transactions*, **38**, 913-920

Strahler, A. N. (1958). Dimensional analysis applied to fluvially eroded land forms. *Bulletin of the Geological Society of America*, **69**, 279-300.

Vandevelde, T., Legendre, L., Therriault, J.-C., Demers, S. & Bah, A. (1987). Subsurface chlorophyll maximum and hydrodynamics of the water column, *Journal of Marine Research*, **45**, 377-396.

The control of gas-vacuolate cyanobacteria

A. E. Walsby

Department of Botany, University of Bristol, Woodland Road, Bristol, BS8 1UG, England

The cyanobacteria that cause problems in water supply are principally the colonial forms that are buoyed up by gas vesicles. The success of these organisms is due, in part, to their gas vesicles, which enable them to perform vertical migrations or to maintain themselves in the euphotic zone. The gas vesicles are also the root cause of the problems. In calm periods they cause the cyanobacteria to float to the water surface forming noxious scums, and they may prevent the colonies from sedimenting in water treatment plants.

Gas vesicles are hollow, gas-filled structures; they are rigid but can be collapsed by the application of pressure. Their critical collapse pressure is influenced by their dimensions, which vary in different organisms. Gas vesicles are formed by the assembly of two types of protein, which determine their mechanical and physical properties.

Methods for collapsing gas vesicles in natural populations of cyanobacteria will be considered. They may have application to the control of cyanobacteria in water supply.

Introduction

Cyanobacteria are true bacteria and have a type of cell organisation that is quite different from any of the eucaryotic algae (Whitton & Carr 1982). They are usually considered to form part of the phytoplankton community, however, because their photosynthetic mechanism is functionally similar to that found in the algae, and they therefore compete for light with the eucaryotic micro-algae in the plankton.

At certain times of year cyanobacteria may come to dominate the phytoplankton in lakes and reservoirs (Carmichael 1981; Gibson & Smith 1982). It may be that one or more of their different bacterial attributes gives them a competitive edge over the eucaryotic members of the phytoplankton. For example, some cyanobacteria are able to fix atmospheric nitrogen and under nitrogen-deficient conditions this may enable them to outcompete the other algae. Not all cyanobacteria fix nitrogen, however, and when nitrogen is available non-fixing forms, such as species of *Microcystis* and *Oscillatoria*, commonly dominate. There are undoubtedly other features of cyanobacteria that determine which species dominate in particular lakes or at particular times, but do not explain the overall success of this group.

The dominance of cyanobacteria in the plankton is partly due to their greater success in harvesting the light that falls on the aquatic system, and there are two ways in which they achieve this: (1) they are able to grow and survive at lower photon irradiances than other algae and (2) they are able to return to the upper euphotic layers, after periods of mixing, by floating up. If we understand these processes we may be in a better position to control the development of cyanobacteria in lakes and reservoirs.

Cyanobacteria and microalgae in competition for light

Photosynthetic systems in all organisms exhibit a biphasic response to irradiance. In the light-limiting phase, the rate of gross photosynthesis is proportional to the irradiance. At higher irradiances the system enters the light-saturating phase at which the photosynthetic rate reaches a maximum value imposed by some other factor such as temperature, which controls the rate of the dark reactions. In a graph of the photosynthetic rate plotted against irradiance the gradient of the curve in the light-limiting phase gives an indication of the efficiency with which light

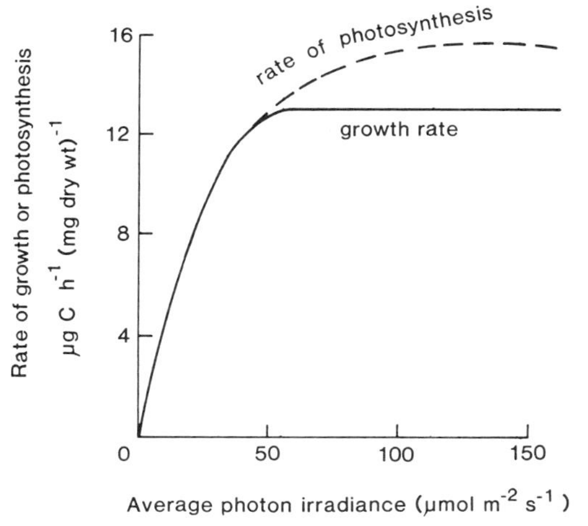

Figure 1. Growth rate and rate of photosynthesis of the cyanobacterium *Anabaena flos-aquae* as a function of photon irradiance. (Redrawn from Van Liere & Walsby (1982) with original data from Shear & Walsby (1975) and W. Zevenboom, unpublished).

energy is converted into chemical energy by the photosynthetic system (Fig. 1). The growth rate of the organism in turn depends on the efficiency of this conversion, and it varies in different organisms.

At the same time as an organism is accumulating chemical energy by photosynthesis it is also dissipating energy through catabolic processes; the overall growth rate depends on the balance of anabolism and catabolism.

Growth rates analysed by photostat culture.

Mur and his colleagues at Amsterdam pioneered the analysis of how growth rate responds to irradiance by growing cyanobacteria and algae in steady-state continuous cultures, which they operated in a parallel fashion to chemostats. In the light-limited chemostat, or photostat, nutrients are supplied at non-limiting concentrations. Again, the inoculated microorganism, growing photosynthetically under illumination from a constant light source, grows rapidly at first, but as its concentration rises the light passing through the culture becomes attenuated by pigments and the average irradiance within the culture decreases until it falls within the light limiting values for photosynthesis. The growth rate decreases until it is equal to the dilution rate of the culture and at this point a steady state is achieved. The cultures were generally operated in such a way that all the light entering the culture was absorbed by the relatively thick layer of cells and in this way it was possible to determine the specific rate of energy uptake (q_E) and to compare it with the growth rate (μ) (Van Liere & Mur 1979). The relationship was found to vary in different photosynthetic organisms.

In summary, their results showed that the greatest increases in growth rate for a given rise in light uptake rate was achieved by the green alga *Scenedesmus*, and at relatively high irradiances this organism would grow faster than the cyanobacteria *Oscillatoria* and *Aphanizomenon*. At low irradiances, however, the cyanobacteria grew more quickly (Fig. 2). The reason for this seems to be that they have a lower maintenance energy requirement. Photosynthesis can therefore compensate for respiration at a lower irradiance in cyanobacteria than in other algae.

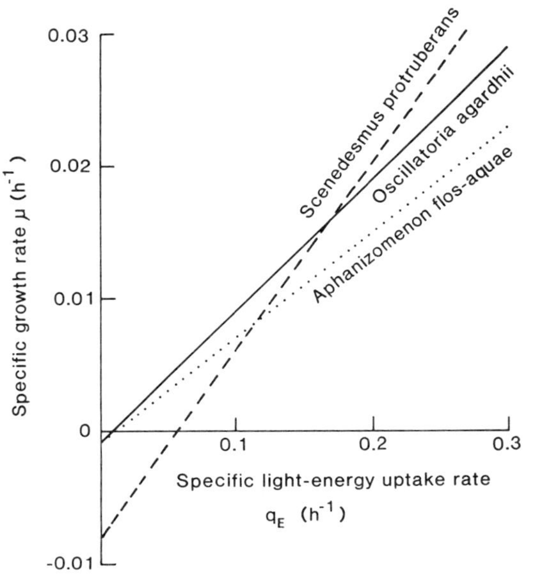

Figure 2. The variation in specific growth rate (μ) with specific light-energy uptake rate (q_E) for the green alga *Scenedesmus* sp., the cyanobacterium *Oscillatoria* sp. and the heterocystous cyanobacterium *Aphanizomenon* sp. (which has an additional energy requirement for making heterocysts and fixing nitrogen). The intercepts of the horizontal axis indicate the maintenance energy requirements of the different organisms at zero growth rate. (After Van Liere & Walsby (1982), with original data of Gons & Mur (1980), Van Liere & Mur (1979) and Zevenboom & Mur (1980)).

Competition between green algae and cyanobacteria in culture

The faster growth of cyanobacteria at low irradiance had been clearly demonstrated by the experiments of Mur et al. (1978) in which the green alga *Scenedesmus protuberans* and the cyanobacterium *Oscillatoria agardhii* were allowed to compete with one another by growing them together in a photostat continuous culture. When the culture was illuminated with a low incident irradiance (1 W m^{-2}, equivalent to 4.5 μmol m^{-2} s^{-1}) the green alga grew so slowly that its rate of division was less than the dilution rate of the culture, and after 25 days it became washed out; the concentration of the cyanobacterium increased 17-fold over the same period (Fig. 3a).

Of more interest is the outcome of an experiment in which a higher incident irradiance (25 W m^{-2}, equivalent to 113 μmol m^{-2} s^{-1}) was used. The green alga initially grew faster than the cyanobacterium, but after 10 days the increased biomass had attenuated light within the culture to such a degree that the average irradiance within the culture was insufficient to allow the green algae to increase faster than the rate at which the culture was being diluted. Again, the cyanobacterium, which continued to grow faster than the dilution rate, won the competition (Fig. 3b). Mur and his colleagues have pointed out that such competition can account for the succession of green algae to cyanobacteria that occurs in many lakes during the summer. This may be summarised in the following way.

Competition in lakes between green algae and cyanobacteria

Imagine the same two organisms growing during the early summer in a clear, shallow but well-mixed lake, with a plentiful supply of nutrients. The average irradiance in the water column is high and the green alga, which grows more rapidly, produces the dominant

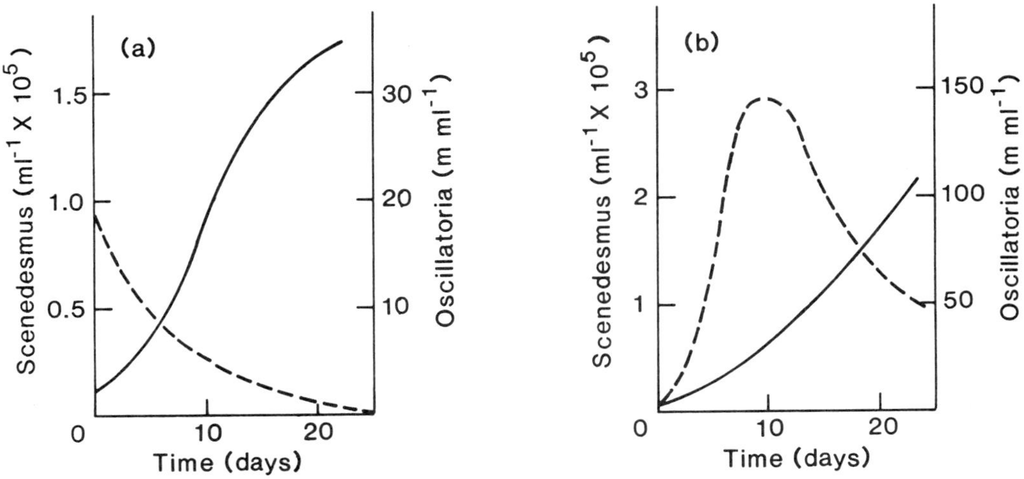

Figure 3. Competitive growth in a continuous culture of the green alga *Scenedesmus protuberans* (- - -) and the cyanobacterium *Oscillatoria agardhii* (—) at incident irradiances of (a) 1.0 W m^{-2}, dilution rate 0.01 h^{-1}; (b) 25 W m^{-2}, dilution rate 0.03 h^{-1}. In each case the cyanobacterium wins the competition. (Redrawn from Van Liere & Walsby (1982), with original data from Mur *et al.* (1978)).

population. As the alga grows it absorbs more of the light entering the system and the average irradiance experienced by the cells circulating within the water column falls (Reynolds 1984). Eventually the average irradiance decreases to a value at which the cyanobacterium grows faster then the green alga, and the cyanobacterium takes over as the dominant phytoplankton organism. The same will happen within the surface mixed layer of a deeper lake. This sort of succession can explain why cyanobacteria succeed green algae in well-mixed lakes.

Advantages of buoyancy provided by gas vesicles

In more stable lakes that stratify for part of the time, buoyant cyanobacteria may derive further advantages from floating up towards the water surface. They then become distributed within a shallower layer with a higher mean irradiance; moreover, in such a position the cyanobacterial population shades out the green algal population. The stable conditions need not persist for very long to give this advantage. All that is required is that the cyanobacteria move upwards in the vertically decreasing light gradient during periods of calm, so that on average they experience a higher irradiance than other non-buoyant cells, which remain dispersed throughout the mixed layer (Humphries & Lyne 1988).

This redistribution of buoyant cyanobacteria towards the surface of a periodically mixed lake has been demonstrated by Ibelings *et al.* (1991) who showed that buoyant *Microcystis* colonies were dominant in the upper 3m of the water column of Vinkeveen, a shallow fen lake, whereas non-buoyant colonies of *Scenedesmus* were distributed towards the lower layers. The vertical distribution of the cyanobacterium was more uniform in the well-mixed expanses of the Ijsselmeer, but even here the colony concentration was much less near the lake bottom where, owing to the weakening of turbulence in the basal boundary layer, the buoyant colonies could escape upwards by floating; the non-buoyant *Scenedesmus*, in contrast, was more abundant in this deep layer.

Planktonic cyanobacteria may also derive advantages over other phytoplankton by regulating their buoyancy. By a number of physiological mechanisms they are able to gain buoyancy at low irradiance and lose buoyancy in high irradiance (Konopka *et al.* 1987 a,b; Walsby 1987).

In this way they may become buoyant and float up in a lake during the night-time and sink during the day. Because of their small size, separate cells and filaments do not move up or down very quickly, but in stable lakes they may stratify, forming layers in the metalimnion (Reynolds *et al.* 1987). The large colonial forms move more quickly, however, and are able to perform vertical migration over depths of 10 metres or more in a day. Such movements may enable these organisms to make excursions into the hypolimnion at night, where they may absorb nutrients that support their growth nearer the water surface during the day (see Reynolds & Walsby 1975, for a review). The quantitative aspects of such movements have been analysed by computer modelling (Walsby 1989; Kromkamp & Walsby 1990).

Controlling cyanobacteria by lake mixing

Artificial mixing is one of the few methods that has been used to discourage the growth of cyanobacteria in lakes, and it can be successful (see Steinberg, this volume) . The general idea behind mixing is that if the phytoplankters can be circulated through a layer that is deep enough, none of them will receive sufficient irradiance for net growth. For the reasons given above, in relatively shallow lakes mixing may be more effective in reducing the growth of green algae than growth of cyanobacteria. Thus, while the overall size of the standing crop of phytoplankton may be reduced, mixing may actually increase the proportion of cyanobacteria.

If a lake could be mixed continually the buoyant cyanobacteria might be maintained in circulation so that they would be unable to accumulate towards the top of the euphotic zone. The problem is, however, to achieve uniform mixing of a large water mass. It may be difficult to maintain mixing velocities over the whole of the lake that exceed the floating velocities of the cyanobacterial colonies, and the colonies would therefore start to accumulate over unstirred areas. Moreover, the buoyancy of the cyanobacteria may increase in response to the lower mean irradiance they experience when stirred through the water column, so that they then float up more rapidly.

An alternative approach to complete and sustained lake mixing is intermittent mixing. The idea behind this is that the phytoplankton cells experience changing conditions that alternately benefit different organisms; the hope is that no one set of conditions exists for long enough to favour one particular species. However, changing conditions may actually benefit certain cyanobacteria, and in particular the colonial *Microcystis* species. They are often found in lakes that are subjected to intermittent mixing, and in the calm periods float upwards very rapidly into the euphotic zone (Walsby & McAllister 1987). During the period of calm they produce surface waterblooms. Some of the colonies may perform diel migrations to and from the water-blooms while others may be able to withstand periods of high irradiance at the water surface.

Control by gas vesicle collapse

The buoyancy of cyanobacteria is provided by gas vesicles. There are ways of destroying gas vesicles which might be applied in lakes to remove the advantages of buoyancy. Gas vesicles are specific to prokaryotic microorganisms and do not occur in any of the eukaryotic phytoplankton algae. Treatments that do no more than destroy gas vesicles might, therefore, selectively remove cyanobacteria. I provide a brief description of the gas vesicle and its properties before discussing possible means of their destruction.

The structure and properties of gas vesicles.

The gas vesicle is a hollow structure formed from proteins. It has the shape of a hollow tube closed at each end by a hollow cone. The tube and cones are formed by ribs of protein running perpendicular to the long axis. Gas vesicles grow by an assembly process from small biconical initials, usually attaining a length of about 0.5 μm (Walsby 1972). The radius of the cylinder

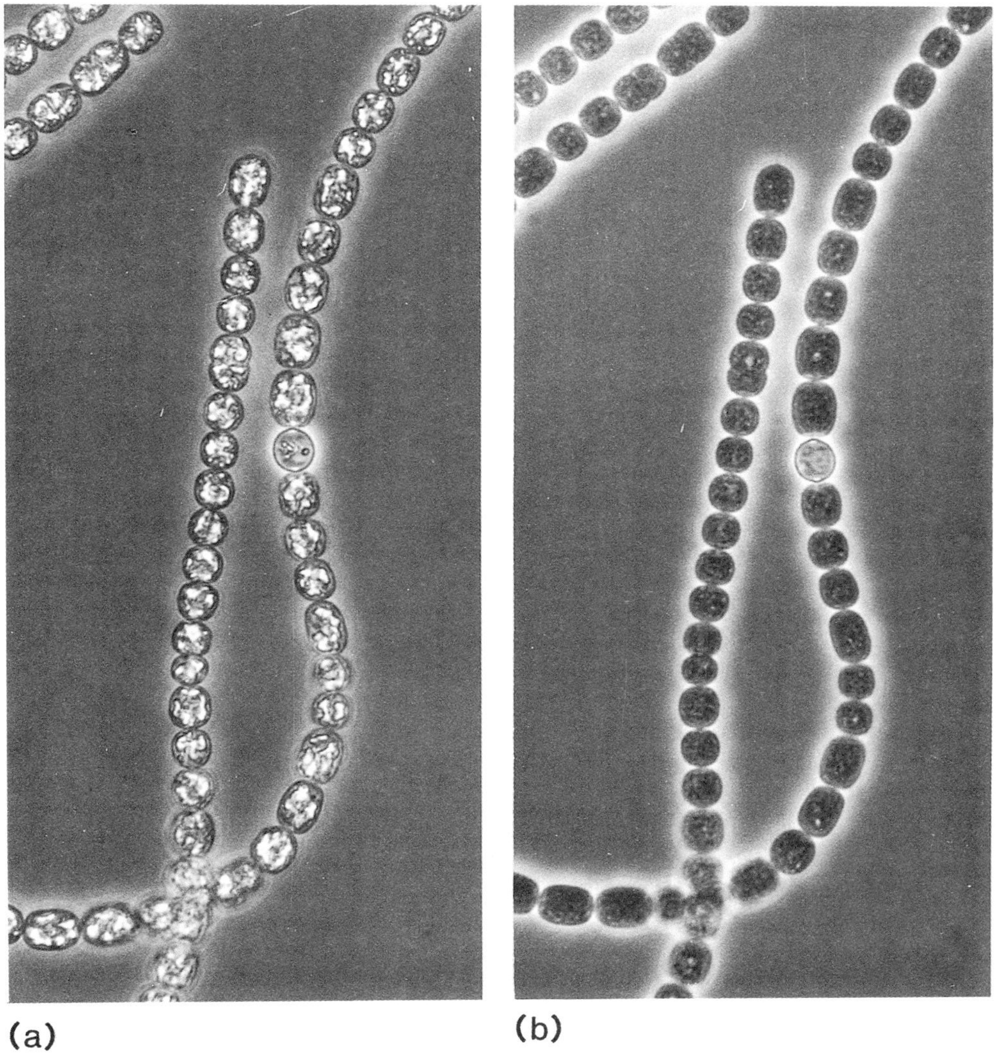

Figure 4. Filaments (of width 5 μm) of *Anabaena flos-aquae* (a) before and (b) after collapse of their gas vacuoles by pressure. The gas vacuoles, seen as bright refractile areas in (a), are aggregates of gas vesicles.

varies between 45 nm and 110 nm in different species but is fairly uniform within each species (Walsby & Bleything 1988). The gas vesicle is therefore too small to be seen under the light microscope but aggregates of gas vesicles, called gas vacuoles, are visible as bright refractile granules in the cells. They are distinguished from other inclusions by their disappearance under pressure (Fig. 4).

The protein wall of the gas vesicle is rigid, so that the structure has a volumetric compressibility of less than one part in a thousand per bar. The outer surface is hydrophilic, minimising the forces of surface tension, but the inner surface is highly hydrophobic, excluding liquid water. The wall is highly permeable to gases which, diffusing in and out in less than a microsecond, are in constant equilibration with the gases dissolved in the surrounding water. The gas is therefore usually air at atmospheric pressure. Further details of the features and properties of the gas vesicle can be found in reviews by Walsby (1972, 1989) and Walsby & Hayes (1989).

The protein of the gas vesicle can be destabilised by extremes of pH, acidity, salt concentration, detergents, and organic solvents (Buckland & Walsby 1971), but it is difficult to see how any of these agents could be applied on a scale that would permit the destruction of gas vesicles in cyanobacteria distributed through large volumes of lake water. Gas vesicles can, however, be destroyed by application of pressure (Fig. 4), and this might be used on a sufficiently large scale.

Collapse of gas vesicles by pressure

When gas vesicles are subjected to sufficient hydrostatic pressure they collapse flat. The cylinder folds down to a rectangular envelope and the cones become partially detached as they collapse to sectors of circles (Walsby 1971). This form of collapse is typical of the failure of a thin-walled cylinder that undergoes instability failure following buckling (Dehart 1969). The pressure at which the structure fails, known as the *critical pressure* (p_c) is determined by the stiffness of the wall material and the ratio of the wall thickness (t) to cylinder radius (r). The wall of the gas vesicle is thought to be of constant thickness in different organisms (which have very similar gas vesicle proteins) but the cylinder radius varies considerably (Walsby & Bleything 1988). Critical pressures of gas vesicles can be measured by turbidimetric techniques, using a pressure nephelometer (Walsby 1973). From a survey of eight genera of cyanobacteria (Walsby & Bleything 1988; see Fig. 5) the relationship between critical pressure and cylinder radius was found to be $p_c = 275(r/\text{nm})^{-1.67}$MPa.

There seems to be a correlation between the strength of the gas vesicles that form on different cyanobacteria and the pressures they normally have to withstand. Thus, in shallow brine pools, where the cells have low turgor pressure and the hydrostatic pressure is small, there appears to have been natural selection for weak gas vesicles that are wider and therefore more efficient at providing buoyancy. Conversely, cyanobacteria from deep freshwater lakes (Fig. 5) possess much stronger gas vesicles that are narrower and therefore less efficient in providing buoyancy (Walsby & Bleything 1988). The range of mean critical pressure encountered is 0.3 MPa in brine pools, 0.5 to 1.0 MPa in freshwater lakes, and up to 3.7 MPa in the deep ocean (1 MPa = 10 bar; a column of freshwater 10 metres deep generates a pressure of 0.98 bar).

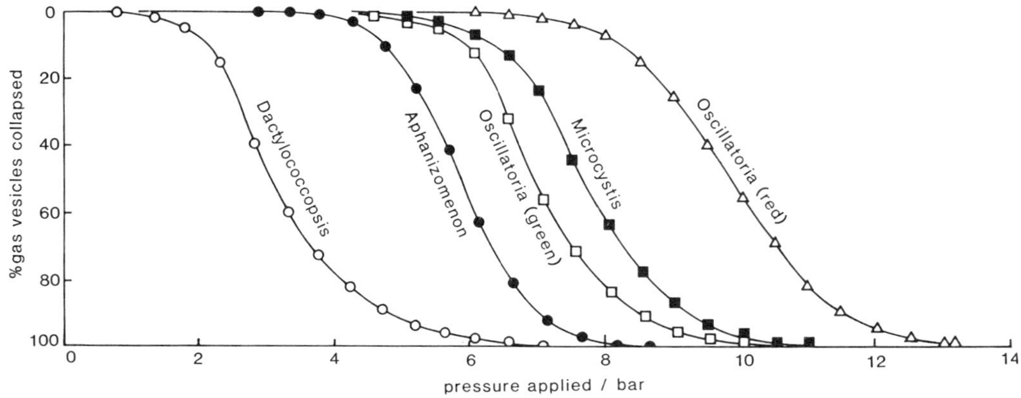

Figure 5. Critical pressure distributions of gas vesicles from various cyanobacteria showing the range in *Dactylococcopsis salina* (○) from a shallow brine lake, and species of *Aphanizomenon* (●), *Microcystis* (■) and *Oscillatoria* (□ green, △ red) from freshwater lakes of different depths. (From Walsby & Bleything 1988).

Within each species both the width and strength of the gas vesicles vary about the mean values referred to above. Thus in *Anabaena flos-aquae*, for example, where the mean p_c is about 0.6 MPa, 10% of the gas vesicles are collapsed at pressures below 0.5 MPa and 10% collapse above 0.7 MPa. In a strain of *Oscillatoria agardhii* from a deep Norwegian lake the mean p_c is 1.0 MPa but the strongest 10% of the gas vesicles require pressures exceeding 1.2 MPa to collapse them. The comparable figures for a strain of *Microcystis* sp. are 0.8 and 0.9 MPa, respectively, though in other strains of this cyanobacterium lower critical pressures are found. Inside healthy cells of the cyanobacteria the gas vesicles are subjected to a turgor pressure, which is usually 0.2 to 0.4 MPa (Fig. 6), and this contributes towards the total pressure required to collapse the structures (Walsby 1971).

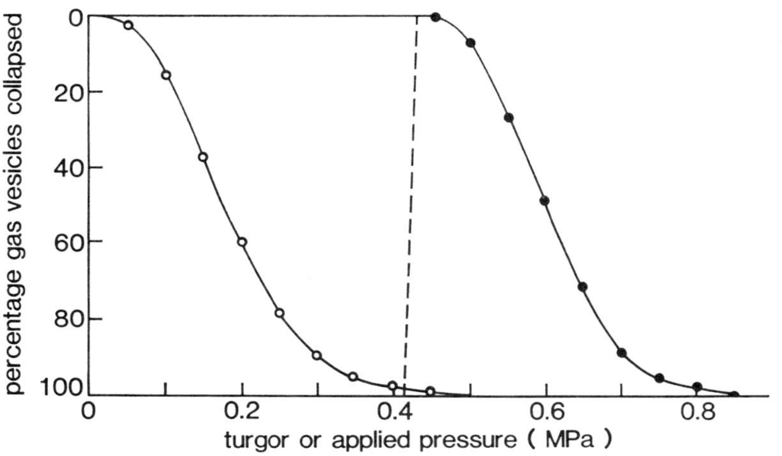

Figure 6. The critical pressure distributions of gas vesicles in cells of *Anabaena flos-aquae* (○) in turgid cells suspended in fresh water, and (●) cells without turgor, suspended in 0.5 M sucrose. The broken line (which gives the difference between the two curves) indicates the turgor pressure of the cells in fresh water. (After Walsby 1980).

The relationship between buoyancy and gas vesicle content

Cyanobacterial cells without gas vesicles usually have a buoyant density within the range 1030 to 1060 kg m^{-3}, depending on the content of dense substances such as carbohydrate. In order to float the cell must have sufficient gas vesicles to decrease the density below that of water, which is 998 kg m^{-3} for fresh water at 20°C. Sometimes the cells have a high gas vesicle content that give densities as low as 942 kg m^{-3}, and it may be necessary to collapse more than 50% of the gas vesicles to make the cells sink (Walsby *et al.* 1991).

When gas vesicles are collapsed the cells initially decrease in volume; much (but not quite all) of the gas volume lost is replaced by water entering the cell (Walsby 1980). The relationship between the volume of gas vesicles collapsed and the change in cell density is approximately linear; for the precise relationship see equation (3) in Walsby *et al.* (1991).

From Stokes's law, the sinking velocity of a cell or colony varies as the excess density, defined as the difference between the density of the object and the density of the suspending medium (e.g., water). It follows that the sinking velocity of the cell or colony will increase (almost linearly) with the proportion of the gas vesicles collapsed.

Methods of applying pressure to collapse gas vesicles

In small containers the hydrostatic pressure in a suspension of cyanobacteria can be raised in a number of ways. The first account of gas vesicle collapse is that of Klebahn (1985) who generated the required pressure by filling a stout glass bottle to overflowing with the cyanobacterial suspension and inserting a cork which he drove in with the blow from a hammer. The pressure can be applied in a more controlled manner by filling a syringe, closing the outlet and pressing on the plunger; this can be scaled up with a purpose-made cylinder and piston. The pressure can also be applied as a gas pressure above the cyanobacterial suspension and by the deceleration of a column of water (Walsby 1971). Variations of these methods are useful for investigating gas vesicle collapse in the laboratory (Walsby 1971) but they are hardly applicable for treating large volumes of cyanobacteria in lake water.

There are two basic approaches to collapsing gas vesicles in cyanobacterial populations in lakes. One is to attempt to apply pressure in the lake itself and the other is to apply the pressure in a pipe through which the cyanobacteria are circulated. The latter approach can be applied to water abstracted from a lake or reservoir before entering a treatment works, or to water recirculated back to the lake.

The pressure required in these applications will vary with the species of cyanobacterium present, as discussed above. It will also depend on the turgor pressure of the cells at the time of treatment (see Fig. 6); this is affected by irradiance and nutrient availability (Walsby 1971, 1980). Some cells may lose turgor pressure through photooxidation in waterblooms (Walsby *et al.* 1991) and the full critical pressure must then be applied in order to collapse their gas vesicles. The proportion of the gas vesicles that must be collapsed to cause the cells to sink will depend on their state of "over buoyancy". In some cases over 50% of the gas vesicles must be collapsed to remove buoyancy (Walsby *et al.* 1991). The sedimentation velocity of the colonies will then depend on the proportion of the gas vesicles remaining and to obtain the maximum velocity it will be necessary to collapse all of the gas vesicles. Although the ideal solution would be to apply a pressure that exceeds the maximum critical pressure of any organism likely to grow in the lake (about 1.3 MPa), in practice there must be a "trade-off" between pressure and cost. The critical pressure curves (Figs 5 and 6) provide the information needed in making these decisions.

Ultrasonic radiation

In 1957 an attempt was made by D. A. Hill and R. F. Packman of the Water Research Association, in association with K. B. Clarke of the Lowestoft Water Company, to treat water heavily laden with *Oscillatoria*. The water was passed through a pipe containing an ultrasonic transmitter delivering pulses of ultrasonic radiation. The discussion section of a paper by Brook (1959) indicates that the method was not successful in collapsing all the gas vesicles or in destroying the buoyancy of the filaments. It has been shown in laboratory experiments, however, that ultrasonic radiation can cause complete collapse of gas vesicles in cyanobacteria (Walsby 1969). A possible problem in scaling up the process may arise from the fact that the intensity of the radiation falls off as the square of distance, so that it requires considerable energy to treat a large volume of water effectively.

A gas vesicle crusher employing a deep concentric pipe

A method requiring little energy to operate has been developed by the East Anglian Water Company (now Suffolk Water Company). Water abstracted from a lake is circulated through a long concentric pipe sunk deep in the ground. At the bottom of the pipe the water experiences a hydrostatic pressure $p = hdg$, where h is the depth (m), d is the density of water (998 kg m^{-3}), and g is the gravitational acceleration (9.8 m s^{-2}) (Clarke & Walsby 1988). The depth required depends on the critical pressure distribution, as discussed above. An 86 m deep pipe has been installed at the Lound Works of Suffolk Water Company. Gas vesicles were completely removed from colonies of *Microcystis aeruginosa* that passed through the pipe, and the colonies lost their buoyancy. Trials are in progress to evaluate the efficacy of the gas vesicle crusher in aiding the settlement of cyanobacteria in water treatment. It is also proposed that, by circulating water from one end of a lake through the crusher and returning it at the other end, the gas vesicles could be removed from the population of cyanobacteria, so that the colonies lose their buoyancy and are removed from suspension (Clarke & Walsby 1988).

Generating pressures in lakes by explosions

When I first started investigating gas vesicles I made a number of devices for collapsing them, and I was curious to see if they could be collapsed by explosions. In one experiment I floated a small firework on a sheet of paper on a waterbath containing a sample of *Microcystis aeruginosa* taken from one of the Metropolitan Waterboard Reservoirs. The firework was detonated and the shockwave from the explosion collapsed the gas vesicles in the *Microcystis* colonies, which quickly sank out of suspension.

In a brief report of this demonstration (Walsby 1968) I suggested that explosions detonated on a larger scale at a lake surface might succeed in sinking out a cyanobacterial waterbloom. This idea was followed up by Menday & Buck (1972) who investigated the possible use of explosives in controlling cyanobacteria. Using Cordtex charges submerged in a flooded quarry they were able to demonstrate that shockwaves from underwater explosions would generate the pressure necessary to collapse gas vesicles as much as 30 metres from the source. Line charges could be laid that would treat bands of water on either side. They analysed the amounts of explosives that would be needed to treat a given area of water, and they also investigated the adverse effects on fish and engineering structures. As they pointed out, "while the technique appears drastic it does provide a quick and cheap control method when a blue-green algal bloom is having a serious effect on a water supply; it is not suggested that explosives should be used as a regular means of algal control".

Reformation of gas vesicles

The cells themselves are not adversely affected by the act of collapsing the gas vesicles; they will, for example, photosynthesize at the same rate before and after gas vesicle collapse (Shear & Walsby 1975). Cells will make new gas vesicles after application of pressure that causes collapse of the existing ones (Lehmann & Jost 1971). The collapsed gas vesicles cannot be reinflated with gas because the membranes are too permeable (Walsby 1969) and because they are anyway split at the edges (Walsby 1971). There is evidence, however, suggesting that some of the protein material in the collapsed gas vesicles may be re-incorporated into the new ones that form, and that the rate of gas vesicle formation does increase after gas vesicle collapse (Hayes & Walsby 1984). The new gas vesicles start as small bicones which then slowly elongate (Waaland & Branton 1969). It takes about 6 hours for an individual gas vesicle to reach the average length of 0.5 μm (Lehmann & Jost 1971), but it takes much longer than this before the cell accumulates sufficient gas vesicles to float.

Deacon & Walsby (1990) found that at the optimal light irradiance for gas vesicle reformation (30 $\mu mol\ m^{-2}\ s^{-1}$) in *Microcystis* sp., it took about 3 days before the cells had made just sufficient gas vesicles to provide buoyancy. Over the same period, cells incubated at low irradiance (<5 $\mu mol\ m^{-2}\ s^{-1}$) or high irradiances (> 80 $\mu mol\ m^{-2}\ s^{-1}$) made less than a quarter of the gas vesicles required to regain buoyancy. Cells that had previously been incubated at high irradiances formed some new gas vesicles when kept in the dark, but after 4 days they also had less than a quarter of the amount required to make them float (Fig. 7).

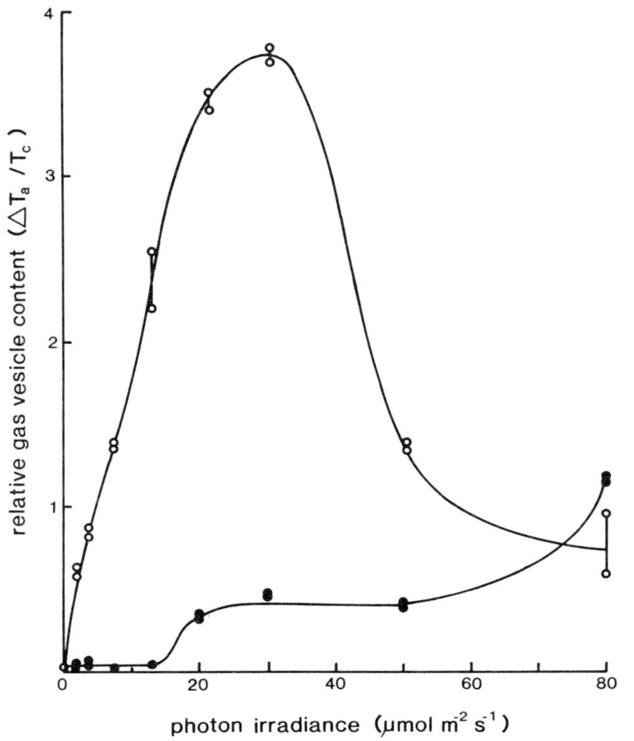

Figure 7. Gas vesicle reformation (indicated by pressure-sensitive turbidity, ($\Delta T_a/T_c$) in *Microcystis* cells after 3 days in the light (○) or 4 days in the dark (●) following collapse of gas vesicles by pressure. The "dark" cells were preincubated at the irradiances shown for 3 days before gas vesicle collapse. Of the cells with the highest gas vesicle content (after 3 days in 30 μmol m^{-2} s^{-1}) only 11% were floating.

These results are relevant to the fate of cyanobacteria that have been subjected to pressures that collapse their gas vesicles in lakes. Without their gas vesicles the colonies will sink to the lake bottom where, except in very shallow, clear waters, the irradiance is likely to be too low to permit the recovery of buoyancy. Buoyancy recovery has not yet been investigated directly in lakes, however, and *in situ* measurements are needed.

Concluding remarks

If colonies are unable to recover their buoyancy they will be lost from circulation in the water, and the gas-vacuolate cyanobacteria will be removed from the competition for light by the phytoplankton. Their place will almost certainly be taken by other microalgae, some of which will doubtless present other problems. However, few other algae form surface scums, cause such visual discolouration and noxious odours, and present such a threat or toxicity as waterbloom-forming cyanobacteria (Carmichael 1981). Gas vesicle collapse offers the prospect of controlling these organisms with a method that is entirely physical and does not entail the application of any chemical substances.

References

Brook, A. J. (1959). The waterbloom problem. *Proceedings of the Society for Water Treatment and Examination*, **8**, 133-137.

Buckland, B. & Walsby, A. E. (1971). A study of the strength and stability of gas vesicles isolated from a blue-green alga. *Archiv für Microbiologie*, **79**, 327-337.

Carmichael, W. W. (editor) (1981). *The Water Environment. Algal Toxins and Health.* Plenum Press, New York.

Clarke, K. B. & Walsby, A. E. (1988). *Improvements in the Treatment of Water to Remove Gas Vacuolate Cyanobacteria.* U.K. Patent Application 2 196 330 A.

Deacon, C. & Walsby, A. E. (1990). Gas vesicle formation in the dark, and in light of different irradiances, by the cyanobacterium *Microcystis* sp. *British Phycological Journal*, **25**, 133-139.

Dehart, R. C. (1969). External-pressure structures. In *Handbook of Ocean and Underwater Engineering.* (eds J. J. Myers, C. H. Holm & R. F. McAllister), pp. 9-3 to 9-20. New York: McGraw-Hll.

Gibson, C. E. & Smith, R. V. (1982). Freshwater plankton. In *The Biology of the Cyanobacteria* (eds N. G. Carr and B. A. Whitton), pp. 463-489. Blackwell Scientific Publications, Oxford.

Gons, H. J. & Mur, L. R. (1980). Growth rate and light uptake rate in light-limited continuous cultures of *Scenedesmus protuberans* Fritsch. *Archives of Microbiology*, **125**, 9-17.

Hayes, P. K. & Walsby, A. E. (1984). An investigation into the recycling of gas vesicle protein derived from collapsed gas vesicles. *Journal of General Microbiology*, **130**, 1591-1596.

Humphries, S. E. and Lyne, V. D. (1988). Cyanophyte blooms: the role of cell buoyancy. *Limnology and Oceanography*, **33**, 79-91.

Ibelings, B. W., Mur, L. R. & Walsby, A. E. (1991). Diurnal changes in buoyancy and vertical distribution in populations of *Microcystis* in two shallow lakes. *Journal of Plankton Research*, **13**, 419-436.

Klebahn, H. (1895). Gasvakuolen, ein Bestandteil der Zellen der Wasserblütebildenden Phycochromaceen. *Flora (Jena)*, **80**, 241-282.

Konopka, A., Kromkamp, J. and Mur, L. R. (1987). Regulation of gas vesicle content and buoyancy in light- or phosphate-limited cultures of *Aphanizomenon flos-aquae* (Cyanophyta). *Journal of Phycology*, **23**, 70-78.

Konopka, A., Kromkamp, J. and Mur, L. R. (1987). Buoyancy regulation in phosphate-limited cultures of *Microcystis aeruginosa. FEMS Microbiology Ecology*, **45**, 135-142.

Kromkamp, J. & Walsby, A. E. (1990). A computer model of buoyancy and vertical migration in cyanobacteria. *Journal of Plankton Research*, **12**, 161-183.

Lehmann, H., & Jost, M. (1971). Kinetics of the assembly of gas vacuoles in the blue-green alga *Microcystis aeruginosa* Kuetz. emend. Elekin. *Archiv für Microbiologie*, **79**, 59-68.

Menday, D. C. & Buck, A. A. (1972). A shock-wave technique to collapse the vacuoles of blue-green algae. *Water Research*, **6**, 279-284.

Mur, L. R., Gons, H. J. & van Liere, L. (1978). Competition of the green algae *Scenedesmus* and the blue-green alga *Oscillatoria. Mitteilungen der Internationale Vereinigung für Theoretisch und Angewandte Limnologie*, **21**, 473-479.

Reynolds, C. S. (1984). *The Ecology of Freshwater Phytoplankton*, Cambridge University Press, Cambridge.

Reynolds, C. S. & Walsby, A. E. (1975). Water-blooms. *Biological Reviews (Cambridge)*, **50**, 437-481.

Reynolds, C. S., Oliver, R. L. & Walsby, E. A. (1987). Cyanobacterial dominance: the role of buoyancy regulation in dynamic lake environments. *New Zealand Journal of Marine and Freshwater Research*, **21**, 379-390.

Shear, H. & Walsby, A. E. (1975). An investigation into the possible light-shielding role of gas vacuoles in planktonic blue-green alga. *British Phycological Journal*, **10**, 241-251.

Van Liere, L. & Mur, L. R. (1979). Growth kinetics of *Oscillatoria agardhii* Gomont in continuous culture, limited in its growth by the light energy supply. *Journal of General Microbiology*, **115,** 153-160.

Van Liere, L. & Walsby, A. E. (1982). Interactions of cyanobacteria with light. In *The Biology of the Cyanobacteria* (eds N. G. Carr & B. A.Whitton), pp. 9-45. Blackwell Scientific Publications, Oxford.

Waaland, J. R., and D. Branton. (1969). Gas vacuole development in a blue-green alga. *Science*, **163**, 1339-1341.

Walsby, A. E. (1968). An alga's buoyancy bags. *New Scientist*, **40**, 436-437.

Walsby, A. E. (1969). The permeability of blue-green algal gas-vacuole membranes to gas. *Proceedings of the Royal Society of London*, **B173**, 235-255.

Walsby, A. E. (1971). The pressure relationships of gas vacuoles. *Proceedings of the Royal Society of London*, **B178**, 301-326.

Walsby, A. E. (1972). Structure and function of gas vacuoles. *Bacteriological Reviews*, **36**, 1-32.

Walsby, A. E. (1973). A portable apparatus for measuring relative gas vacuolation, the strength of gas vacuoles, and turgor pressure in planktonic blue-green algae and bacteria. *Limnology and Oceanography*, **18**, 653-658.

Walsby, A. E. (1980). The water relations of gas-vacuolate prokaryotes. *Proceedings of the Royal Society of London*, **B208**, 73-102.

Walsby, A. E. (1987). Mechanisms of buoyancy regulation by planktonic cyanobacteria with gas vesicles, In *The Cyanobacteria* (eds P. Fay & C. Van Baalen), pp. 377-392. Elsevier Science Publishers, Amsterdam.

Walsby, A. E. (1989). The gas vesicle: a stable gas-filled structure in bacteria. In *Supersaturation and Bubble Formation in Fluids and Organisms* (eds A. O. Brubakk, B. B. Hemmingsen & G. Sundnes), pp. 69-103. Tapir Publishers, Trondheim, Norway.

Walsby, A. E. & Bleything, A. (1988). The dimensions of cyanobacterial gas vesicles in relation to their efficiency in providing buoyancy and withstanding pressure. *Journal of General Microbiology*, **134**, 2634-2645.

Walsby, A. E. & Hayes, P. K. (1989). Gas vesicle proteins. *Biochemical Journal*, **264**, 313-322.

Walsby, A. E., Kinsman, R., Ibelings, B. W. & Reynolds, C. S. (1991). Highly buoyant colonies of the cyanobacterium *Anabaena lemmermannii* form persistent surface water blooms. *Archiv für Hydrobiologie*, **121**, 261-280.

Walsby, A. E. & McAllister, G. K. (1987). Buoyancy regulation by *Microcystis* in Lake Okaro. *New Zealand Journal of Marine and Freshwater Research*, **21**, 521-524.

Whitton, B. A. & Carr, N. G. (1982). Cyanobacteria: current perspective. In *The Biology of the Cyanobacteria* (eds N. G. Carr & B. A. Whitton) pp. 1-8. Blackwell Scientific Publications, Oxford.

Zevenboom, W. & Mur, L. R. (1980). Nitrogen-fixing cyanobacteria: why they do not become dominant in Dutch hypertrophic lakes. In *Hypertrophic Ecosystems* (eds J. Barica & L. R. Mur), pp. 123-130. Junk, den Haag.

Physical measures to inhibit planktonic cyanobacteria

CHRISTIAN E. W. STEINBERG[1] AND ELKE GRUHL

[1] *GSF – Forschungszentrum für Unwelt and Gesundheit, Institut für ökologische Chemie, Ingolstädter Landstr. 1, D-W-8042 Neuherberg, Germany*

[2] *Obere Bergstrasse 16, D-W-8311 Eching-Berghofen, Germany*

In a small lake, intermittent destratification was installed after several other physico-chemical and physical in-lake therapy measures (phosphorus immobilization, permant destratification) had been tested without great success. If an aerobic sediment-water interface can be maintained, intermittent destratification removes cyanobacteria and prevents optimal development of other members of the photoautotrophic plankton. During growing seasons, increasing abundances of small-bodied herbivores (*Bosmina*) and *Daphnia* may have accounted for relatively low phytoplankton biomass as well.

Intermittent destratification is a very fast-working in-lake measure and seems to be applicable even in relatively shallow lakes (< 15 m), in which permanent destratification seems to be risky.

Introduction

Cyanobacterial dominance is assumed to be a clear sign of eutrophication (e.g. Vollenweider 1968, also see Table 1; Wetzel 1983). In temperate climates, cyanobacteria are the most prominent producers of (extracellular) organic substances, which can negatively affect other aquatic organisms and cause changes in the trophic interactions as well as impair raw-water qualities by, for instance, off-flavours (e.g. Berglind *et al.* 1983). The two compounds most frequently identified as causing off-flavours in water are geosmin and 2-methylisoborneol.

Furthermore, strains of all dominant cyanobacteria genera (*Anabaena*, *Aphanizomenon*, *Microcystis*, *Oscillatoria/Limnothrix*) are known to be capable of toxin production. These toxins can cause illness or death in almost any mammal, bird or fish which ingests enough of the toxic cells or extracellular toxins (for details see Carmichael *et al.* 1985; Berg *et al.*, 1987; Steinberg 1990). Toxin-producing cyanobacteria are widely spread. Whereas instances of toxicity to fish, livestock and human consumers were, at one time, believed to be largely associated with the warmer climates at low latitudes, the application of recently introduced methods of assaying for cyanobacterial toxins (Berg *et al.* 1987) testifies to the widespread occurrence of toxic strains among populations in lakes and reservoirs of temperate climates as well. For instance, in Scandinavian lakes, Berg *et al.* (1986) found that about 50% of the examined blooms were toxic to mice. They also found that neurotoxins of cyanobacteria caused deaths of three cows.

These examples serve to illustrate that the prevention of cyanobacterial blooms in lakes is one of the most important issues in applied limnology and – at least in some regions – in drinking-water supplies.

In this paper, we want to demonstrate that certain physical and physico-chemical in-lake management measures can be more or less successfully applied in reducing planktonic cyanobacterial dominance. Our experience is based on several whole-lake experiments, which have been carried out in a small Bavarian lake.

Table 1. *Minimum phosphorus requirement per unit cell volume for several algae common to lakes of progressively increasing productivity (compiled by Vollenweider 1968).*

Algae	Minimum P requirement (μg mm^{-3} cell volume)
Asterionella	<0.2
Fragilaria	0.2-0.35
Tabellaria	0.45–0.6
Scenedesmus	>0.5
Oscillatoria	>0.5
Microcystis	>0.5

Study site, material and methods

Fischkaltersee is a head-lake of the Iffeldorf Lake District, located south of the Starnberger See (Upper Bavaria, Germany). It has the following morphometric and hydrological features (Table 2).

Table 2. *Morphometric and hydrological data for Fischkaltersee.*

Surface area	3.4 ha
Maximum depth	11.2 m
Mean depth	5.7 m
Volume	180,000 m^3
Drainage area	30 ha
Theoretical retention time	143 days

The water of the lake has a bicarbonate content of 5 mequiv. l^{-1}. In the phytoplankton, filamentous cyanobacteria (Oscillatoriales) were dominant prior to in-lake management measures (Steinberg 1983). The lake is fed by several small groundwater sources and two surface inflows, of which one drains from an agricultural area. During winter this inflow loads the lake with de-icing salts from an adjacent street junction. As a consequence the lake, which formerly had been holomictic (Melzer 1976), became meromictic in the late 1970s.

The second inflow drains from a nearby settlement, which was connected to a sewage diversion in 1981. However, the aquifer still desorbs appreciable amounts of phosphorus (Fig. 1). The decreasing phosphorus content can be described by the relationship: $[P] = 541\ e^{-0.1y}$, $r^2 = 0.92$, where y is time in years.

Except for winter months, dissolved or soluble reactive phosphorus (SRP, assumed to be equal to phosphorus available to algae) in this second inflow was immobilized by partly hydrolyzed aluminium chloride, between 1982 and 1987 (Fig. 2). On an annual average, about 50–70% of the dissolved phosphorus was immobilized by this external management measure during 1982 to 1986. Since 1987, dissolved phosphorus is clearly increasing (Fig. 1).

From April 1980 to the fall of 1984, the lake was permanently destratified. In April 1985, the permanent destratification was changed into an intermittent mode. By fluorimetric tracing, it was determined that during destratification the turnover time of the water was 2.7 days.

Shortly before changing from quiescent to mixing conditions and from mixing to quiescent conditions, respectively, water samples were taken by means of a Schindler sampler at 1-m intervals. Immediately after sampling, temperature, pH and base capacity (to pH 8.2 as a measure of free carbon dioxide) were determined. All other parameters were analyzed in the laboratory. All determinations followed German Standard Methods or equivalents. Phytoplankton samples were taken by an integrating sampler after Schröder (Züllig & Bärlocher Ltd., Rheineck, Switzerland). They were fixed with Pfeiffer's solution. The counting was done with an inverted microscope (according to Utermöhl 1958).

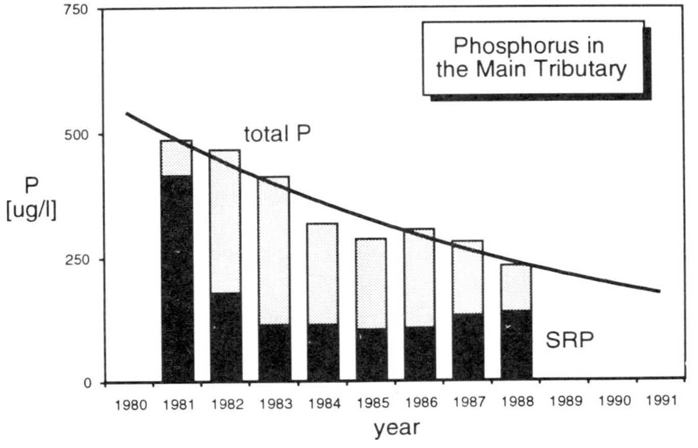

Figure 1. Mean annual phosphorus concentrations in the main tributary of Fischkaltersee. SRP is dissolved or soluble reactive phosphorus.

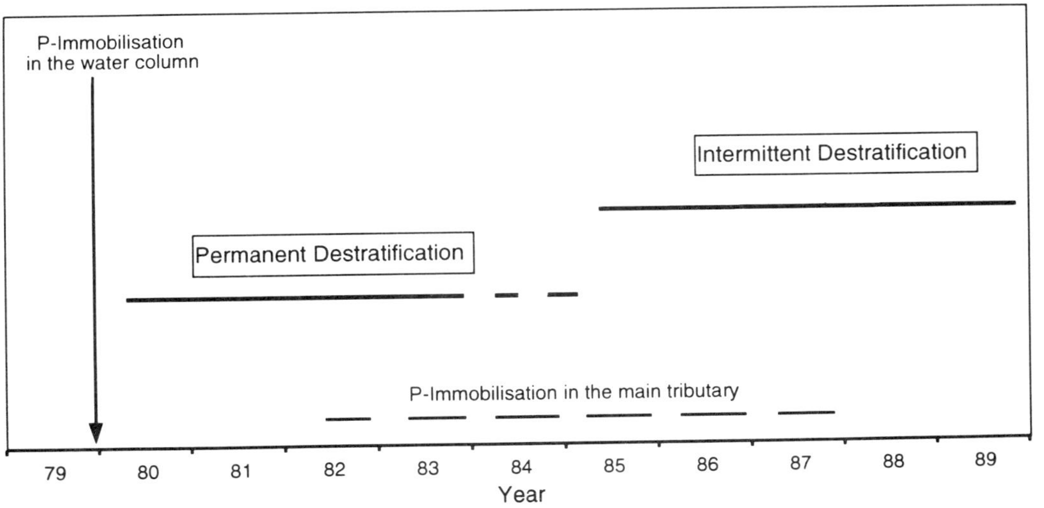

Figure 2. Chronology of internal and external lake management measures in/at Fischkaltersee.

First attempt at cyanobacterial inhibition: in-lake phosphorus flocculation

Prior to the in-lake management measures employed in our investigations, the phytoplankton community in Fischkaltersee was dominated most of the time by filamentous cyanobacteria, *Limnothrix redekei* (van Goor) Meffert, and *Oscillatoria rubescens* D.C. If cyanobacterial dominance has to be regarded as a direct consequence of nutrient enrichment (eutrophication), as was shown in a classical compilation by Vollenweider (1968; see Table 1), nutrient reduction in the water column or reduction of nutrient availability should lead to the decreasing presence of cyanobacteria.

Following this rationale, in December 1979 (prior to all destratification experiments), phosphorus in the lakewater was immobilized by flocculation with aluminium hydroxide, treating the various layers of the lake according to their specific phosphorus contents (Steinberg & Buchsteeg 1980). The molar ratio of A1:P was about 22, being ten times larger than the ratio used in successfully operating tertiary chemical sewage treatment plants. As a consequence of this measure, inactivation of dissolved phosphorus was about 70% in the tropholytic zone and almost 50% in the trophogenic layer immediately after the treatment procedure. The decrease of dissolved phosphorus was up to 65% in the lowermost zones and nearly 100% in the uppermost zones. The phosphorus content of the lake remained at the low post-treatment level from December 1979 to April 1980. Thus the adsorbing capacity of the aluminium layer in the sediment-water interface compensated for both allochthonous phosphorus-loading and phosphorus-release by the anaerobic sediments. The values may elucidate the chemical effect of the aluminium treatment. Whereas the phosphorus content in the uppermost layers increased slightly, the content in the sediment–water interface clearly decreased (Table 3).

Table 3. *Whole-lake changes of total phosphorus content (kg) in Fischkaltersee due to aluminium treatment in December 1979.*

Compartment	29 Nov 1979	8 Jan 1980	1 Apr 1980
0 – 6 m depth	11.6	7.2	8.1
6 – 8 m depth	3.0	1.5	1.8
8 – 11 m depth	24.1	6.3	4.8
Whole-lake totals	38.7	15.0	14.7

Phytoplankton biomass was reduced up to 90% by co-precipitation directly after the flocculation treatment. However, after a few days, *Limnothrix redekei* reached its former abundances, probably due to at least two facts. (1), This cyanobacterium re-invaded the pelagial by formerly co-precipitated trichomes. These microorganisms carried phosphorus from the monimolimnion into the trophogenic zone. Thus on a long-term basis the phosphorus fractions available to photoautotrophic microorganisms did not significantly fall below 50 $\mu g\ l^{-1}$. (2), From more recent experience it is known that the phosphorus content has to be reduced below 10 to 15 $\mu g\ l^{-1}$ (not 50 $\mu g\ l^{-1}$), in order to prevent cyanobacterial development in lakes on a long-term basis. In meromictic (sheltered) lakes with high water-column stability, cyanobacteria development, such as *Oscillatoria*/*Limnothrix* and *Aphanizomenon*, can be observed above this threshold (Steinberg & Hartmann 1988).

In other studies of phosphorus precipitation and control of sediment-phosphorus release, much higher A1:P ratios have been applied. Kennedy & Cooke (1982) calculated dosages directly from the alkalinity of the water to be treated. For the alkalinity of Fischkaltersee, a ratio of about 50 should have been applied to obtain pH 6.0, where insoluble aluminium hydroxide predominates. Cooke *et al.* (1982) treated several lakes according to their recommendations. However, in all lakes the phosphorus concentration remained above 35 µg l^{-1} and several lakes still had dense cyanobacteria populations.

In Medical Lake, Washington, USA, an Al:P ratio of more than thirty times that of Fischkaltersee was applied (EPA 1980). Total phosphorus contents were reduced from 0.45 mg l^{-1} to between 0.07 and 0.05 mg l^{-1}. In the post-treatment year, however, total phosphorus concentrations were again as high as 0.130 mg l^{-1} (Soltero *et al.* 1981).

In White Lough, Northern Ireland – a lake fairly well comparable to Fischkaltersee by size and shape – phosphorus release to the hypolimnion was reduced by 92% as a result of adding ferric aluminium sulphate in the first post-treatment year. In the third year it returned to pre-treatment levels. Overwintering phosphorus concentrations were in the range 31–53 mg P l^{-1} compared to 111–194 mg P l^{-1} in pre-treatment years (Foy 1985).

A reason for the failure of the precipitation treatment in Fischkaltersee is that this lake had become meromictic by the input of de-icing salts prior to the aluminium treatment in December 1979, as mentioned above. This means that phosphorus concentrations had to fall below the extremely low threshold (10–15 µg P l^{-1}) for cyanobacterial absence to be successfully established and maintained via low phosphorus content/availability. However, the actual phosphorus content of Fischkaltersee directly after the immobilization treatment was about three times higher than the threshold level. In general, this negative statement is valid for all precipitation treatment measures for which information is available. Consequently, most of the limno-biological studies report on failures and read like the one for White Lough (Foy & Fitzsimmons 1987):

> "Reduced total phosphorus concentrations in the summer which followed the addition of iron aluminium sulphate to White Lough failed to reduce significantly the phytoplankton, which continued to be dominated by *Oscillatoria agardhii* var. *isothrix* Skuja. This species was present throughout the 4 years studied, forming over 50% of the algal volume in 80% of samples".

Second attempt at cyanobacterial inhibition: permanent destratification

A permanent destratification was installed in Fischkaltersee in order to (1), destroy meromixis (= decrease water column stability); (2), oxidise the sediment-water interface to avoid phosphorus release from the sediment and to fix phosphorus from the water column onto the sediment; (3), reduce net light availability to phytoplankton.

Since April 1980 (Fig. 2), we have aerated the lake at the deepest point by means of a membrane compressor with a maximum pumping capacity of 275 litres of air min^{-1}. To prevent the diffusor from sinking into the relatively soft sediment, it was mounted on a rack. The destratification process was started on low power to avoid a sudden distribution of reduced chemical compounds and oxygen deficiency, which could cause severe fish kills (cf. Danecker 1971).

At the beginning, the effects of this in-lake measure were generally positive. Although the biomass of the chlorophytes and diatoms nearly doubled (Steinberg 1983), the population of cyanobacteria was removed almost completely. However, considering long-term effects, this measure was not successful. Although total phosphorus concentrations were reduced from about 90 to 50 g l^{-1} and dissolved phosphorus was almost undetectable, *Limnothrix* species reappeared in the third year of this in-lake management measure. The cyanobacteria peaked at the end of the fourth year of permanent destratification, the biomass being much higher than during the pre-destratification period (Steinberg & Tille-Backhaus 1990).

On a whole ecosystem scale, the findings at Fischkaltersee confirm laboratory as well as limnetic enclosure studies. For example, from enclosure experiments, several species of the *Oscillatoria/Limnothrix* group are known to be turbulence-tolerant or even turbulence-preferring forms (Reynolds 1986, 1989). Whether or not Oscillatoriales can survive or even dominate under mixing conditions, depends on mixing depth, euphotic depth and the ratio of both, provided that biotic factors, such as changes in trophic features of the zooplankton, can be excluded. In Fischkaltersee during the period under question, no clear change in the zooplankton structure was observed. Large-bodied filter-feeders were low in abundance all the time.

Actually, *Limnothrix redekei* developed in the shade of very dense populations of coccal greens and cryptophytes. As an additional mechanism, which may have favoured cyanobacterial re-occurrence, it is conceivable that these dense populations of eukaryotic algae depleted concentrations of dissolved carbon dioxide, as discussed among others by Shapiro (1990). During this phase of cyanobacterial re-occurrence, however, carbon dioxide concentrations were still well above detection limits, thus the carbon depletion mechanism may be of only minor importance. Similar findings have been reported by Spencer & King (1987). Possible mechanisms will be discussed in more detail later on.

Third attempt at cyanobacterial inhibition: intermittent destratification

In April 1985, the permanent stratification in Fischkaltersee was changed into an intermittent mode with the following objectives. (1), To repeat the positive success obtained at the beginning of the previous aeration, i.e. a decrease of cyanobacterial biomass and presence. (2), To aggravate the conditions of competition and development for all phytoplankton species in order to decrease the biomass of the algae, or (3), to elevate biomass losses via sedimentation but prevent phosphorus release from the sediment during quiescent periods. (4), To improve underwater light climates as a consequence of diminished phytoplankton biomass in order to avoid future cyanobacterial re-occurrence.

In the first year of intermittent destratification the rhythm was approximately 21 days of mixing to 7 days of quiescence. Since the second year the rhythm has been approximately 14:14 days. In order to increase phytoplankton biomass losses via sedimentation, the quiescent periods were slightly prolonged during later years.

Even during the prolonged quiescent periods, there was no significant phosphorus release from the sediments, not even during periods of low oxygen content in the sediment–water interface (Fig. 3). The regression line has almost no slope, indicating that the phosphorus concentration in the sediment–water interface appears to be redox-independent.

In 1984, the permanent destratification was in operation for less than 6 months due to technical problems and the lake restratified during the late growing season, with subsequent sediment-release of nutrients. With respect to intermittent mixing, 1984 may serve as a reference for pre-treatment conditions.

Total phosphorus decreased from 60 (1984) to 30 $\mu g\ l^{-1}$ in 1988; $[P] = 60 - 5.65 \log t$, (t = units of 7.6 days), $r = -0.582$ (Fig. 4). This concentration, however, can still support cyanobacterial growth under conditions of high water stability (see above).

From 1984 through 1988, on an equivalence basis, N:P ratios are increasing; $N{:}P = 20.9 + 5.07 \log t$, ($t$ = units of 7.6 days), $r = 0.29$ (Fig. 5), which in turn may prevent invasion of hetero-cysteous cyanobacteria (Steinberg & Hartmann 1988).

Due to rapidly decreasing silica concentrations ($[Si] = 5.58 - 0.86 \log t$, ($t$ = units of 7.6 days), $r = -0.70$) (Fig. 6), the Si:P ratio declines from about 120 to 40; $Si{:}P = 122 - 16.3 \log t$,

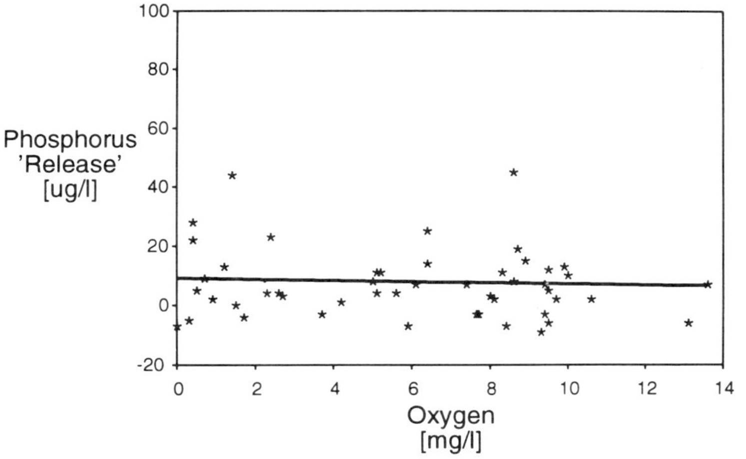

Figure 3. Differences in phosphorus content of the 11-m and 10-m depths, as an estimate of phosphorus release from the sediments, plotted against oxygen concentrations during quiescent periods in Fischkaltersee.

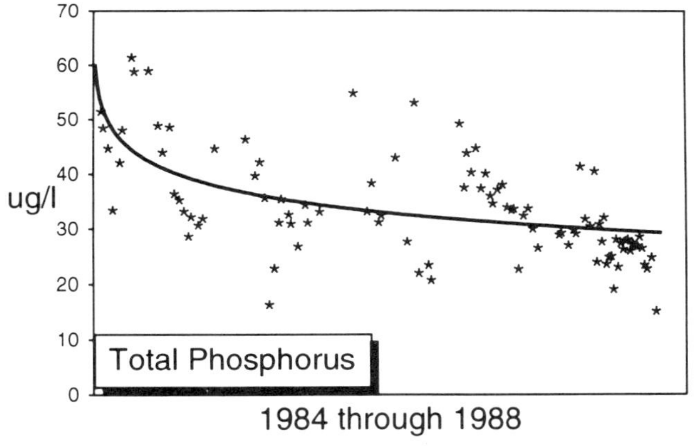

Figure 4. Mean total phosphorus concentrations in the water column of Fischkaltersee in 1984 through 1988.

(t = units of 7.6 days), $r = -0.53$ (Fig. 7). According to Tilman (1982), such changes in Si:P ratios cause shifts in the diatom composition from species with a high silica requirement, for instance *Synedra* spp., to those with lower silica requirements, such as *Asterionella formosa* and some centric diatoms (*Cyclotella meneghiniana*, *Stephanodiscus minutus*). Actually, this trend can be detected in the Fischkaltersee plankton as well. During the years of intermittent destratification, *Synedra acus* was gradually replaced by *Asterionella* and several centric diatoms (Figs 8–11).

Changes in biotic parameters

As expected, the initial success caused by permanent destratification was repeated. Thus immediately after the onset of intermittent destratification, *Limnothrix redekei* peaked only for a few days and was finally out-competed by chlorophytes and diatoms (Fig. 8). Disregarding the fact that there was no stable pattern of phytoplankton succession in the following years, the

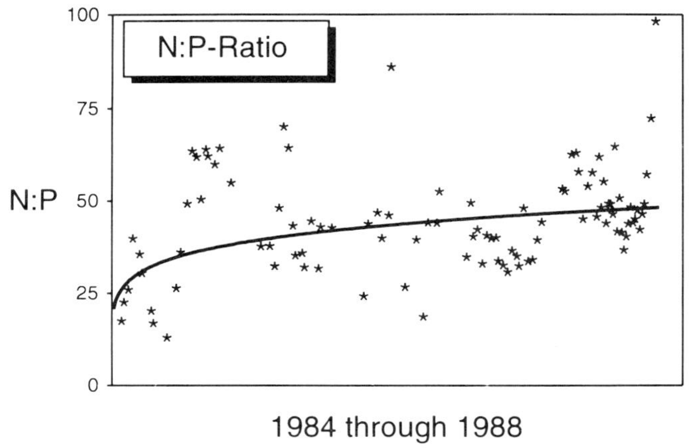

Figure 5. The ratios of inorganic nitrogen and total phosphorus (equivalence basis) in the water column of Fischkaltersee in 1984 through 1988.

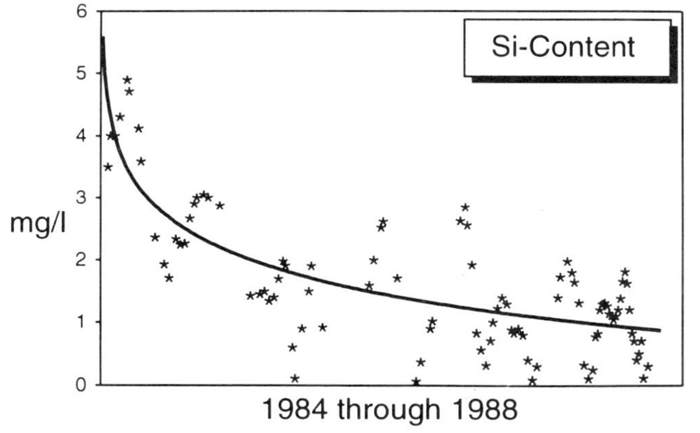

Figure 6. Mean dissolved reactive silica concentrations in the water column of Fischkaltersee in 1984 through 1988.

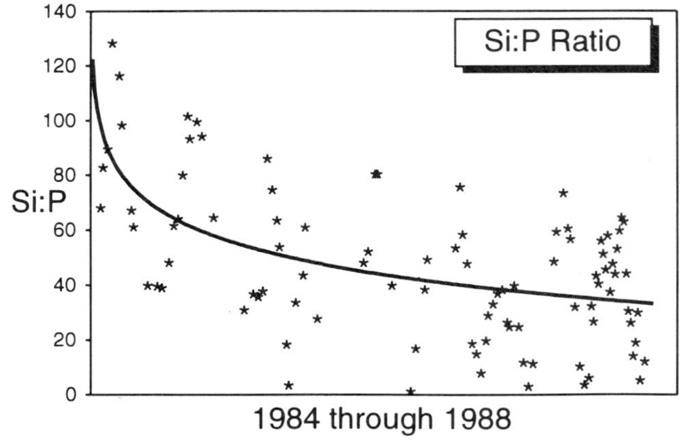

Figure 7. The ratios of dissolved reactive silica and total phosphorus (equivalence basis) in the water column of Fischkaltersee in 1984 through 1988.

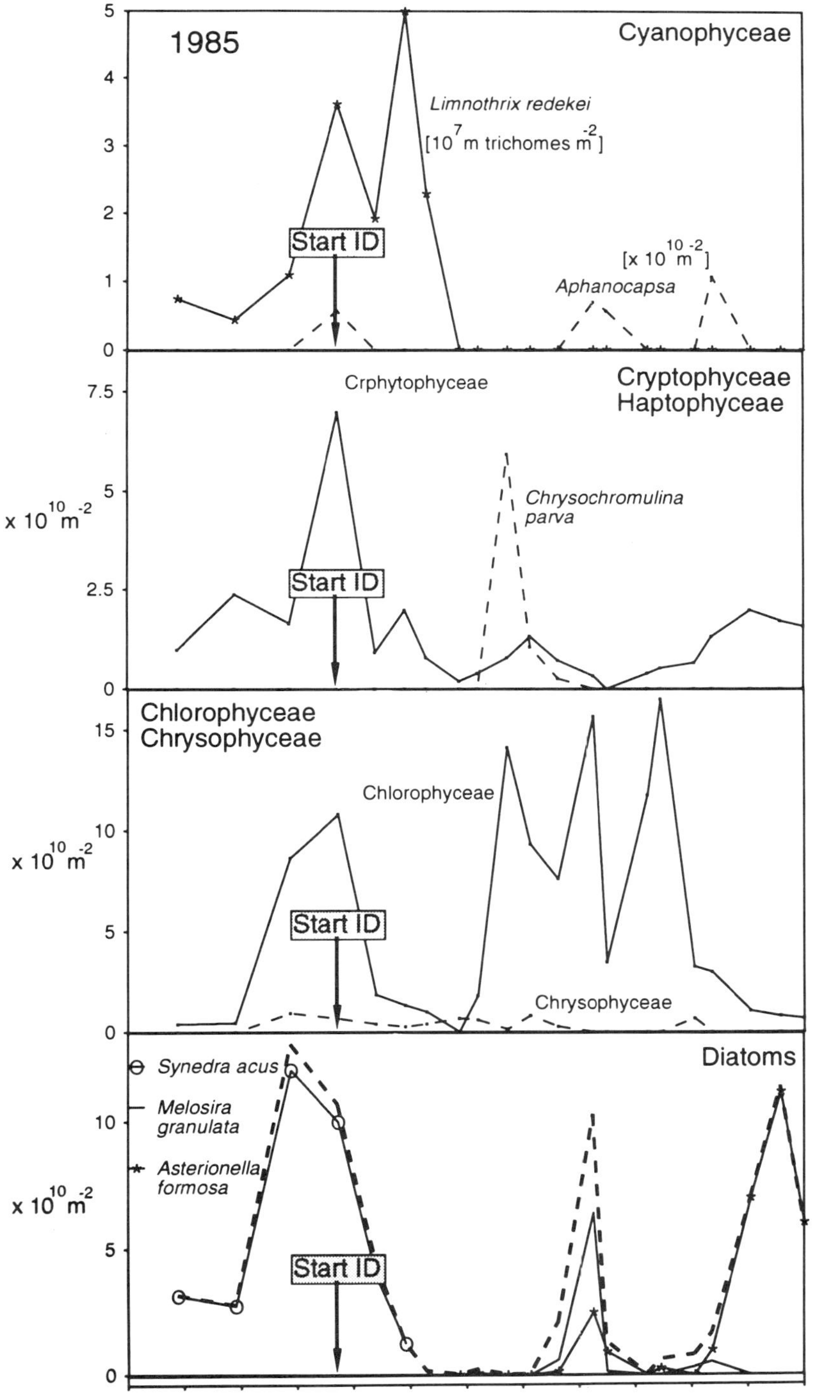

Figure 8. Phytoplankton succession in the first year of intermittent destratification (=ID) in Fischkaltersee, in 1985.

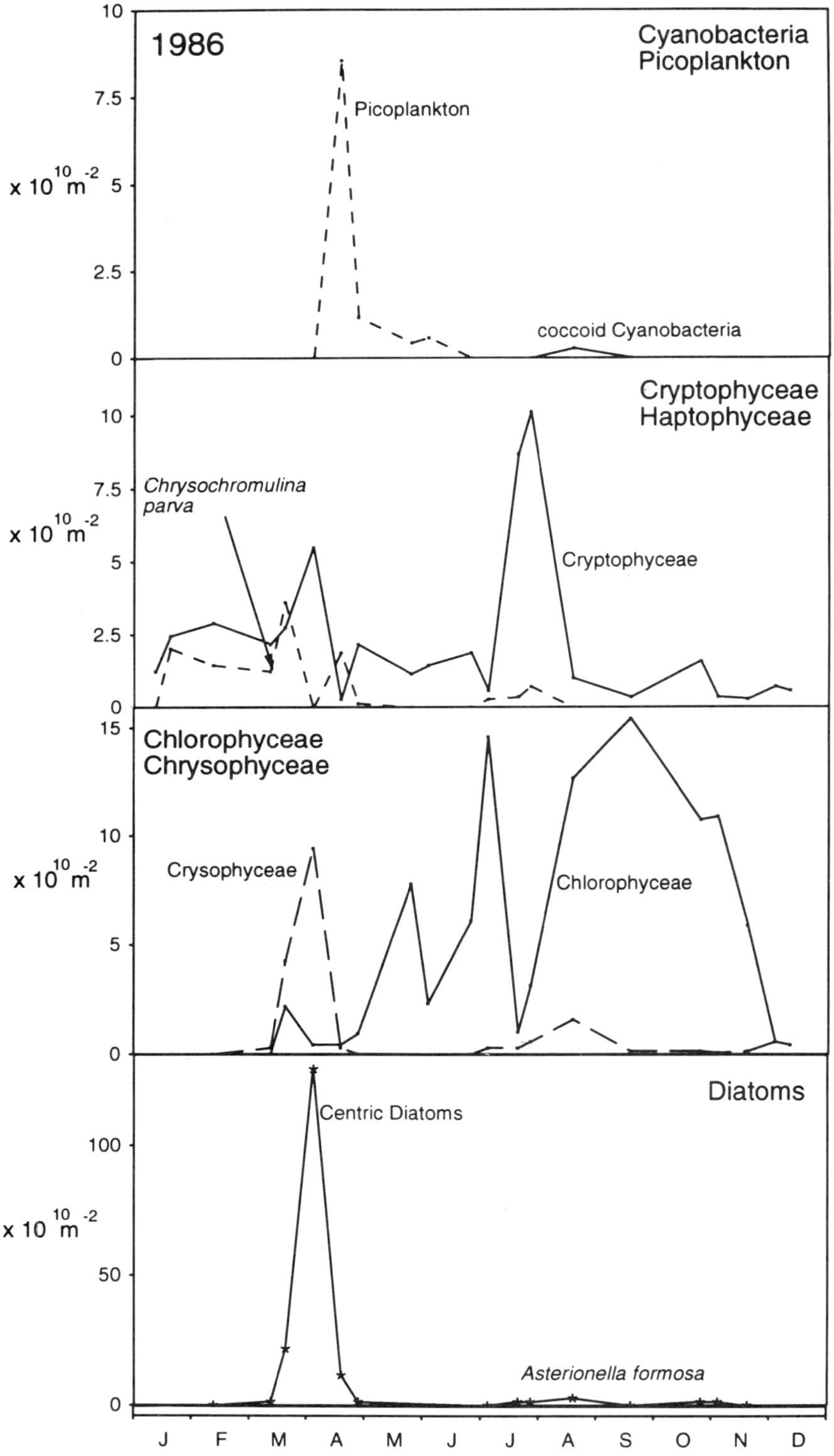

Figure 9. Phytoplankton succession in the second year of intermittent destratification (=ID) in Fischkaltersee, in 1986.

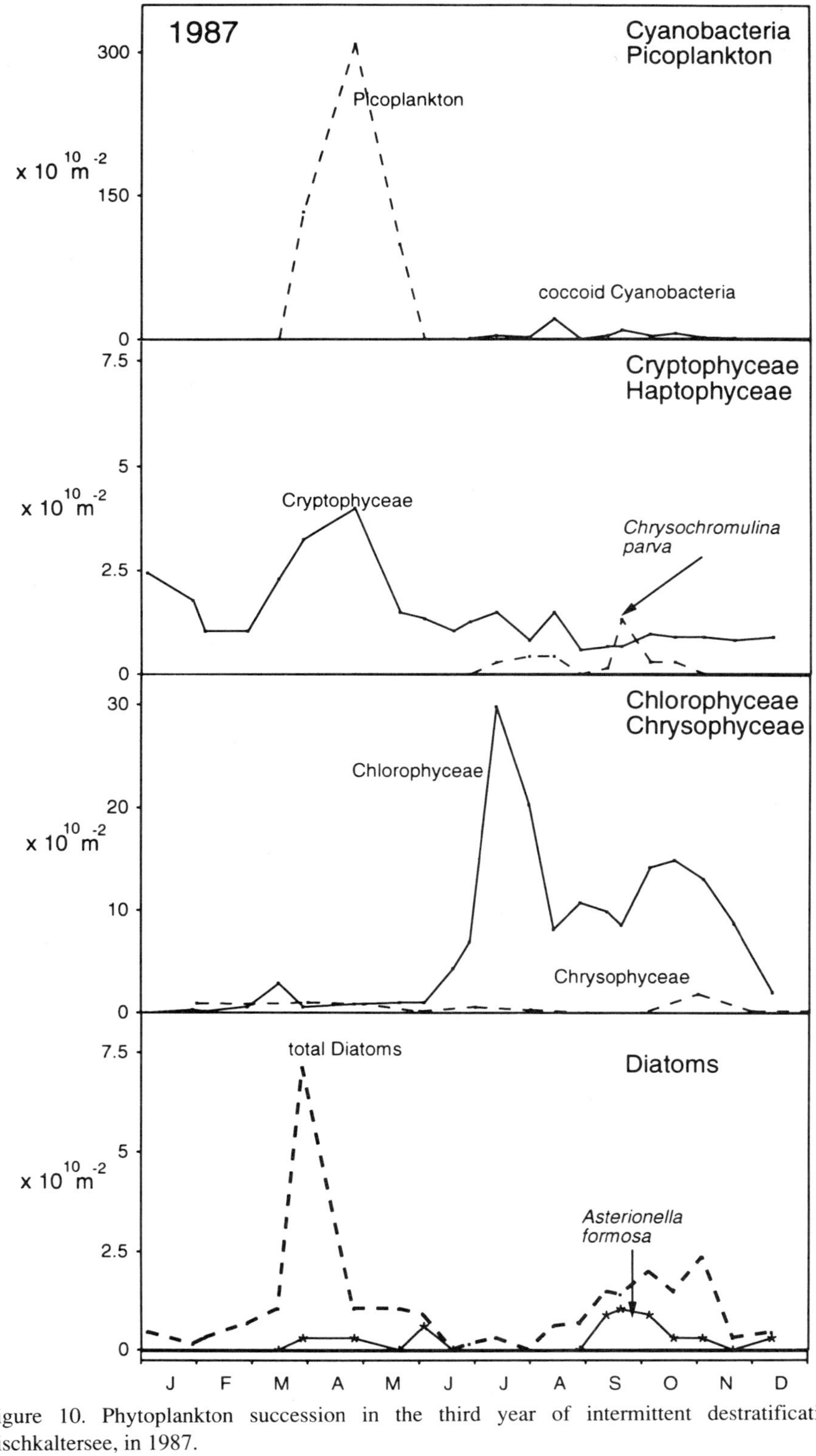

Figure 10. Phytoplankton succession in the third year of intermittent destratification (=ID) in Fischkaltersee, in 1987.

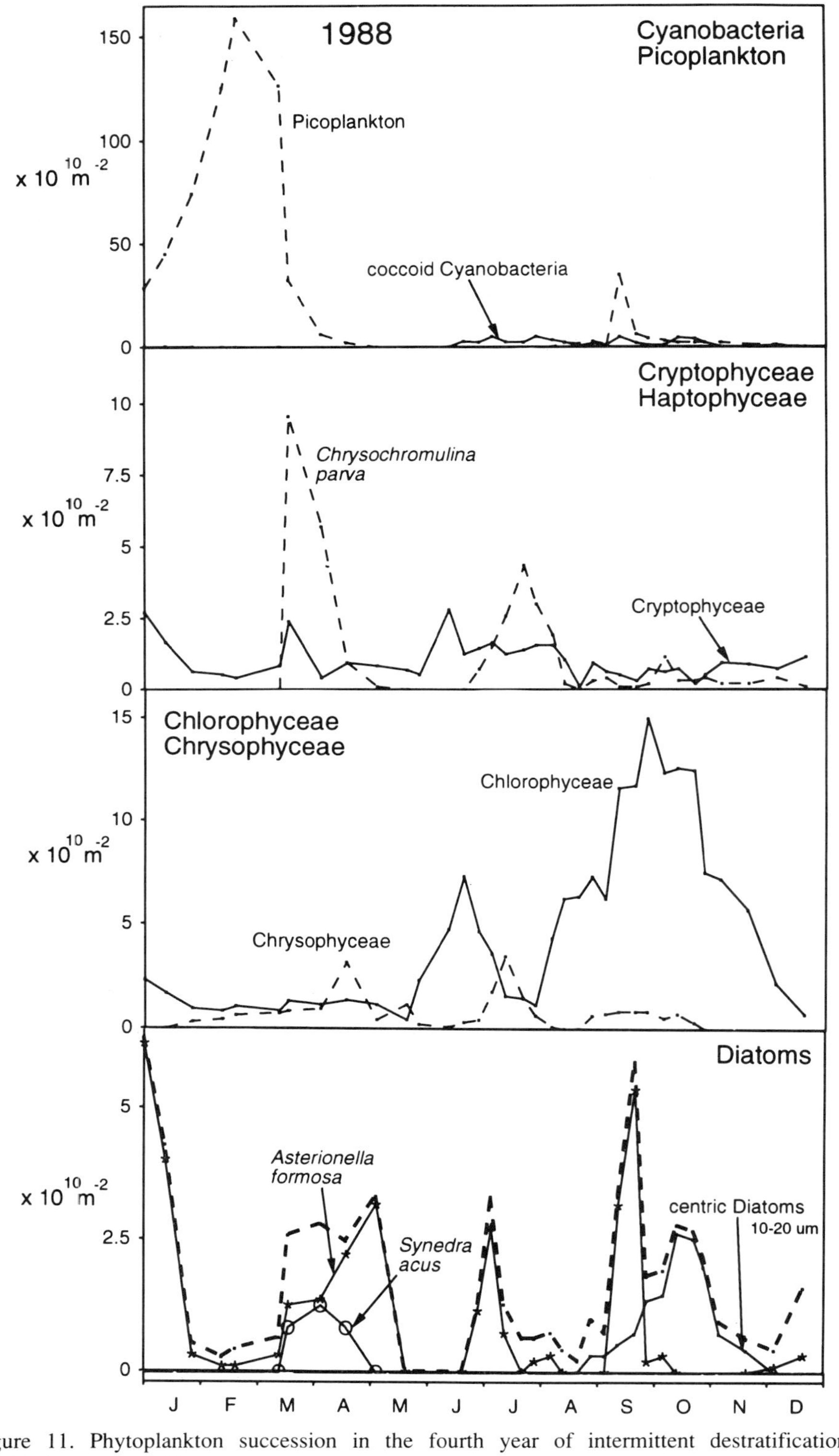

Figure 11. Phytoplankton succession in the fourth year of intermittent destratification (=ID) in Fischkaltersee, in 1988.

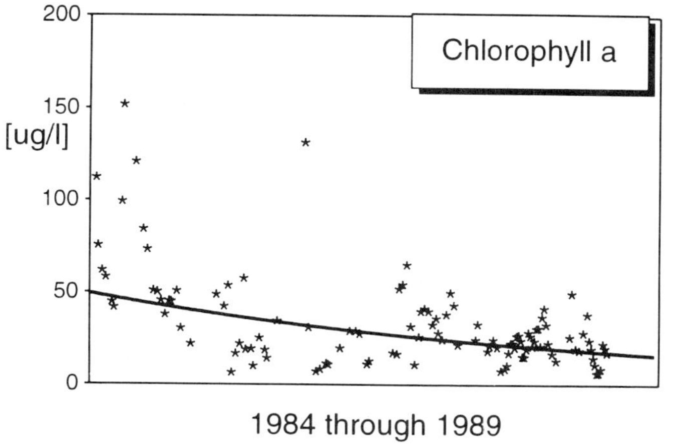

Figure 12. Mean chlorophyll-*a* concentrations in the water column of Fischkaltersee in 1984 through 1989.

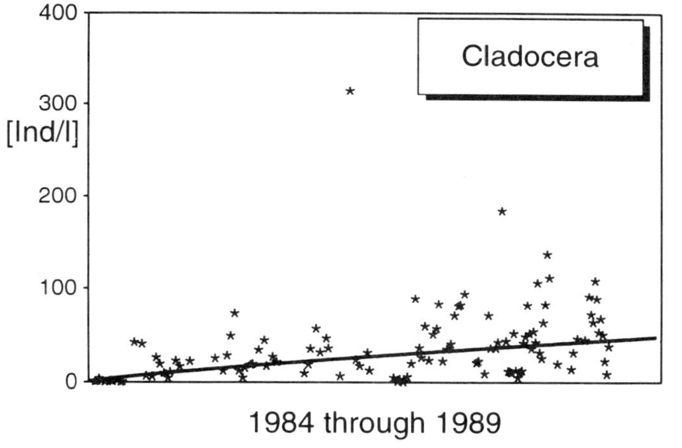

Figure 13. Abundance (individuals l^{-1}) of cladocera in Fischkaltersee in 1984 through 1989.

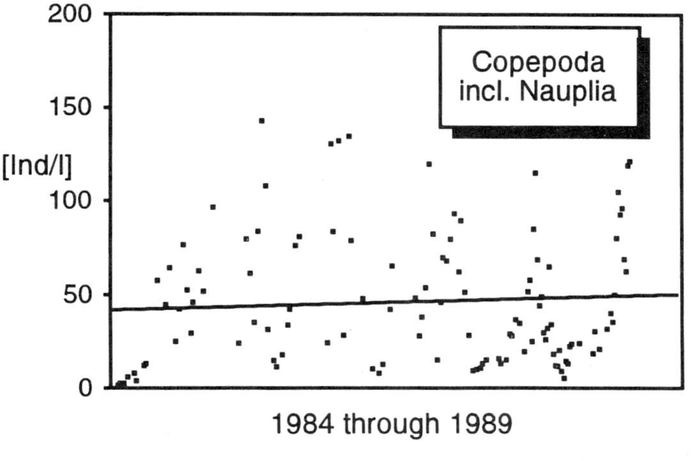

Figure 14. Abundance (individuals l^{-1}) of Copepoda (including nauplia) in Fischkaltersee in 1984 through 1989.

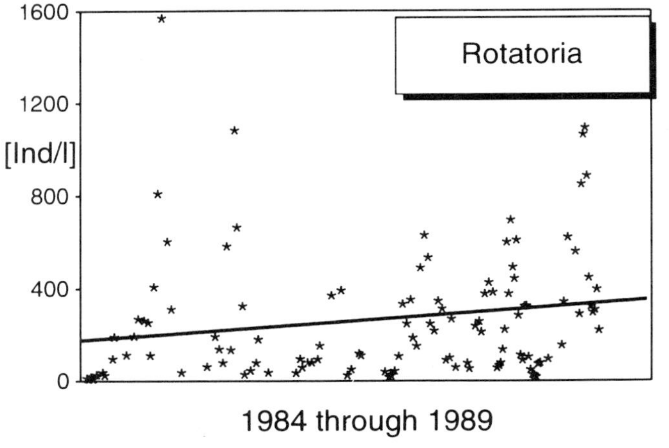

Figure 15. Abundance (individuals l^{-1}) of Rotatoria (Rotifera) in Fischkaltersee in 1984 through 1989.

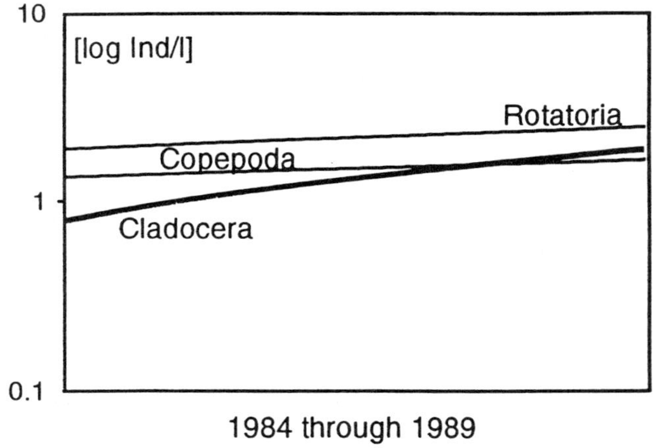

Figure 16. Comparison of regression lines (see Table 4) for abundances (log individuals l^{-1}) for components of the zooplankton in Fischkaltersee in 1984 through 1989.

Table 4. *Increases of zooplankton populations in Fischkaltersee during intermittent destratification, as described by regression equations.*

$\log_{10} Ind\ l^{-1} = a + bt$

t = units of 7.6 days; n = 115; r = correlation coefficient

Zooplankton group	a	b	r
Rotatoria	1.89	0.002	0.315
Copepoda incl. nauplia	1.35	0.001	0.069
Cladocera	0.77	0.004	0.542

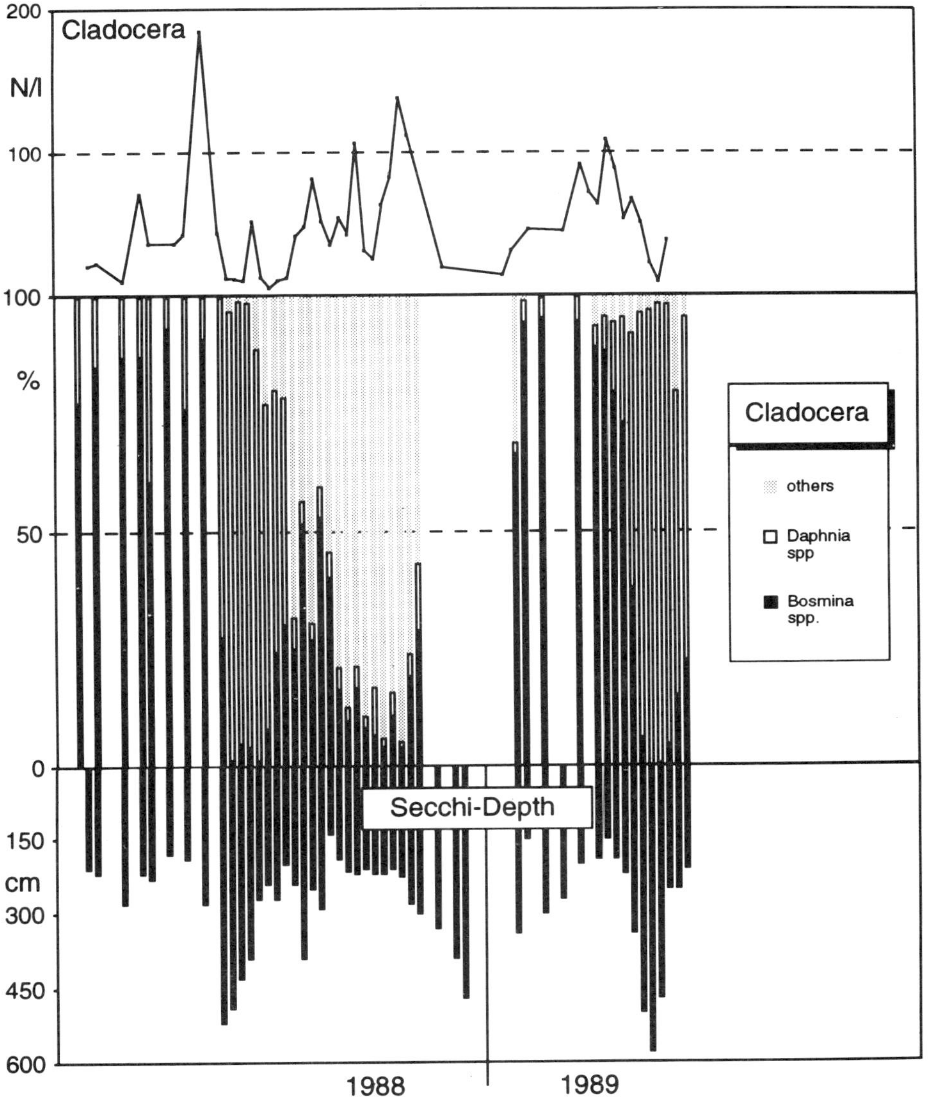

Figure 17. Cladoceran abundances (number l^{-1}), cladoceran composition (%) and Secchi disc readings during 1988 and early 1989.

general statement, however, remains valid throughout the total period of intermittent destratification. Filamentous planktonic cyanobacteria were traded-off completely (Figs 9–11). In several years, the occurrence of picoplankters has been monitored. Since neither the taxonomic problems nor the ecological significances of these forms are sufficiently resolved, a detailed assessment of their importance is not possible.

In contrast to permanent mixing, possible nuisance cyanobacteria did not re-appear after a certain period of adaptation. Again in contrast to permanent mixing, the biomass of planktonic algae clearly and continuously decreased as indicated by chlorophyll-*a* (Fig. 12). Mean chlorophyll-*a* concentrations fell from 50 µg l^{-1} (pre-treatment period) to less than 25 µg l^{-1} in

1989. Furthermore, the annual oscillation of chlorophyll-*a* concentrations clearly declined, and the planktonic system appeared to have a more stable status.

Probably due to better food conditions (replacement of filamentous cyanobacteria by chlorophytes and diatoms), the herbivorous zooplankton – cladocerans, copepods (mainly nauplia) and rotifers – increased in abundance (Figs 13–15). The turbulent conditions in the lake did not affect their development, since all of the studied groups reached densities which are common in slightly eutrophic lakes and reservoirs (e.g. Wetzel 1983). Increases for all zooplankton groups are statistically significant ($P < 0.1$); regressions are given in Table 4 and depicted in Figure 16.

The highest increases in abundances can be observed with Cladocera, the most efficient herbivores among the zooplankton (Fig. 16). Among grazing zooplankton, *Bosmina* species dominate by numbers (Fig. 17). However, *Daphnia* is about 1000 times more effective than *Bosmina* is at grazing algae (Geller 1991), so in Fischkaltersee the clear-water phases appear to be caused by both zooplankton genera. As a result of sedimentation during quiescent periods and grazing by zooplankton, Secchi-disc readings exceeding 4.5 m in depth are common in the spring. In other words, Secchi-disc readings of less than 2 m depth, which occurred frequently during the period of permanent mixing (Steinberg 1983, Steinberg & Zimmermann 1988), have become rare during the later period of intermittent destratification (Fig. 17). As an overall consequence of the improved light climate, it appears most likely that cyanobacteria are unable to re-adapt and gain dominance by their ability of buoyancy regulation under low light conditions.

The establishment of zooplankton, functionally dominated by highly efficient cladoceran grazers, may exert an additional stabilizing "top–down" influence on the total plankton community. *Vice versa*, however, in the absence of considerable populations of large-bodied cladocerans, early season blooms of diatoms and green algae reduce light availability, thus creating conditions favourable for increased buoyancy and bloom formation by cyanobacteria. Similar to Fischkaltersee, in the system studied by Spencer & King (1987), algal buoyancy and surface bloom formation were enhanced markedly by decreased light intensity, and only to a lesser extent by decreasing availability of carbon dioxide. Furthermore, the appearance of cyanobacterial blooms could be prevented by a reduced density of planktivorous fish, which allowed development of dense cladoceran populations (also see Benndorf, this volume).

Discussion

In general, turbulence appears to have a diverging influence on phytoplankton species. For instance, during induced intermittent artificial mixing, several diatoms, desmids, and the cyanobacteria *Oscillatoria* and *Limnothrix*, achieved rapid rates of growth during mixing which then declined or were overtaken by loss rates during quiescent, restratifying episodes. *Eudorina* and *Sphaerocystis* grew well during quiescence but they declined during mixing episodes. During mixing phases, *r*-strategists (e.g. *Rhodomonas* and *Ankyra*) reached maximum "true" growth rates, not exceeeding 0.75 $days^{-1}$. The growth rates of other "summer species" (*Anabaena*, *Ceratium*, *Volvox*, *Microcystis*) were depressed during mixing, though they generally maintained their existing biomass. By thus controlling the biomass of *r*-selected species and delaying the crop maxima of the *K*-selected "summer species", the total standing crop of phytoplankton was below the mean nutrient-carrying capacity (Reynolds *et al.* 1984).

Our findings are at least partly in good agreement with the general statements of Reynolds *et al.* (1984). Discrepancies appear to be mainly attributable to differences in the study seasons. Whereas Reynolds *et al.* (1984) studied summer species, we calculated changes of net growth rates of dominant phytoplankters during two occasions in spring 1988 and 1989. Results are given in Table 5. With the diatoms, only *Fragilaria crotonensis* achieved clearly elevated

Table 5. *Changes in net growth rate during destratification of Fischkaltersee*
(0 = no significant change; + = slight increase in growth during mixing; ++ = increase in growth during mixing; – = slight decrease in growth during mixing; – – = decrease in growth during mixing.

Species	1988 19 April to 16 May	1989 4 April to 1 May
Fragilaria crotonensis		++
Asterionella formosa	0	0
Synedra acus	(–)	
centric diatoms>20μm	0	0
centric diatoms>10-20μm		(+)
centric diatoms>10μm		0
Cryptomonas ovata	0	(–)
Rhodomonas minuta	+	0
Katablepharis ovalis		–
Dinobryon cylindricum		0
Dinobryon divergens	– –	–
Dinobryon sociale		0
Dinobryon sertularia		(+)
Uruglena americana		0
Gymnodinium helveticum	0	
Chrysochromulina parva		–
Closterium spp.	0	
Ankistrodesmus spp.	0	
Coelosphaerium spp.		–
Crucigenia tetrapedia		+
Scenedesmus spp.	0	0
Tetraedron minimum		0

growth rates, whereas there were only slight changes in the growth rates of medium-sized centric diatoms (a small increase) and *Synedra acus* (a small decrease). Also consistent with Reynolds *et al.* (1984) is the finding that *r*-strategists (here: *Rhodomonas*, *Crucigenia*) are favoured and *K*-strategists (here: *Dinobryon divergens*) are not favoured in growth under mixing conditions.

As demonstrated, sufficiently strong turbulence, which changed the conditions for competition between phytoplankton species, can be artificially induced by permanent as well as intermittent destratification. At least for certain periods of time, the development of filamentous cyanobacteria was disturbed or even prevented by both types of destratification. As depicted in Figure 8 and by Steinberg (1983), they were out-competed by chlorophytes and diatoms. The structural changes can be explained by changes in the chemical and/or physical environment so that (1) chlorophytes and diatoms gained advantage in competition over cyanobacteria, or (2) cyanobacteria were at a disadvantage.

However, in the third year of permanent destratification, the filamentous cyanobacteria (*Limnothrix redekei*) reappeared and reached maximum densities far above those of the untreated lake (Steinberg & Tille-Backhaus 1990). So it has to be discussed, whether or not the explanations above are sufficient to explain not only the change from filamentous cyanobacteria to chlorophytes and diatoms, but also the return of *Limnothrix* in the third and fourth year of permanent destratification.

Concerning the first hypothesis, the destratification of Fischkaltersee changed the chemical environment of the production zone. This was shown, for instance, by the carbon dioxide increase in the first year (1980) of permanent destratification (Steinberg 1983). This fact is considered advantageous for non-cyanobacteria (Shapiro 1990). The resulting structural changes are reproducible and can even be found under alkaline conditions (Benndorf 1988). There are two explanations for such shifts. Firstly, due to low carbon dioxide compensation points and low photorespiration, cyanobacteria can assimilate at lower concentrations of carbon dioxide than eukaryotic algae can and are even capable of directly utilizing bicarbonate (Paerl 1988). Secondly, even a slight decrease of pH stimulates cyanophages and leads to lysis of cyanobacteria (Shapiro 1984).

At the beginning of the permanent destratification in 1980, this carbon dioxide hypothesis appears to be a reasonable explanation for the phytoplankton community changes in favour of eukaryotic photosynthetically active organisms, since the availability of carbon dioxide increased from concentrations below detection limits to 40 to 100 μequiv. l^{-1}. However, the inverse deduction is not valid, because this mechanism does not sufficiently explain the re-occurrence of *Limnothrix redekei* during the third and fourth year of permanent destratification. The cyanobacterial re-occurrence took place when carbon dioxide levels clearly ranged above detection limits (50 to 100 μequiv. l^{-1}, Steinberg & Tille-Backhaus 1990) which, according to Shapiro's (1984, 1990) hypothesis, should have favoured eukaryotic photoautotrophs. The hypothesis which most convincingly explains the re-appearance of *Limnothrix* during permanent destratification is the ability of this cyanobacterium to maintain growth even under conditions of resource poverty (rather low P concentration) and of severe light limitation through mixing (Reynolds 1989) as well as through shading by other photoautotrophs. (Reynolds (1989) reports on a generation time of 5 days for *Oscillatoria* at 30 μg P l^{-1} even at daily perceived light integrals of only 0.01 mol photon. This ability is obviously lacking in other cyanobacteria and in oligotrophic eukaryotes, such as several chrysophyte species.)

Furthermore, the above-mentioned carbon dioxide hypothesis does not explain the shifts caused by intermittent destratification since 1985. Since then, non-cyanobacteria dominated for periods of several months when the free carbon dioxide concentrations in the water were below detection levels (Steinberg & Zimmermann 1988 and unpublished data). If, during the intermittent destratification, the carbon dioxide effect worked at all it was overlaid by other stronger mechanisms.

Concerning the second hypothesis, from other field studies which are summarized by Fogg *et al.* (1973) and Reynolds & Walsby (1975), it is known that cyanobacteria need zones of special physical conditions (e.g. light climate) and chemical conditions (e.g. elevated concentrations of organic and inorganic nutrients) for optimal growth. The reason appears to be their limited metabolic flexibility (Carr, referred to by Fogg *et al.* 1973). Apart from the carbon dioxide hypothesis, this limited metabolic flexibility also seems to sufficiently explain the described decrease of cyanobacteria at the beginning of the permanent destratification (Steinberg 1983).

Oscillatoriales appear to be able to adapt even to turbulent systems, but they exhibit very long lag phases. Steinberg & Tille-Backhaus (1990) documented the re-invasion of *Limnothrix redekei* during permanent destratification after two years of absence.

Permanent destratification

Compared to untreated lakes with thermal stratification, *permanent destratification* produced unnatural light–dark rhythms for the photoautotrophs, which the filamentous cyanobacteria could not adapt to as fast as eukaryotic algae could. The ability of cyanobacteria to regulate buoyancy, and thus more or less actively migrate into convenient light climates, no longer counted in the competition with eukaryotic algae. The continued mixing, however, produced relatively constant mixing patterns in the lake. Reynolds *et al.* (1983) discovered that good conditions for growth of *Oscillatoria (Planktothrix) agardhii* exist when the ratio of mixing depth (Z_{mix}) to Secchi depth (Z_{sec}) is greater than 2. In Fischkaltersee, Z_{mix} was about 11 m and Z_{sec} varied between 60 and 300 cm in 1983. If the results with *O. agardhii* can be transferred to other Oscillatoriales, for instance *Limnothrix* spp., reasonably good growth-supporting conditions should have existed for these cyanobacteria in Fischkaltersee in 1983, when *Limnothrix* started to re-occur. During the questionable period, the ratio of Z_{mix}:Z_{sec} varied from 3.5 to a maximum of 17.5. This may, at least in part, explain the re-appearance and subsequent dominance of filamentous cyanobacteria. Nevertheless, *Limnothrix redekei* needed more than a 2-year period to adapt to the new conditions, disturbed by permanent mixing.

Intermittent destratification

In contrast to permanent mixing, during *intermittent destratification* the uniform movement of the water masses suddenly stops when air pumping is switched off. It was shown that during quiescence, slight thermal and chemical stratifications began to proceed (Steinberg & Zimmermann 1988). As a consequence, sedimentation could take place with subsequent improved light climates. With Secchi depths of more than 6 m, the ratio Z_{mix}:Z_{sec} fell well below 2, producing unfavourable conditions for cyanobacterial competition.

What are the reasons for the success of intermittent mixing as an in-lake measure against cyanobacterial (mass) development? There are at least three reasonable explanations why cyanobacteria cannot adapt fast enough to the intermittently disturbed conditions.

(1). Species with vacuoles (like *Limnothrix redekei*) are not able to adjust the size of their gas vacuoles fast enough (Oskam 1978; Reynolds 1986; Reynolds & Reynolds 1985). The rate of photosynthesis of cyanobacteria – as well as of all photosynthetic organisms – diminishes when they are transported into low light zones. So the turgor in the cells drops. The vacuoles, however, do not collapse and are not even reduced. When the water movement stops, the cells ascend and probably die at the water surface because of the high light intensities to which they are not adapted (Reynolds 1984). As a proof of this statement, during the onset of intermittent mixing, many *Limnothrix redekei* trichomes with incomplete or even with defect vacuoles have been observed.

(2). The chemical microscale conditions change rapidly and the environment of the cells is best renewed during circulation. The microscale conditions turn drastically worse during quiescence, when the zones around the cells become nutrient-depleted. Obviously, cyanobacteria cannot adapt fast enough to these conditions of alternating disturbance and quiescence. As mentioned above, this is more probably due to a lack of rapid enzymatic adaptation. In stratified lakes, cyanobacteria "search" for feasible zones by varying their specific weights. In Fischkaltersee, under conditions of intermittent mixing, this capability is no longer a competitive advantage, but is an obstacle for survival.

(3). Ward & Wetzel (1980) found that adaptation to varying light climates by selected bloom-forming cyanobacteria, was not completed even within 48 hours. In Fischkaltersee, with a mixing rhythm of one turnover in less than 2 days, cyanobacteria should not have been able to adapt to strongly changing light levels, even twice within *c.* 48 hours.

All three mechanisms result in the fact that the long time-period required by cyanobacteria for their adaptation to environmental conditions, appears to be the most important limiting factor in competing with eukaryotic algae. As a consequence, when optimizing the rhythm of intermittent mixing as a fast-working and effective tool of in-lake management, the following targets should be addressed.

(i). Periods of disturbance (intermittent mixing) should be kept at a minimum, and sedimentation of decaying phytoplankton biomass (bio-seston) should reach a maximum.

(ii). Periods of quiescence must not result in internal fertilization by nutrient release from the sediments. Microbial nutrient-release from settling bio-seston or from the sediments may be significantly enhanced by elevated temperatures in the sediment–water interface!

(iii). Artificial mixing should be applied in short but strong disturbing pulses, which all photoautotrophs have to adapt to with specific lag-periods. The photoautotrophs should still be in the period of (re)-adaption when, after a period of quiescence, the next disturbance occurs. An optimal mixing frequence should maintain phytoplankton growth within the lag phase or, if not achievable, far below optimal growth.

The optimal rhythm of intermittent mixing is a result of these three mechanisms and has to be determined specifically for each lake or reservoir under consideration.

Regarding the long-term operation of this in-lake management measure against cyanobacterial dominance, intermittent mixing must be carried out for as long as external inputs of nutrients affect the lake or reservoir. In this case, intermittent destratification is a treatment of symptoms, to prevent cyanobacterial blooms in the water column. However, as has been demonstrated in the Biesbosch reservoirs (see next section), mixing measures cannot avoid benthic cyanobacterial developments with their adverse effects on raw water quality (off-flavour by geosmin and 2-methylisoborneol). In the Biesbosch reservoirs, mixing improved the light climate and this promoted or even caused the development of benthic cyanobacteria, due to the artificially constructed littoral zones which only allow biofilms to grow. However, it is also most likely that in natural lakes, improved light climates would favour the invasion and growth of macrophytes, if their resting stages are present, and if the littoral zones are not physically hostile as they are in most reservoirs.

In the case of Fischkaltersee, two different allochthonous inputs have to be stopped: phosphorus from desorbing resources in an aquifer and de-icing salts from a street junction. These inputs have caused meromixis and indirectly favoured cyanobacterial dominance. Intermittent mixing must also be continued until nutrient concentrations and availability have decreased to levels which do not support cyanobacterial growh. Concerning phosphorus, this threshold is 10 to 15 $\mu g\ l^{-1}$ (Steinberg & Hartmann 1988). Moreover, in the specific case of Fischkaltersee, intermittent destratification has to be maintained for as long as the winter-time salt loadings, which cause meromictic conditions in the lake, are not significantly reduced.

Advantages of intermittent destratification compared with permanent destratification

The time-scale of the therapy success in Fischkaltersee may be compared with that of the Biesbosch project (see the article by Oskam & van Breemen, this volume). This project, in the Netherlands, includes permanent destratification in artificial reservoirs. Nutrient-rich water of the River Meuse is gathered in considerably deep reservoirs to yield raw water for drinking-water purposes. The in-lake management measures of the Biesbosch reservoirs are also carried out to reduce the biomass of planktonic algae and to extinguish cyanobacteria. Since permanent destratification is regarded as one of the fastest-working therapy measures, and because it is one of the few measures which show positive results – not only in terms of geochemistry but also in limnobiology – we take this management measure as a standard of reference.

Permanent destratification needs a minimum depth. The artificial Biesbosch lakes are significantly deeper than Fischkaltersee. The greater depth results in partial light limitation for photoautotrophs and a subsequent decrease in phytoplankton biomass. If permanent destratification, however, is applied to relatively shallow lakes, clear increases in phyto-plankton biomass must be anticipated, as demonstrated for Fischkaltersee phytoplankton in the first years of permanent destratification (Steinberg 1983). On the other hand, *intermittent destratification* appears to be applicable to shallow lakes. However, it should not be applied to shallow water-bodies in which, during periods of quiescence, sedimentation and (more or less permanent) sediment accumulation do not occur or in which, during mixing periods, re-suspension overcompensates sedimentation during quiescent periods.

Both permanent and intermittent destratification can reduce the biomass of photoautotrophs. Whereas with permanent destratification the prevailing mechanisms are light limitation and grazing of herbivorous zooplankton (Oskam 1978), with intermittent destratification the sedimentation of (decaying) phytoplankton biomass is an additional advantage. Compared to permanent mixing, which is a relatively inexpensive method, intermittent destratification is even less costly (Table 6).

Table 6. *The specific costs (DM per m³ of lakewater) of each physical measure tested in Fischkaltersee*

Measure	Installation costs	Operation costs
Phosphorus precipitation		
Chemicals	–	0.10
Employees	–	0.10
Permanent destratification	<0.02 (1980)	0.006 (1980 – 1984)
Intermittent destratification	<0.02 (1980)	0.003 (1985 – 1989)

Parts of this study have been funded by the German Environmental Agency, Berlin, by contract No. 102 04 411, and parts of this paper are based on the final report of this contract. The authors are grateful to the Bavarian Water Board, Munich, for having provided excellent working facilities and to Dr L. Ballhorn for improving the wording.

References

Benndorf, J. (1988). Destratifikation zur Blaualgenbekämpfung in der Talsperre Bautzen. Zuarbeit zur Forschungsaufgabe NVe 32/88 der Wasserwirtschaftsdirektion Obere Elbe – Neisse, Dresden (*Manuscript*).

Berg, K., Skulberg, O. M. & Skulberg, R. (1987). Effects of decaying toxic blue-green algae on water quality – a laboratory study. *Archiv für Hydrobiologie*, **108**, 549-563.

Berg, K., Skulberg, O. M., Skulberg, R., Underdal, B. & Willen, T. (1986). Observations of toxic blue-green algae (Cyanobacteria) in some Scandinavian lakes. *Acta Veterinaria Scandinavica*, **27**, 440-452.

Berglind, L., Holtan, H. & Skulberg, O. M. (1983). Case studies on off-flavours in some Norwegian lakes. *Water Science and Technology*, **15**, 199-209.

Carmichael, W. W., Jones, C. L. A., Mahmood, N. A. & Theiss, W. C. (1985). Algal toxins and water-based diseases. *CRC Critical Reviews on Environmental Control*, **15**, 275-313.

Cooke, G. D., Heath, R. T., Kennedy, R. H. & McComas, M. R. (1982). Changes in lake trophic state and internal phosphorus release after aluminium sulfate application. *Water Resources Bulletin*, **18**, 699-705.

Danecker, E. (1971). Die künstliche Durchmischung des Krottensees. *Carinthia II*, **31**, 41-62.

ENVIRONMENTAL PROTECTION AGENCY (EPA) (1980). Restoration of Medical Lake. *Capsule Report*, **EPA-625/2-80-025**.

Fogg, G. E., Stewart, W. D. P., Fay, P. & Walsby, A. E. (Editors) (1973). *The Blue-Green Algae*. Academic Press, London.

Foy, R. H. (1985). Phosphorus inactivation in a eutrophic lake by the direct addition of ferric aluminium sulphate: impact on iron and phosphorus. *Freshwater Biology*, **15**, 613-629.

Foy, R. H. & Fitzsimmons, A. G. (1987). Phosphorus inactivation in a eutrophic lake by direct addition of ferric aluminium sulphate: changes in phytoplankton populations. *Freshwater Biology*, **17**, 1-13.

Geller, W. (1991). Die Planktongemeinschaft als dynamisches System. In *Ökologie der Oberbayerischen Seen*, pp. 31-47. Rundgespräche der Kommission Ökologie der Bayerischen Akademie der Wissenschaften, Pfeil Verlag, Munich.

Kennedy, R. H. & Cooke, G. D. (1982). Control of lake phosphorus with aluminium sulfate: dose determination and application techniques. *Water Resources Bulletin*, **18**, 389-395.

Melzer, A. (1976). Makrophytische Wasserpflanzen als Indikatoren des Gewässerzustandes oberbayerischer Seen. *Dissertationes Botanicae*, **34**, 1-195.

Oskam, G. (1978). Die Vorausberechnung der Algenbiomasse in den Biesbosch Speicherbecken. Theorie und Praxis. In: *Verminderung der Algenentwicklung in Speicherbecken und Talsperren* (eds G. Oskam & H. Bernhardt). *DVGW-Schriftenreihe Wasser*, **16**, 90-107.

Paerl, H. W. (1988). Nuisance phytoplankton blooms in coastal, estuarine, and inland waters. *Limnology and Oceanography*, **33**, 823-847.

Reynolds, C. S. (1984). Artificial induction of surface blooms of cyanobacteria. *Verhandlungen der Internationalen Vereinigung für Theoretische und Angewandte Limnologie*, **22**, 638-643.

Reynolds, C. S. (1986). Experimental manipulations of the phytoplankton periodicity in large limnetic enclosures in Blelham Tarn, English Lake District. *Hydrobiologia*, **138**, 43-64.

Reynolds, C. S. (1989). Relationships among the biological properties, distribution, and regulation of production by planktonic cyanobacteria. *Toxicity Assessment*, **4**, 229-255.

Reynolds, C. S. & Walsby, A. E. (1975). Water-blooms. *Biological Reviews*, **50**, 437-481.

Reynolds, C. S. & Reynolds, J. B. (1985). The atypical seasonality of phytoplankton in Crose Mere, 1972: an independent test of the hypothesis that variability in the physical environment regulates community dynamics and structure. *British Phycological Journal*, **20**, 227-242.

Reynolds, C. S., Wiseman, S. W. & Clarke, M. J. O. (1984). Growth- and loss-rate responses of phytoplankton to intermittent artificial mixing and their potential application to control of planktonic algal biomass. *Journal of Applied Ecology*, **21**, 11-39.

Reynolds, C. S., Wiseman, S. W., Godfrey, B. M. & Butterwick, C. (1983). Some effects of artificial mixing on the dynamics of phytoplankton in large limnetic enclosures. *Journal of Plankton Research*, **5**, 203-234.

Shapiro, J. (1984). Blue-green dominance in lakes. The role and management significance of pH and CO_2. *Internationale Revue der Gesamten Hydrobiologie*, **69**, 765-780.

Shapiro, J. (1990). Current beliefs regarding dominance by blue-greens: The case for the importance of CO_2 and pH. *Verhandlungen der Internationalen Vereinigung für Theoretische und Angewandte Limnologie*, **24**, 38-54.

Soltero, R. A., Nichols, D. G., Gasperion, A. F. & Beckwith, M. A. (1981). Lake restoration: Medical Lake, Washington. *Journal of Freshwater Ecology*, **1**, 155-165.

Spencer, C. N. & King, D. L. (1987). Regulation of blue-green algal buoyancy and bloom formation by light, inorganic nitrogen, CO_2, and trophic level interactions. *Hydrobiologia*, **144**, 183-192.

Steinberg, C. E. W. (1983). Effects of artificial destratification on the phytoplankton population in a small lake. *Journal of Plankton Research*, **5**, 855-864.

Steinberg, C. E. W. (1990). Alterations of organic substances during eutrophication and effects of the modified organic substances on trophic interactions. In *Organic Acids in Aquatic Ecosystems* (eds E. M. Perdue & E. T. Gjessing), pp. 189-208. J. Wiley & Sons, Chichester.

Steinberg, C. E. W. & Bucksteeg, K. (1980). Testing of aluminium precipitation as a lake restoration method. *Vom Wasser*, **55**, 47-61 (in German).

Steinberg, C. E. W. & Hartmann, H. M. (1988). Planktonic bloom-forming cyanobacteria and the eutrophication of lakes and rivers. *Freshwater Biology*, **20**, 279-287.

Steinberg, C. E. W. & Tille-Backhaus, R. (1990). Re-occurrence of filamentous planktonic cyanobacteria during permanent artificial destratification. *Journal of Plankton Research*, **12**, 661-664.

Steinberg, C. E. W. & Zimmermann, G. (1988). Intermittent destratification: A therapy measure against cyanobacteria in lakes. *Environmental Technology Letter*, **9**, 337-350.

Tilman, D. (1982). *Resource Competition and Community Structure*. Princeton University Press.

Utermöhl, H. (1958). Zur Vervollkommnung der quantitativen Phytoplanktonmethodik. *Mitteilungen der Internationalen Vereinigung für Theoretische und Angewandte Limnologie*, **9**, 1-38.

Vollenweider, R. A. (1968). Scientific Fundamentals of the Eutrophication of Lakes and Flowing Waters, with Particular Reference to Nitrogen and Phosphorus as Factors in Eutrophication. Paris, Report of the Organisation for Economic Cooperation and Development, **DAS/CSI/68.27**, 192 pp.; Annex, 21 pp.; Bibliography, 61 pp.

Ward, A. K. & Wetzel, R. G. (1980). Photosynthetic responses of blue-green algal populations to variable light intensities. *Archiv für Hydrobiologie*, **90**, 129-138.

Wetzel, R. G. (1983). *Limnology*. 2nd edition. Saunders College Publishing, Philadelphia. 753 pp. + appendices.

Active reservoir management: a model solution

J. HILTON, A. E. IRISH AND C. S. REYNOLDS

Institute of Freshwater Ecology, Windermere Laboratory, Far Sawrey, Ambleside, Cumbria LA22 0LP, England

Predictions from a functional model of algal growth capacity (maximum achievable algal biomass) and algal species succession for Chew Valley Lake (UK) were compared with observed data for 1989. Initial simulations grossly underestimated algal growth from the late spring onwards. However, inclusion of a pulsed input of phosphorus from the sediment, at a level consistent with field observations, resulted in very good predictions. The model was then applied to a number of hypothetical operational scenarios as examples of the use of such a model as a tool for reservoir management. An example is also given of an extension of the model to predict chemical properties of abstracted and treated water.

Introduction

Several presentations in this conference have outlined the difficulties of using both steady-state and statistically-fitted models when predicting the effects of remedial actions on algal growth. Steady-state procedures, of their very nature, cannot deal with dynamic situations. Statistical models require extensive calibration, and predictions often have to be made for environmental conditions which are often outside the original calibration conditions. In addition, the calibration requirement makes them difficult to transfer to other lakes. To date, no computer programs have been developed which will successfully predict changes in species of algae.

The obvious solution to these limitations is to apply our limnological knowledge to the problem and develop functional models, so reducing the requirement for such rigorous calibration. Reynolds has proposed a model (see later), based on fundamental principles of algal response to environmental events, which has successfully recreated the maximum observed biomass, the timing of events and a fair simulation of the species succession in several lakes. A forerunner of this model was developed jointly with Welsh Water under contract to Messrs. Wallace Evans and Partners, for use in the Cardiff Bay Barrage study. Reynolds (1989a) was able to authenticate this model against a data-set obtained in respect of Slapton Ley, Devon, which like the proposed barrage lake is shallow and extremely discharge sensitive. In this paper we will test a much developed form of this original model against a more complex data-set and, using a simple example, show how it can be applied as an aid in the choice of management strategy for the reduction of problems caused by eutrophication. Some further developments of the model are indicated.

Model description

The mathematical bases of this model have been presented elsewhere (Reynolds 1984, 1989a). A schematic diagram of the operation of the model, in its simplest form, is given in Figure 1. Daily hydraulic flow rates (F_{in}), metabolisable phosphorus concentrations (P_{in}, usually approximated as soluble reactive P) and nitrogen concentrations (N_{in}, usually approximated by the nitrate concentration) are fed into the computer model for each stream, pumped input, sewage discharge or industrial discharge. The model then calculates the phosphorus and

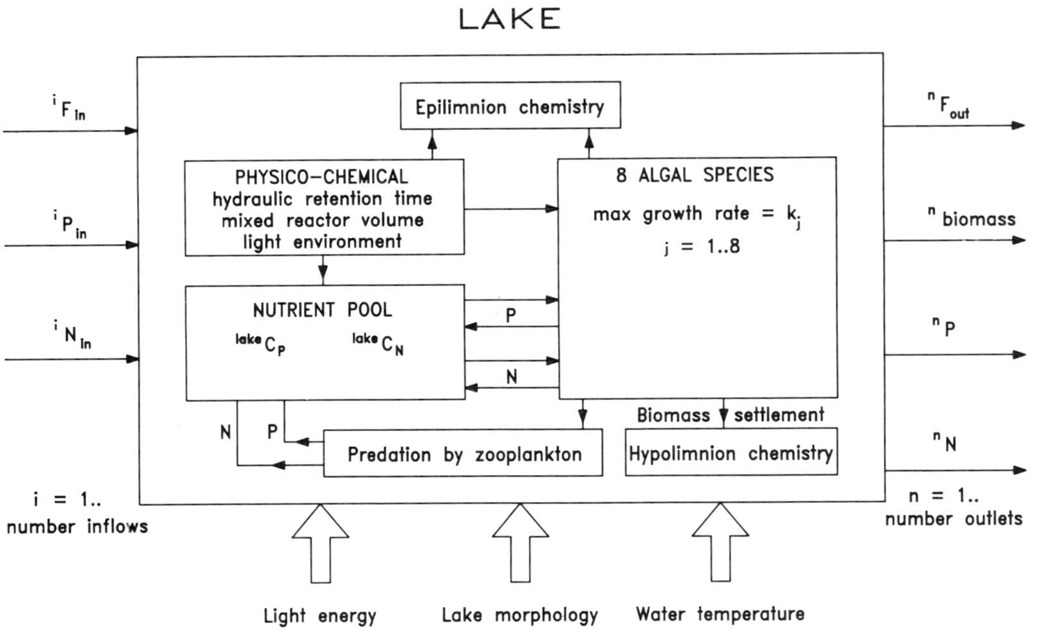

Figure 1. A schematic diagram of the algal growth capacity model. Explanation in the text.

nitrogen concentrations in the lake (C_p and C_n), assuming no utilisation by algae. Using input data identifying the mixed layer volume, depth and mean depth, the light climate in the water column is estimated from the incoming insolation (input). Assuming no limitation by nutrients, inocula of up to eight algal species, characteristic of different types found in the water-body, are then allowed to grow at their own individual maximum rates, adjusted according to the temperature and light conditions in the water. There is linkage between nutrient concentrations and the growth rate only when the availability falls below the growth rate saturating concentration. The model checks to see if there are sufficient nutrients to sustain the calculated amount of growth during the day, assuming that a Redfield ratio of nutrients (Redfield 1934) holds within the cell. Provision is made for more efficient use of nutrients when external concentrations fall. If no limitation is encountered the biomass estimate is retained and the appropriate amount of phosphorus and nitrogen is removed from the nutrient pool. If there is insufficient phosphorus or nitrogen to sustain this amount of growth then the biomass is increased by the maximum amount allowed by the limiting nutrient. The rate-limiting nutrient concentration is reduced to zero and the other nutrient pool is reduced accordingly.

Certain types of algae can be eaten by zooplankton. The model allows the existing population of zooplankton to feed and increase in mass, according to the temperature and the extent to which food concentration meets the minimum survival requirement or saturates its maximum. New terms are calculated each day. As a result of consumption of algae by zooplankton, phosphorus and nitrogen are assumed to be released instantaneously and the appropriate amount is added back into the available nutrient pools. Algal cells, phosphorus and nitrogen in solution are washed out of the system at a rate proportional to the rate of water abstraction and/or compensation flow. The computer program then cycles to calculate values for the next day.

As a result of this approach, in particular the use of a maximum growth rate at a given temperature and light environment, the resulting biomass estimates indicate the maximum level (or capacity) of algal biomass a lake or reservoir can recruit. As has been stated earlier (Reynolds, this volume) this maximum level may not always be observed in reality. Unaccounted natural occurrences, such as infection by chytrids or limitation by an unmodelled nutrient, such as silica, may not allow the populations to achieve their theoretical levels.

Comparison of model simulations with real events

Because the model is based on fundamental ecophysiological processes and the growth rates etc. are fixed, the model needs no calibration to adjust the model parameters. This makes it easy to transport from lake to lake. However, the model can only make reliable predictions in those situations where all the relevant processes are included within the algorithm. The predictive ability of the model is tested by its ability to simulate existing conditions and the resulting responses of populations. In this paper we present real data from Chew Valley Lake and a simulation as an example of the model's capabilities. There are three reasons for this choice.

(a). The first model fit was not completely realistic. As a consequence, the example shows some of the pit-falls of uncontrolled application of the model, and the way in which the model itself can assist the search for improved authenticity in the simulations.

(b). A full chemical data set was available for the modelled year (Hilton *et al.* in prep.).

(c). The model was successfully applied by Bristol Water, as an aid to its decision-making on some proposed operational changes to the lake, operated as a reservoir.

However, it must be said that the examples of possible management options given at the end of this paper are purely exemplary and in no way constitute any part of a commission brief.

Chew Valley Lake

The lake (reservoir) was built in 1953 by the Bristol Waterworks company. It has a surface area of 4.86 km^2 at top water level, a mean depth of 4.27 m and a maximum depth of 11.5 m. Wilson *et al.* (1975a) showed that, under natural conditions, it would stratify intermittently, during appropriate episodes in the summer. At these times, anoxia rapidly develops in the hypolimnion, significantly reducing the quality of abstracted water. As a result of Wilson's work, destratification was introduced to maintain chemical water quality to the treatment works. This equipment is still in continuous operation during the summer. The biology of the lake has been described by Hammerton (1959) and Wilson *et al.* (1975b).

Data on the lake's morphology, volumes of water in the inflowing streams, the volumes abstracted for treatment, the quantity of compensation flow, and nutrient concentrations in the inflows, were supplied to the model for 1989. The model was run with the assumption of a completely mixed lake, producing the predicted algal capacity shown in Figure 2. Comparison with algal biomass measurements (Fig. 2, initial prediction – expressed as chlorophyll-*a*) made in the lake during 1989 showed that the initial spring peak of chl*a* (60 $\mu g\ l^{-1}$) was well reproduced in both timing and maximum concentration. In the year in question the diatoms became limited by a lack of silica (Hilton *et al.* in prep.) and growth by green algae was handicapped by low water temperatures. The real algal biomass did not achieve the predicted maximum capacity. For the rest of the year, the first model run suggested that biomass levels would be low, not exceeding about 15 μg chl*a* l^{-1}. In reality, a significant peak of algae occurred in the autumn.

Although it is quite possible for the model to predict a maximum capacity which is not achieved in reality, the converse, i.e. observed algal levels higher than predicted levels, should

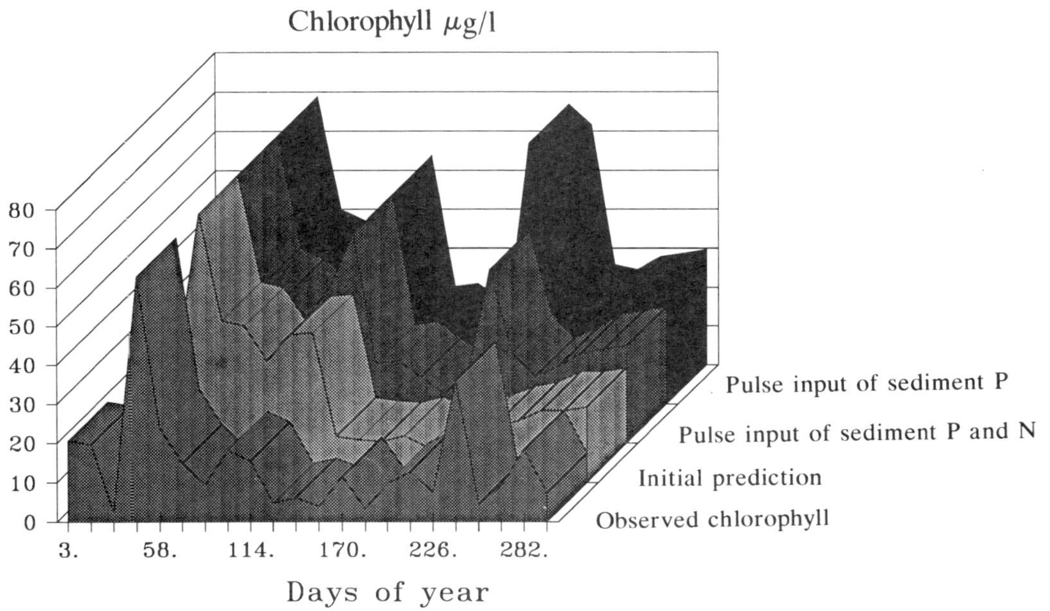

Figure 2. A comparison of observed chlorophyll-*a* concentrations (µg l^{-1}) with algal capacities predicted with different amounts of sedimentary nutrient input to Chew Valley Lake.

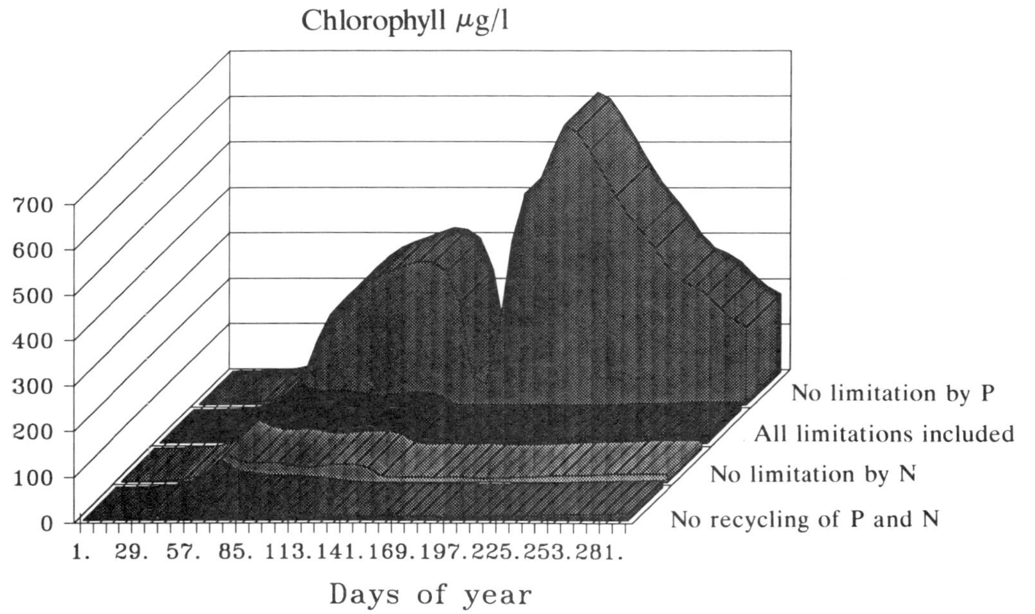

Figure 3. A comparison of predicted algal capacities (chlorophyll-*a*, µg l^{-1}) in Chew Valley Lake after removal of selected limitations.

not be possible if the model is viable. In order to find possible causes of the model's failure to predict accurately, the limitations on algal growth imposed by each of the three constraining variables, (1) phosphorus concentrations, (2) nitrogen concentrations and (3) zooplankton nutrient recycling, were removed one at a time. The resulting predicted levels are shown in Figure 3. Removal of either the capacity limitation by nitrogen or the lack of nutrient recycling by zooplankton had only a small effect on standing crops, in both cases reducing biomass predictions, compared to the case of full limitation. However, removal of capacity limitation by phosphorus allowed the predicted capacity to rise to over 600 $\mu g\ l^{-1}$ of chl*a*, with a significant peak in algal concentration at the end of summer. Hence, a possible cause of the poor predictive ability at the end of the summer in the original simulation could have been an unrecognised source of phosphorus.

Trial simulations (not shown) using a range of constant input rates of phosphorus throughout the summer period suggested that an amount equivalent to $3\mu g\ P\ l^{-1}$ in the whole lake, added each day, would be sufficient to give the observed chlorophyll levels in the autumn. Comparison of these phosphorus loads with measured sources showed that underestimation of the input load from streams was unlikely and that roosting birds were present in insufficient numbers to generate the required influx of phosphorus (Hilton, unpublished). By deduction, release from the sediments was implicated. The simulations of biomass, assuming an extra, constant summer input of phosphorus, also suggested that the missing input probably occurred in a pulse or series of pulses, rather than at a constant rate.

A study of the phosphorus concentrations measured in the lake water (Hilton *et al.* in prep.) showed that in May there was a rapid increase of about 200 $\mu g\ l^{-1}$ in both soluble reactive phosphorus and total phosphorus in the water, which was maintained until the autumn. By an iterative process it was found that if 50% of this released phosphorus was made available to the algae for growth, then the late summer peak in algal biomass was reproduced by the model at observed levels (Fig. 2, sediment-P pulsed input; any reference to this sediment input in the rest of this paper assumes that only 50% is utilised). The major increase in phosphorus in the lake water occurred during a storm, following a hot, calm period of several weeks duration and was probably due to release from the sediments.

Having established that the model can simulate the correct quantities of total algal biomass, when given phosphorus inputs which satisfy the observations, we may consider the observed and predicted distribution of species. The changes observed in the dominant algal species in the lake are shown in Figure 4. Notable concentrations of the green alga, *Closterium*, were observed in early January. The spring peak was dominated by a small diatom, *Stephanodiscus hantzschii*, with a small contribution from large diatoms, including *Stephanodiscus astraea*, and *Closterium*. Mid-summer algal populations, although much lower than the spring peak, consisted of cyanobacteria and cryptomonads, with a significant contribution from nitrogen-fixing species in mid to late summer (day 170). The late summer population was dominated by *Stephanodiscus astraea*, with a small contribution from *Fragilaria* and the nitrogen-fixing cyanobacterium, *Anabaena*.

Figure 4 also shows species composition predicted by the model, including a pulsed input of sediment phosphorus. The large *Closterium* population observed at the start of the year is not simulated, as it results from the cumulative conditions of the previous year and input data were only available for 1989. The model predicts a spring peak of diatoms, both large and small species, with a small input contribution from large green algae such as *Closterium*. As stated earlier, the early summer populations of diatoms were, in reality, limited by silica, so other species became dominant, but the present model does not allow for silica limitation. The mid to late summer growth of nitrogen-fixing cyanobacteria is well represented by *Anabaena*.

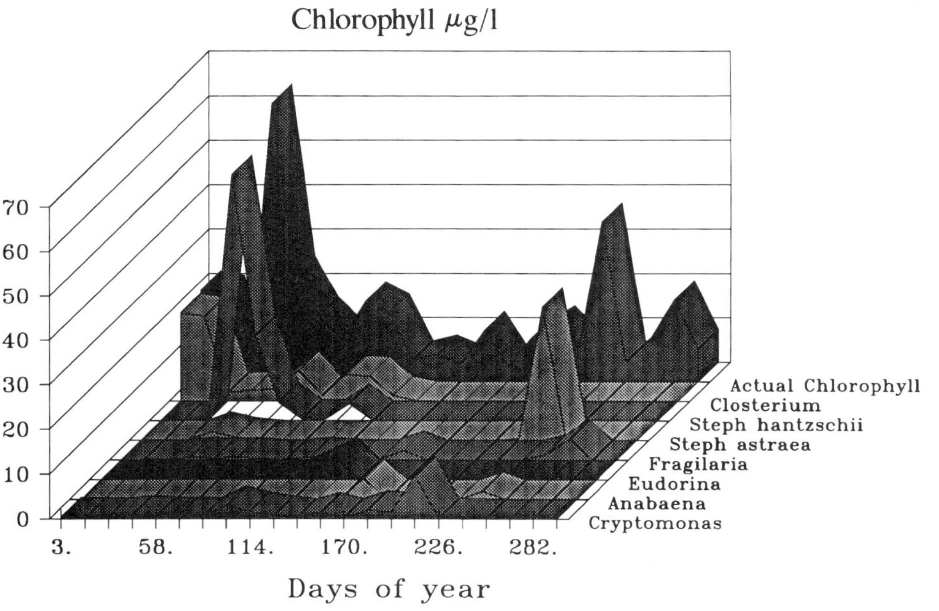

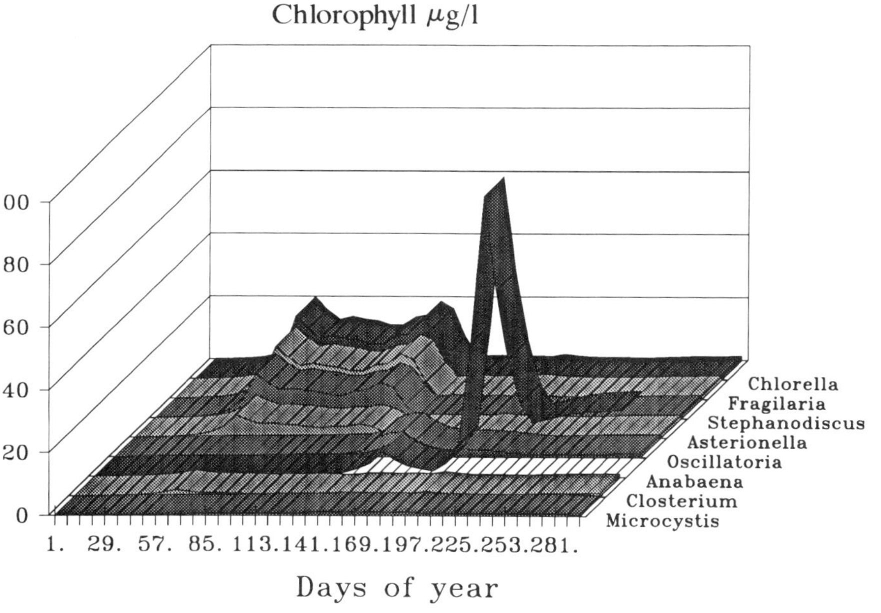

Figure 4. Above: observed composition of algal species in Chew Valley Lake in 1989. Below: predicted composition of algal species from the model, assuming an input of phosphorus from the sediments. Algal biomass is expressed as chlorophyll-*a* ($\mu g\ l^{-1}$)

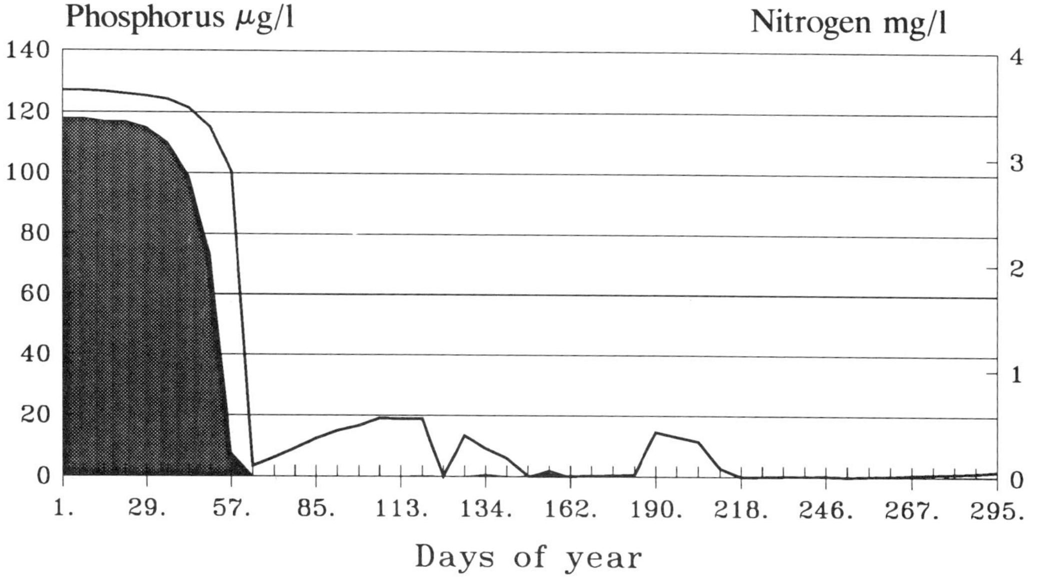

Figure 5. Dissolved nitrate-nitrogen (mg l^{-1}, —) and reactive phosphorus concentrations (µg l^{-1}, shaded areas) predicted by the model, assuming a sedimentary phosphorus input to Chew Valley Lake.

However, the late summer peak is completely dominated by nitrogen-fixing cynaobacteria, rather than the observed diatoms

The appearance of nitrogen-fixing cyanobacteria is symptomatic of a lack of nitrogen in the system. This is borne out in Figure 5 which shows that, under the model conditions, nitrogen levels would have been very low during the periods around day 168 and from day 218 onwards. Hence, although nitrogen was not limiting the total algal biomass, more nitrogen (in addition to the quantities brought in by the inflows and included in the model) must have been introduced into the real system. Otherwise, dominance by the nitrogen-fixers would have occurred in reality as well.

A rapid reduction in nitrate-nitrogen concentrations was observed in the water column. It coincided with the large increase of phosphorus in the water column (around day 155). So the nitrate chemistry was predicted correctly, resulting in the prediction of the dominance of nitrogen-fixing blue-green algae around this period. However, in mid June (approximately day 168), about six weeks after the large release of phosphorus, ammonia levels in the water column were seen to rise rapidly from about 100 µg l^{-1} to about 700 µg l^{-1}. These increased levels were observed for the next six weeks. Again a search of the external inputs could locate no reason for the increase. Other data (Hilton *et al.* in prep.) were consistent with this event also being a release from the sediments.

When an amount of nitrogen equivalent to this release was introduced into the model, the predictions of biomass capacity remained essentially unaltered (some reduction was observed due to the less efficient use of phosphorus at higher nitrogen concentrations) but the species dominance in the late summer changed from nitrogen-fixing cyanobacteria to a mixture of almost all the species introduced in the theoretical inoculum (Fig. 6). Both *Anabaena* and *Fragilaria* were seen in significant quantities in the autumn. Since growth rates of *Stephanodiscus astraea* were not included in the model its dominance could not be predicted directly. However, all the algae which show significant growth in the model are efficient in

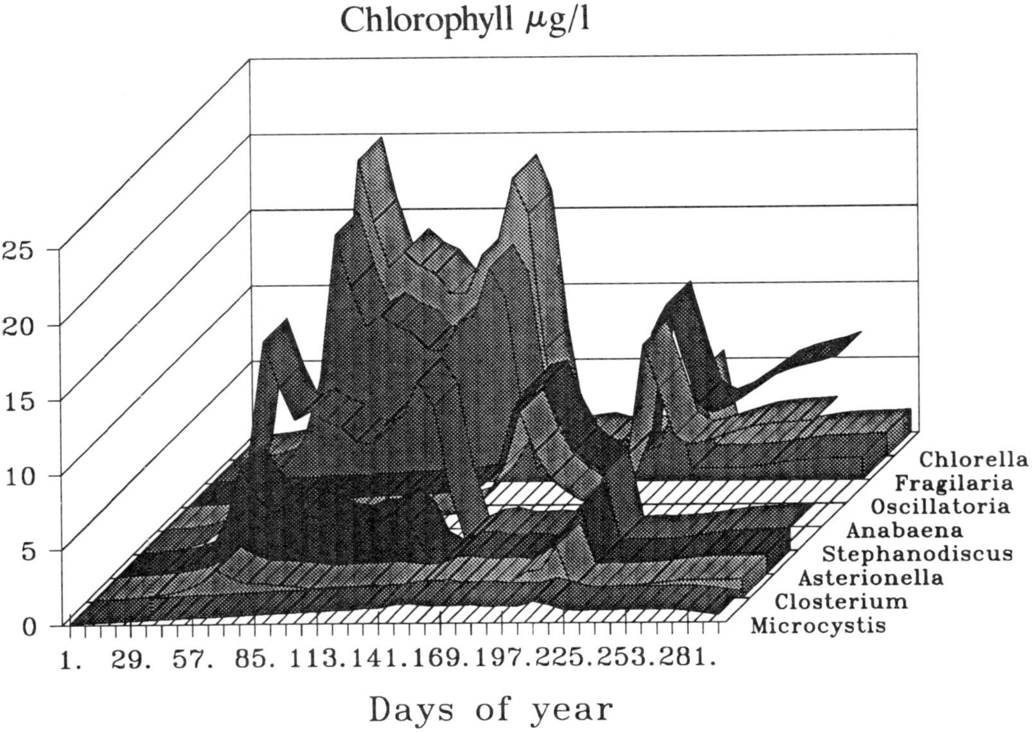

Figure 6. Algal species composition predicted from the model (expressed as chlorophyll-*a*, µg l^{-1}), assuming pulsed inputs of phosphorus and nitrogen from the sediments of Chew Valley Lake.

their use of light (diatoms and *Oscillatoria*) and/or are slow growing efficient users of nutrients. *Stephanodiscus astraea*, a large diatom, is consistent with the model selections.

Hence, when given the correct amounts of nutrient input, the model can correctly predict the levels of algal biomass, the timing of peak events and the changes in species. When the model is used in conjunction with a good chemical and algal data-set it can direct the user to a greater understanding of the lake system, which in itself is an aid to better decision making. Since it would be unrealistic to include growth rates for all algal species in the model, the results will never predict the actual species dominance in all cases. Experience is required to interpret the results in terms of the conditions (low light, low nutrient supply, etc.) which are selecting the dominant species. However a major limitation to the use of the model, without direct comparison with actual data, is our ability to predict phosphorus and nitrogen release from the sediments.

Management strategies

Having checked both the completeness of the input data for the model and the similarity of simulations to real life, the model can then be used in a predictive mode. Suppose that the peak levels of algae in the lake were considered to be unacceptably high, possibly as the result of significant down time of the treatment plant due to filter clogging. Would phosphorus removal from the inflows be a viable policy ? There are seven inflows to Chew Valley Lake, of which the River Chew is by far the largest, constituting at least 90% of the total incoming volume of water. The obvious policy would be to treat water in the River Chew only. If the concentration of phosphorus in the River Chew is reduced to zero in the model, the output shown in Figure 7

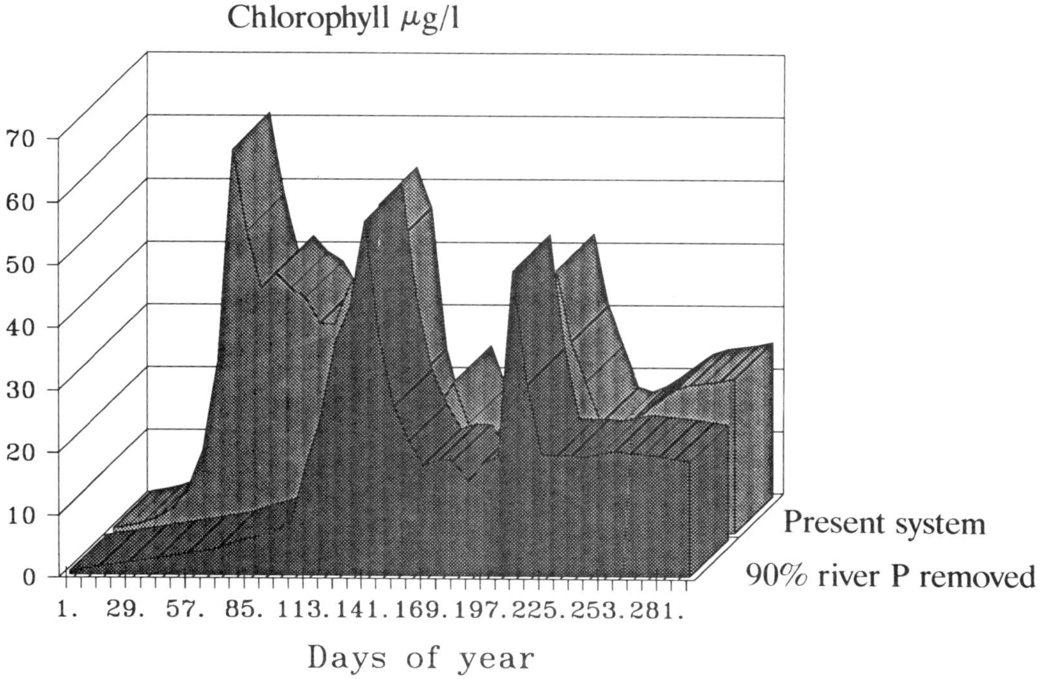

Figure 7. A comparison of predicted algal capacities (chlorophyll-*a*, µg l^{-1}) in Chew Valley Lake with phosphorus input as at present and after its removal from the major inflow.

is obtained. As can be seen, the early spring peak of algal biomass is almost completely removed. Since this is one of the times of peak algal production it would have a beneficial effect. However, removal of phosphorus from the major inflow would give no significant reduction in the algal biomass from late spring onwards, so that no benefit would accrue during this period. The sediment releases, resulting from many years of phosphorus inputs, dominate the system at this time and, as a result, reductions in observed biomass levels during the late spring and summer would take many years to achieve without a programme to remove or stabilise sedimentary phosphorus. They would still be expected to improve after an unspecified number of years, as the internal load reduces, through gradual equilibrium and eventual burial of the phosphorus-rich sediment. Hence, if the treatment problems consistently occurred only in the spring then the phosphorus-removal option may be appropriate (the actual decision will also rely on relative costs). If the problem is typical of the late spring to summer period then alternative or joint removal of phosphorus and sediment-treatment strategies would be required.

As a further example it would be interesting to see if the destratification, which is used to improve hypolimnetic water quality, has any beneficial effect on algal biomass levels. Before destratification was introduced a thermocline often existed at about 7-m depth. If we introduce a permanent thermocline at this depth into the model then the biomass predicted in Figure 8 would probably result. It is not significantly different from the destratified situation. The species composition also shows only minor changes from the destratified situation (Fig. 9). We must conclude that the only benefit (although sufficient in itself) of destratification in this reservoir is the removal of anaerobic conditions in the bottom waters, improving the quality of water abstracted for treatment.

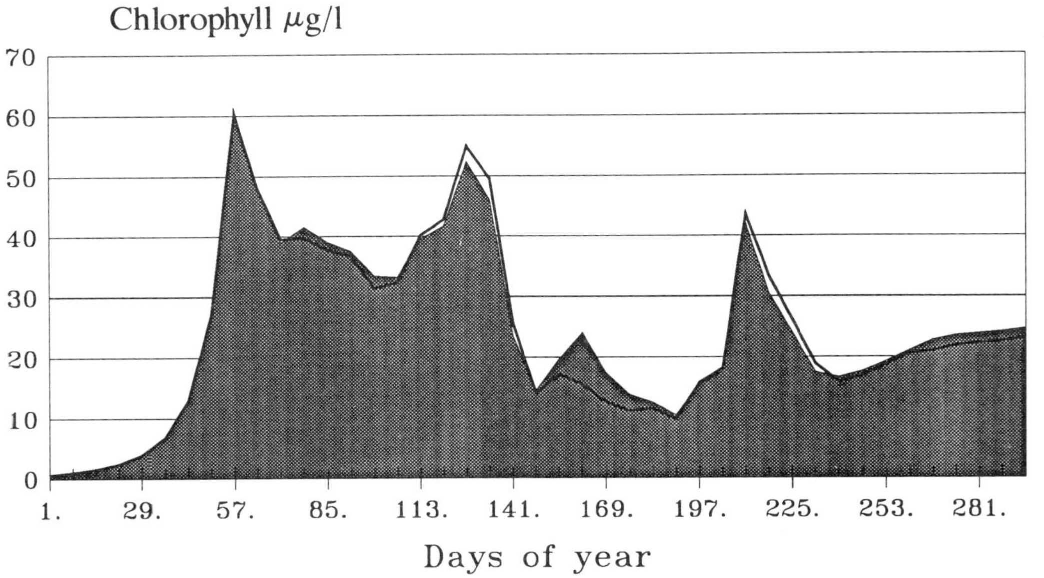

Figure 8. A comparison of (theoretical) predicted algal capacities (chlorophyll-*a*, μg l^{-1}) in Chew Valley Lake with input as at present but assuming either complete mixing (no thermocline; shaded area) or a continuous summer stratification at a depth of 7 m (—).

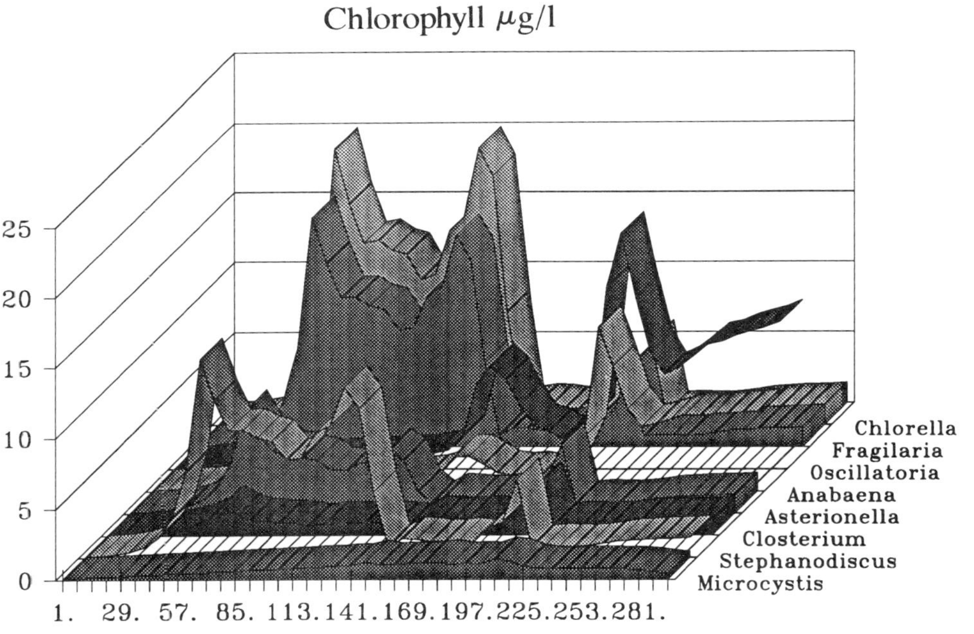

Figure 9. Species composition (chlorophyll-*a*, μg l^{-1}) predicted from the model assuming a pulsed sedimentary input of phosphorus and nitrogen into Chew Valley Lake under conditions of continuous summer stratification.

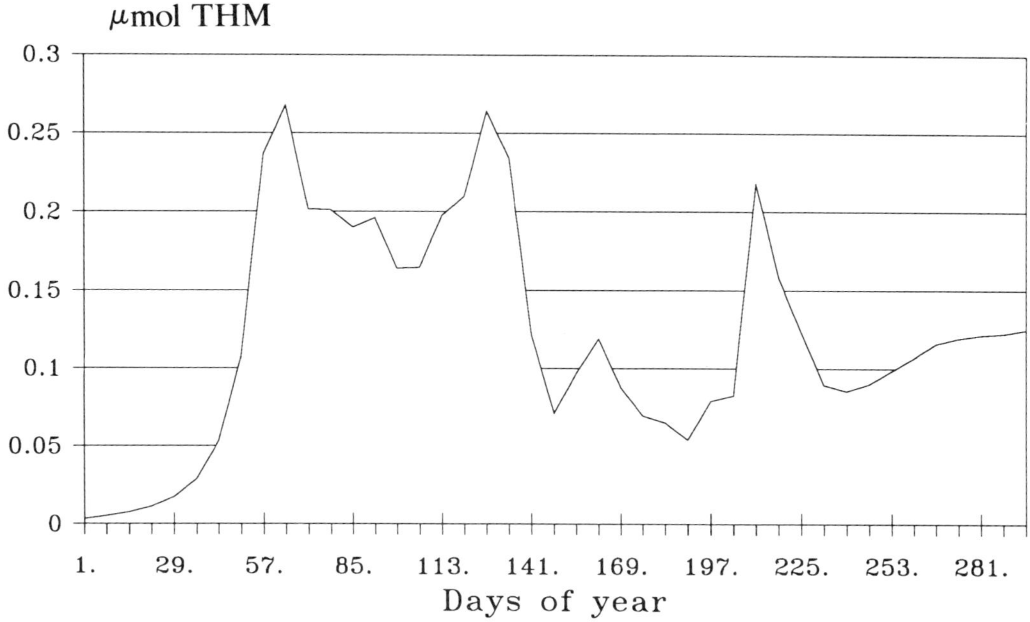

Figure 10. Predicted maximum concentrations of tri halo methanes (μmol l^{-1}) in Chew Valley Lake after treatment, assuming that chlorine is allowed to come into contact with algal cells.

Further developments

Many chemical processes in lakes and reservoirs are either driven by, or strongly influenced by, the algal carbon produced by photosynthesis. If the simulations from our algal growth model are used as a data input to a chemical model it is possible to make reliable predictions of many chemical processes. For example, if algal cells are exposed to chlorine then tri halo methanes (THM) are formed. By using Hoehn *et al.*'s (1980) relationship between algal biomass and THM production it is possible to predict the worst case level of THM production after treatment for the present reservoir (Fig. 10). Using more rigorous chemical relationships (Stumm & Morgan 1981) we have accurately predicted epilimnetic pH levels and the dynamics of hypolimnetic water chemistry changes (Hilton *et al.* in prep.). There is even a good possibility that sediment phosphorus influx could be predicted by combining the hypolimnetic model and the epilimnetic pH model. Similarly reliable prediction of nitrate removal by reservoirs is also within grasp.

Conclusions

The application of a realistic, functionally-based model of algal growth can benefit many areas of limnological research and application. Choices between different management strategies can be made on a more rational basis and, by identifying those areas where the model is still limited, research topics for the future can be identified.

The authors would like to thank Mr E. Rigg for the preparation of some diagrams; staff at Bristol Water, particularly Mr D. Smith and Mr P. Hodge, for assistance with the collection of data; and Bristol Water for supporting the work and for permission to publish.

References

Hammerton, D. (1959). A biological and chemical study of Chew Valley Lake. *Proceedings of the Society for Water Treatment and Examination*, **8**, 87-132.

Redfield, A. C. (1934). On the proportions of organic derivatives in sea water and their relation to the composition of plankton. *James Johnstone Memorial Volume*, pp. 176-192. Liverpool University Press, Liverpool.

Reynolds, C. S. (1984). Phytoplankton periodicity: the interactions of form, function and environmental variability. *Freshwater Biology*, **14**, 111-142.

Reynolds, C. S. (1989a). The FBA-Welsh Water simulation of phytoplankton dynamics in flushed systems: authentication in a coastal lagoon. Reference report of the Freshwater Biological Association, Ambleside.

Reynolds, C. S. (1989b). Physical determinants of phytoplankton succession. In *Phytoplankton Ecology* (ed. U. Sommer), pp. 9-56. Science Publishers, Madison.

Stumm, W. & Morgan, J. J. (1981). *Aquatic Chemistry*. Wiley, New York. 780pp.

Wilkson, R. S., Sleigh, M. A., Maxwell, T. R. A., Mance, G. & Milne, R. A. (1975a). Physical and chemical aspects of Chew Valley and Blagdon lakes, two eutrophic reservoirs in North Somerset, England. *Freshwater Biology*, **5**, 357-377.

Wilson, R. S., Maxwell, T. R. A., Mance, G., Sleigh, M. A. & Milne R. A. (1975b). Biological aspects of Chew Valley and Blagdon lakes, England. *Freshwater Biology*, **5**, 379-393.

Management of Biesbosch Reservoirs for quality control with special reference to eutrophication

GIJSBERT OSKAM AND LAMBERT VAN BREEMEN
Water Storage Corporation Brabantse Biesbosch Ltd, P.O. Box 61, 4250 DB Werkendam, The Netherlands

The three Biesbosch Reservoirs are pumped storage reservoirs, fed with rather polluted and highly eutrophic water from the River Meuse. Air injection at the bottom of the reservoirs prevents thermal stratification, which would otherwise result in serious water quality deterioration. Reservoir mixing also serves as an economic algal control measure; mixing over sufficient depth causes light to play the role of limiting factor and this, combined with zooplankton grazing, keeps the biomass of phytoplankton at acceptable levels. Special problems are caused by benthic, geosmin-producing *Oscillatoria* species growing on the inner embankment. Rooting up the bottom with a harrow is used as the method of control, based on underwater observations by biological staff trained as SCUBA-divers.

With regard to pollutant behaviour the three reservoirs act as a series of fully mixed reactors. This enables the application of kinetic models to describe their behaviour and allows the use of a selective intake policy, e.g. for suspended solids with associated contaminants, ammonia and polynuclear aromatic hydrocarbons. A combination of selective intake and self-purification processes – enhanced by the compartmentalisation of the storage volume in three reservoirs – leads to a striking improvement for many water-quality parameters. Unfortunately this does not apply to some herbicides, like atrazin, which require extra treatment.

Introduction

The Biesbosch Reservoirs are man-made pumped storage reservoirs (Table 1), fed with moderately polluted but highly eutrophic water from the River Meuse. The reservoirs were constructed during the 1970s and serve as the raw water supply for several waterworks in the southern and western parts of the Netherlands, e.g. Rotterdam and its wide surroundings. The delivery of water in 1991 amounted to 166 million m^3 $year^{-1}$, about one-eighth of the total supply in the Netherlands.

The three reservoirs are connected in series as shown in Figure 1. Honderd en Dertig and Petrusplaat are designed as process reservoirs to guarantee a minimum residence time for quality improvement, whereas De Gijster reservoir really serves the storage function of the project and can be nearly emptied, ensuring a safe yield of about 2 months at the present throughput.

Table 1. *Characteristics of the three Biesbosch reservoirs*

Variable	Units	De Gijster	Honderd en Dertig	Petrusplaat
Surface area	ha	305	210	100
Volume	10^6 m^3	40	33	13
Maximum depth	m	27	27	15
Mean depth	m	13	15	13
Retention time	weeks	11	9	4

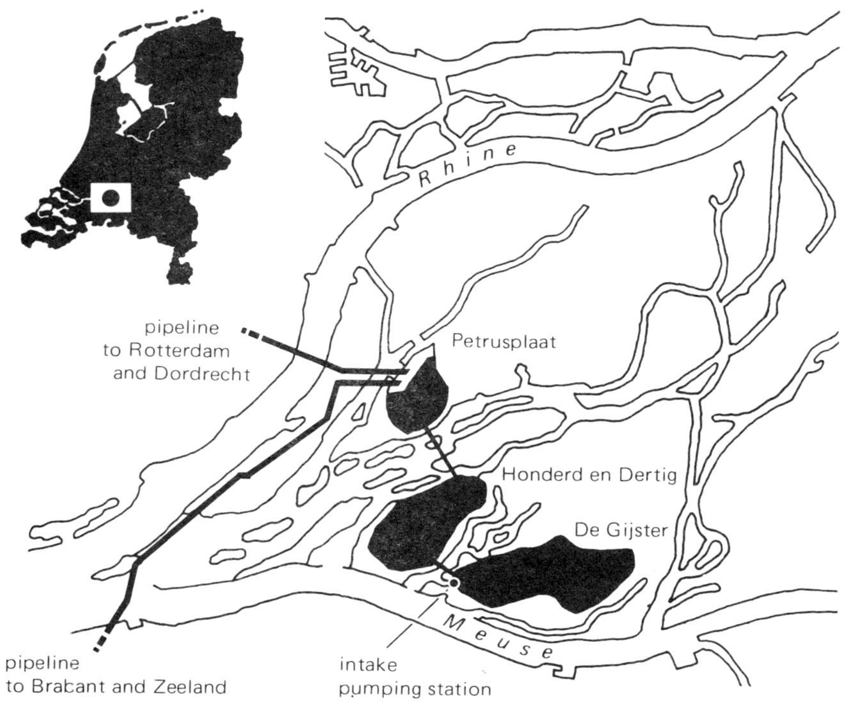

Figure 1. Location of the Biesbosch Reservoirs, in the Netherlands.

Judged by Dutch standards the Biesbosch reservoirs are very deep, varying from 15 m to 27 m locally. Based on the experience with the Thames Valley reservoirs, thermal stratification had to be expected (Ridley *et al.* 1966). When a eutrophic reservoir stratifies, deoxygenation of the hypolimnion water often leads to anaerobic conditions in the bottom layers and this initiates several undesirable processes. Particulate $Fe(OH)_3$ and MnO_2 will be reduced chemically to dissolved Fe^{2+} and Mn^{2+}, also resulting in solution of associated phosphates. Biological processes are denitrification ($NO_3^- \rightarrow N_2$), sulphate reduction ($SO_4^{2-} \rightarrow H_2S$), formation of NH_4^+, and putrefaction of organic substances. As a result the quality of the hypolimnion water becomes unacceptable for treatment to drinking water.

The Meuse water is rich in nutrients, although phosphorus concentrations have decreased because of the replacement of phosphates in detergents. At present the average concentrations of total phosphorus and nitrogen amount to 0.3 and 5 mg l^{-1} respectively. When storing such water the mass development of algae will occur, causing a lot of problems for water supply, like formation of taste and odour compounds, impairment of coagulation, clogging of filters, filter passage of algae, etc. As the objective of storage is also quality improvement and not deterioration, the control of possible adverse effects is the first aim of the management of the reservoirs. In this the air injection system plays an essential role.

Air injection for reservoir mixing

The destratification system, developed for the reservoirs, uses air injection at the bottom for maintaining isothermal conditions. Air injection takes place in so-called injection units, the number of units per reservoir varying from three to six, depending on the size of the reservoir. Figure 2 shows an injection unit, which is very compact and consists of a 1 × 3 metres steel

Figure 2. An air-injection unit.

frame on which six Flygt (polyethylene) foam plastic blocks are installed. The land-based compressors feed the injection units by means of plastic pipes, forcing air through the foam blocks and producing a bubble column of very finely divided air, thus entraining large amounts of water.

The system has several advantages. Maintenance cost is very low, because the foam blocks do not clog. Materials deposited during long periods of non-use (autumn and winter) crumble, because the foam block expands when air injection is started again. Installation is easy and very flexible as the position of the units can be easily altered, if their original configuration does not prove to be optimal. The overall investment is moderate.

Table 2. *Characteristics of the air injection system in the three Biesbosch reservoirs*

Variable	Values
Total surface area	615 ha
Total volume	86 10^6 m^3
Water flow	166 10^6 m^3 $year^{-1}$
Injection units	3–6 per reservoir
Injection depths	15–25 m
Air flow	0.8 m^3 s^{-1}
Compressor power	250 kW
Compressor working hours	2,200 h $year^{-1}$
Annual energy consumption	550,000 kWh $year^{-1}$
Price per kWh	0.10 Hfl

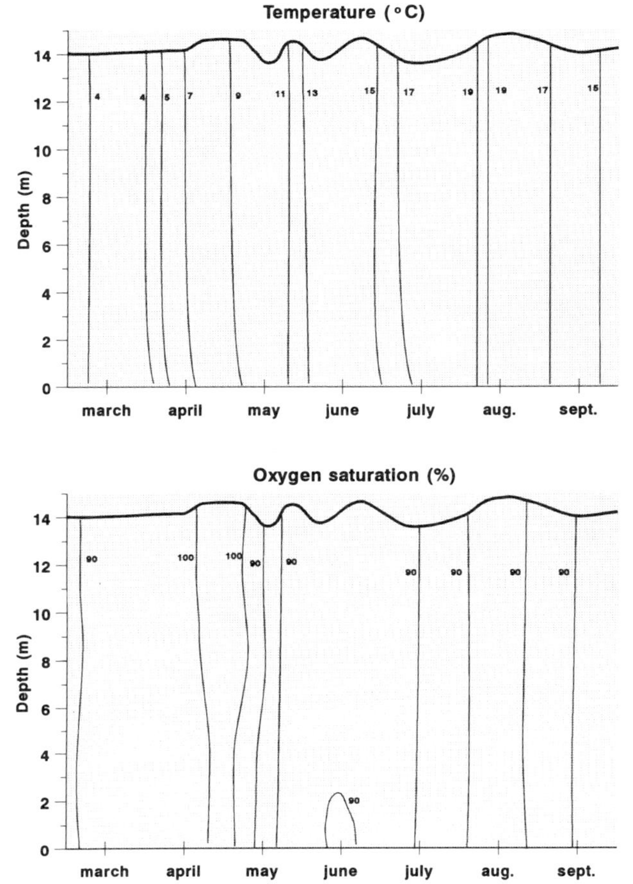

Figure 3. Temperature and oxygen isopleths for Petrusplaat Reservoir.

Table 2 gives the operational characteristics of the air injection systems in the three Biesbosch reservoirs. The total operating cost of the reservoir mixing system amounts to about Hfl. 80,000. The capital cost is Hfl. 150,000 (total investment is Hfl. 1.5 million), so the total cost can be calculated at less than 0.15 cents per cubic metre of delivered water and is thus very moderate.

The air injection is very effective in maintaining homogeneous conditions in the reservoirs. Measurements have shown that under optimal conditions, 1 m^3 of air entrains up to 250 m^3 of water, thus ensuring efficient vertical and horizontal mixing. This is illustrated in Figure 3, where the temperature and oxygen isopleths in Petrusplaat are shown. It can be calculated that the mixing time of the reservoirs is small compared to the average residence time, so the reservoirs can be considered as a series of fully mixed reactors.

Control of planktonic algal growth

Algal control in the Biesbosch reservoirs is based on the principle of light limitation through reservoir mixing over sufficient depth. Artificial mixing as an algal control method was first recognized and applied in the management of the deeper Thames Valley reservoirs, which are circulated by means of jet-type inlet pumps (Steel 1975). Whether or not light can operate as a limiting factor depends mainly on the available mixing depth in relation to the optical properties of the stored water. Simple models of light-limited algal growth have been developed, which consider net algal growth as the balance between gross photosynthesis and respiration. Photosynthesis is confined to the upper layers, whereas respiration occurs in the whole water column. With a large ratio of mixing depth to euphotic depth, respiration will dominate the algal metabolism and thus restrict growth. The algal concentration itself of course affects light penetration. This self-shading phenomenon leads to the concept of a maximum algal concentration for every given situation of mixed depth and extinction of the water, assuming that nutrients and sunlight conditions are optimal.

In very simplified form the expression for the maximum algal biomass (C_{max}, mg Chl m^{-3}) is (Oskam 1978):

$$C_{max} = \frac{1}{\varepsilon_c}\left(\frac{27}{Z_m} - \varepsilon_w\right)$$

where ε_c is specific extinction per unit of algal concentration (m^2 mg Chl^{-1}), ε_w is the extinction coefficient of the water without algae (m^{-1}), Z_m is the mixing depth (m), and 27 is a numerical factor incorporating approximate light intensity and duration.

Figure 4 illustrates the influence of water extinction, ε_w, on the maximum algal biomass for different mixing depths. In the Biesbosch reservoirs, ε_w varies from 0.5 to 1.0 m^{-1} (the corresponding Secchi disc transparency is 2–4 m), so that with average depths of 13–15 m a maximum chlorophyll concentration between 40 and 70 mg m^{-3} can be expected.

Figure 5 shows the course of algal development in Petrusplaat during 1986 to 1991, together with zooplankton numbers. Up to 1988 the pattern was the same every year, with a moderate spring pulse of diatoms (30–60 mg Chl m^{-3}) and small algal crops during summer and autumn. These low levels during most of the growing season were caused by the grazing pressure of the zooplankton, mostly consisting of large cladocerans such as *Daphnia longispina*, *D. magna* and *D. pulex*.

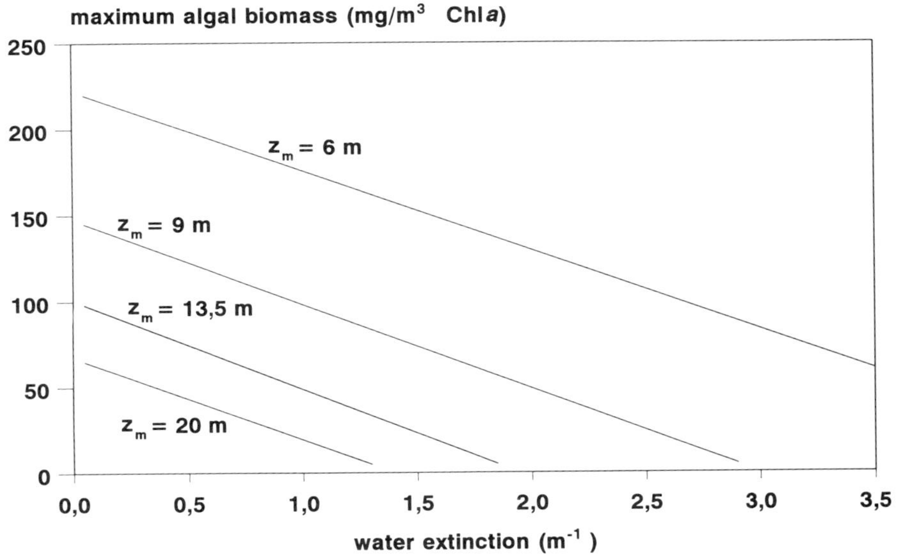

Figure 4. Relationship between maximum algal biomass (mg Chl*a* m^{-3}) and water extinction (m^{-1}) for different mixing depths (specific extinction ε_c = 0.02 m^2 mg Chl^{-1}).

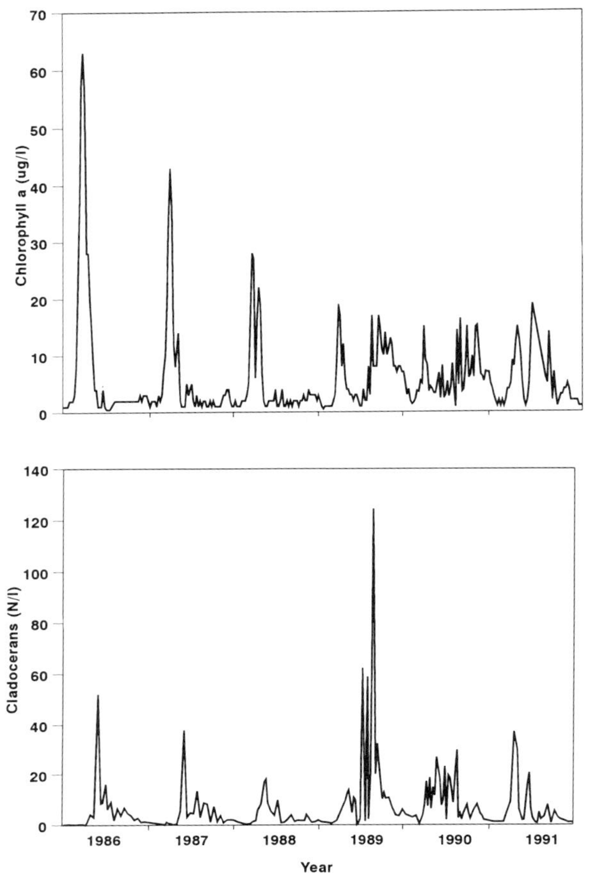

Figure 5. Concentrations of chlorophyll-*a* (mg l^{-1}) and planktonic cladocerans (numbers per litre) in Petrusplaat Reservoir during 1986-1991.

Only in early spring, with water temperatures under 10°C, the grazing pressure of the zooplankton was low and the amount of diatom growth was determined by the light climate. During the rest of the growing season, mixing over sufficient depth retarded the algal growth rate to levels where the zooplankton could "keep up" with the algae, suppressing them to low levels. This controlling effect of the herbivorous zooplankton was possible because predation pressure on the zooplankton was low, as each year the year-class 0 of planktivorous fish was eaten by pike-perch, perch and eels.

Since 1988 the pattern of events seems to have changed. The spring pulse of diatoms decreased, because the mild winters allowed an early appearance of the zooplankton. But also a shift is occurring from predominantly cladoceran to mixed cladoceran and copepod zooplankton populations. The herbivorous copepods are less efficient grazers. Consequently there is a trend to higher algal levels during the summer period. This shift to copepod dominance could well be connected to the presence of the carnivorous zooplankter *Bythotrephes longimanus*. After its first appearance in 1987, it has now become well-established with levels of about 1–2 individuals per litre of water during the summer months. This development gives cause for some concern and merits close attention. Nevertheless, the present levels of algal biomass in the reservoirs are still five to ten times lower than in

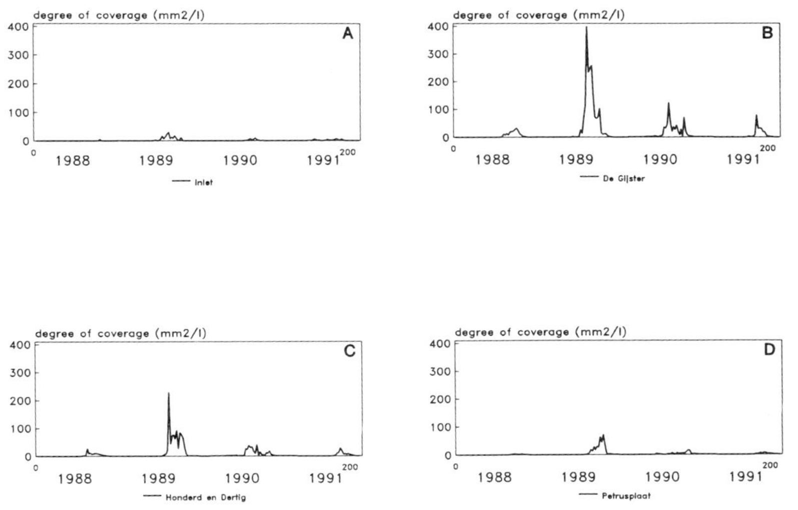

Figure 6. Second distribution of planktonic cyanobacteria expressed as relative aerial units ($mm^2\ l^{-1}$) in the Biesbosch Reservoirs during 1988-1991. A, river water from the Meuse; B, de Gijster Reservoir; C, Honderd en Dertig Reservoir; D, Petrusplaat Reservoir.

comparable water-bodies, but which are shallow and have large populations of herbivorous fish like bream and roach.

In spite of the growth-control by vertical mixing and grazing, some algal species are less easily controlled. These are benthic species and planktonic algae with buoyancy by gas vacuoles, such as the planktonic blue-green algae *Microcystis aeruginosa*, *Aphanizomenon flos-aquae* and *Anabaena* spp. These cyanobacteria reflect a horizontal and vertical inhomogeneity in large water-bodies, induced by wind. Therefore it is necessary to take samples at several locations in the reservoir to get a real impression of the abundance of the cyanobacteria. A mixture of these samples represents the average biomass in the whole reservoir.

Figure 6 shows the abundance of the cyanobacteria in the period 1988–1991, in the inlet and the three reservoirs, expressed as relative aerial units ($mm^2\ l^{-1}$). The small inlet concentrations can be explained by the turbulent and turbid nature of the river water. The growth period in the reservoirs starts in June–July and ends in September–October, with the maximum biomass in August–September. Although biomass varies greatly in different years, the constant pattern is that the highest biomass is found in the first reservoir, De Gijster. Some time-shift can be seen in the starting point of the growth period in each reservoir. This could indicate that the appearance of the planktonic cyanobacteria in the last two reservoirs is caused by successive inoculation.

The comparatively high biomass of cyanobacteria in De Gijster might be explained by its different morphometry. All around its perimeter the reservoir has a shallow area (width 100 m, depth 3–7 m). This has a negative influence on the mixing characteristics of the reservoir and favours the growth of cyanobacteria. *In situ* measurements are being planned to confirm this hypothesis.

It has long been known that cyanobacteria can produce toxins. These substances are held responsible for the deaths of animals (wild and livestock), birds and fish, and have sometimes been associated with human illness after skin contact and ingestion. Especially *Microcystis*, but also *Anabaena*, *Aphanizomenon* and *Oscillatoria* species have been implicated in producing toxins. A recent overview of the problems encountered has been given by Lawton & Codd (1991). From their data it appears that 44 to 75% of the blooms tested showed toxicity in a mouse bioassay. A recent Dutch study showed that the toxins of *Microcystis* were present intracellularly as well as in the water (Hoekstra *et al.* 1991). For this reason the reduction of cyanobacterial numbers in the process reservoirs Honderd en Dertig and Petrusplaat is considered to be very important.

Control of geosmin-producing benthic cyanobacteria

Objectionable tastes and odours in drinking water, especially earthy-musty odours, are a worldwide problem. Two organic compounds, geosmin and 2-methylisoborneol, have been indicated as the main cause of earthy-musty odours in water (Gerber & Lechevalier 1965; Rosen *et al.* 1970). These compounds are produced by actinomycetes (Gerber 1983; Wood *et al.* 1983) and cyanobacteria (Persson 1983).

In April–May 1984 an earthy-musty odour developed in Petrusplaat. Geosmin was detected at a concentration of 33 ng l^{-1}. One of the receiving waterworks had no facilities for sufficient geosmin removal and many consumer complaints were received as the geosmin concentration of the drinking water exceeded the 5 ng l^{-1} threshold odour concentration. The high concentration of geosmin could not be explained by the composition of the biota in the pelagial zone of the reservoir. Algal biomass was low (2 µg Chl*a* l^{-1}) and consisted mainly of diatoms (*Stephanodiscus hantzschii*, *Melosira islandica* and *Asterionella formosa*) which are not recorded as possible geosmin producers (Persson 1983, 1988). Also the number of actinomycetes (maximum 0.4 CFU per ml) had not increased compared to previous years.

Water samples taken adjacent to the inner embankment showed higher taste (dilution) numbers. Therefore an intensive sampling campaign of the bottom was started with an Eckman–Birge mud-sampler. Microscopic examination showed the presence of benthic cyanobacteria, *Oscillatoria limosa*, *O. tenuis*, *O. brevis* and *Pseudoanabaena catenata*. Cyanobacteria of the genus *Oscillatoria* in general, but also benthic *Oscillatoria*, are recorded as producers of geosmin and 2-methylisoborneol (Izaguirre *et al.* 1982; Persson 1983). This information led to the suggestion that these *Oscillatoria* were responsible for the production of geosmin. This was confirmed by *in situ* measurements and laboratory experiments, where pure culture isolates of *Oscillatoria splendida* and *O. limosa* were shown to produce geosmin but not 2-methylisoborneol (van Breemen *et al.* 1991, 1992).

As it was considered impossible to monitor *Oscillatoria* development by conventional sampling it was decided to train biologists as SCUBA-divers, so that *in situ* observations would allow control measures at the right moment (Means *et al.* 1984). In the spring of 1985 a team of three divers was operational. The underwater observations showed that *Oscillatoria* were growing in a narrow band on the inner embankment at a depth of 3–6 m. This confined area of growth offered two possible methods of control: dosing with copper sulphate pellets, as already successfully applied in California (McGuire *et al.* 1983) and a newly-developed method, in which cyanobacteria are stirred up by rooting up the bottom with a harrow-like device (van Breemen *et al.* 1991)

Experiments on controlling the cyanobacteria were started in May 1985. Four areas (each approximately 2,000 m^2) with a benthic cyanobacterial mat were chosen, based on underwater observations. Two locations served as controls and two were experimental areas where copper sulphate was applied in doses of 250 and 500 kg ha^{-1} respectively. Elsewhere, rooting up the bottom was done with a harrow consisting of 21 chains, dragged by a boat. Figure 7 shows the harrow in its original form.

The results of this experiment (Fig. 8) indicated that both methods of control had the same effect: within one week there was a sharp decrease in the area of cyanobacterial growth. After

Figure 7. A harrow used for rooting up the bottom of the reservoirs.

two weeks, recolonisation was observed in the area where the bottom was rooted up and in the area where copper sulphate was dosed at about 250 kg ha^{-1}. In the area where 500 kg ha^{-1} was applied, regrowth was not observed, indicating that this method is successful in eliminating benthic cyanobacteria.

Copper sulphate not only had a great impact on the cyanobacteria, it appeared to be lethal for all living benthos at 500 kg ha^{-1} and killed more than 50% at 250 kg ha^{-1}. Because of this harmful effect of copper sulphate on an essential part of the aquatic ecosystem, it was decided to continue the experiment in 1986 by only rooting up areas of cyanobacterial growth. The results (Fig. 9) show that in the untreated area, cyanobacterial growth was comparable to that in 1985, and the growth period lasted from April to July. In the treated areas, hardly any growth occurred, showing that a frequency of once every fortnight is sufficient to control the growth of benthic cyanobacteria.

In the final reservoir, Petrusplaat, various *Oscillatoria* species produced geosmin from February until June. Underwater observations indicate that growth of benthic cyanobacteria in the first two reservoirs can be neglected. To understand to what extent production or reduction of taste and odour compounds occurs in the Biesbosch reservoirs a survey was started in 1988. River water and outlets of Honderd en Dertig and Petrusplaat were analysed routinely for geosmin and 2-methylisoborneol. The results for the period 1988–1990 are shown in Figure 10.

The concentrations of both compounds in the river, with maxima of 10–50 ng l^{-1}, are higher than after storage in the first two reservoirs. The seasonal variations of geosmin and 2-methylisoborneol in the river water are the result of processes in the river and catchment area; both compounds increase during periods of high river flow. In the first two reservoirs a substantial decrease of geosmin and 2-methylisoborneol concentrations takes place; at the outlet of Honderd en Dertig, maximum values of about 3-6 ng l^{-1} were found.

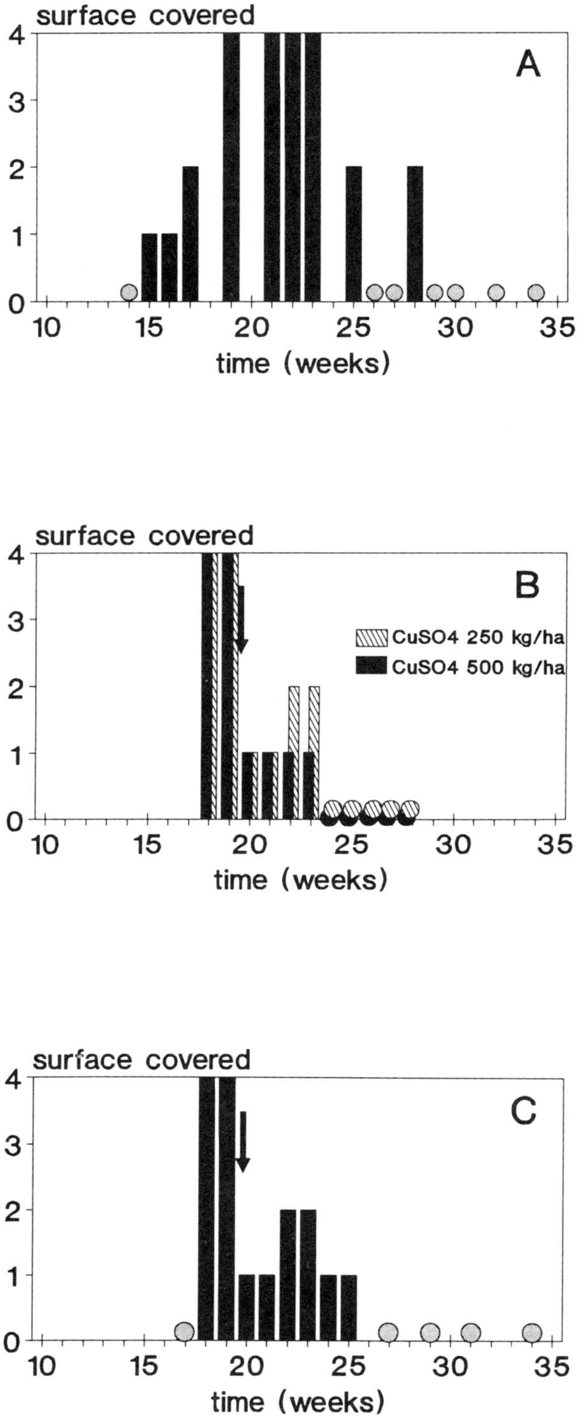

Figure 8. Percentage of the bottom covered with benthic cyanobacteria in the growth-control experiment in Petrusplaat Reservoir in 1985. A, untreated area; B, area treated with $CuSO_4$; C, area rooted up with a harrow. ◎, no cyanobacteria observed, and 1 = 0-25%, 2 = 25-50%, 3 = 50-75%, 4 = 75-100% of the bottom covered with cyanobacteria. ↓:time of $CuSO_4$ dosage or start of the rooting up treatment.

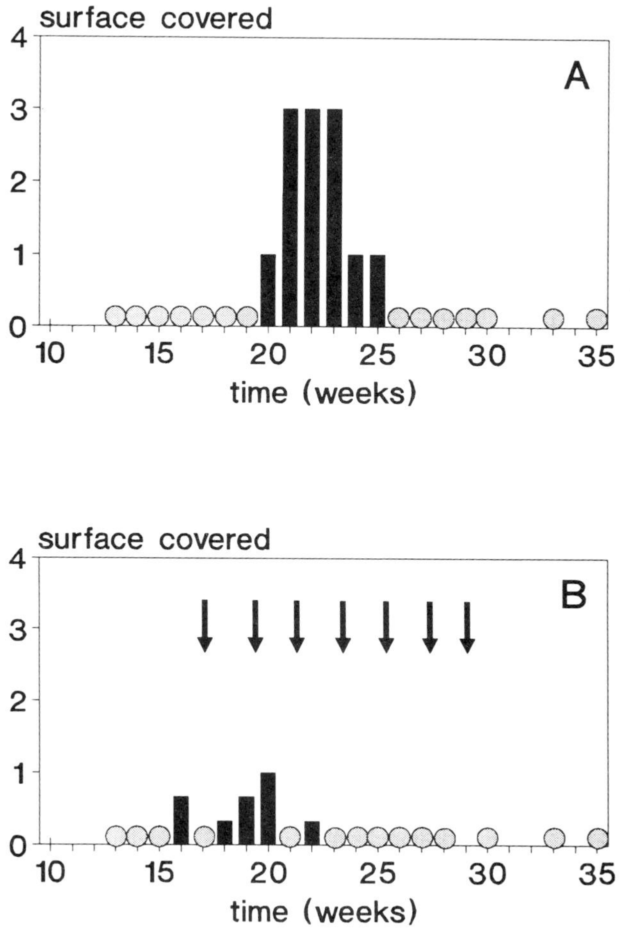

Figure 9. Percentage of the bottom covered with benthic cyanobacteria in the growth-control experiment in Petrusplaat Reservoir in 1986. A, untreated area; B, area rooted up with a harrow. 0-4 on the ordinate represent proportions covered with cyanobacteria, as in Fig. 8. ↓ start of the rooting up treatment.

As a result of cyanobacterial control in Petrusplaat, maximum geosmin concentrations sharply decreased from 33 ng l^{-1} in 1984 to 6.4 ng l^{-1} in 1985, and to 2–3 ng l^{-1} in 1986-1988. In 1989 and 1990, geosmin concentrations in Petrusplaat increased, while those of 2-methylisoborneol did not change. Underwater observations showed that the growth of the benthic cyanobacteria started earlier in 1989 (March) and much earlier in 1990 (January–February) than in previous years. These observations, and the fact that the maximum geosmin concentrations in those years were found 5-8 weeks earlier, indicates that the growth of the benthic cyanobacteria may be influenced by water temperature.

In 1990 cyanobacterial growth started in January. Because geosmin concentrations were low it was decided not to start rooting up the bottom. However, the rather sharp increase of the benthic cyanobacteria in March resulted in an increase of the geosmin concentrations. Although the bottom was then rooted up daily it was not possible to prevent a rise in the geosmin concentrations (maximum of 16.3 ng l^{-1}). This experience in 1990 shows that for a successful control of growth it is necessry to start rooting up the bottom as soon as the

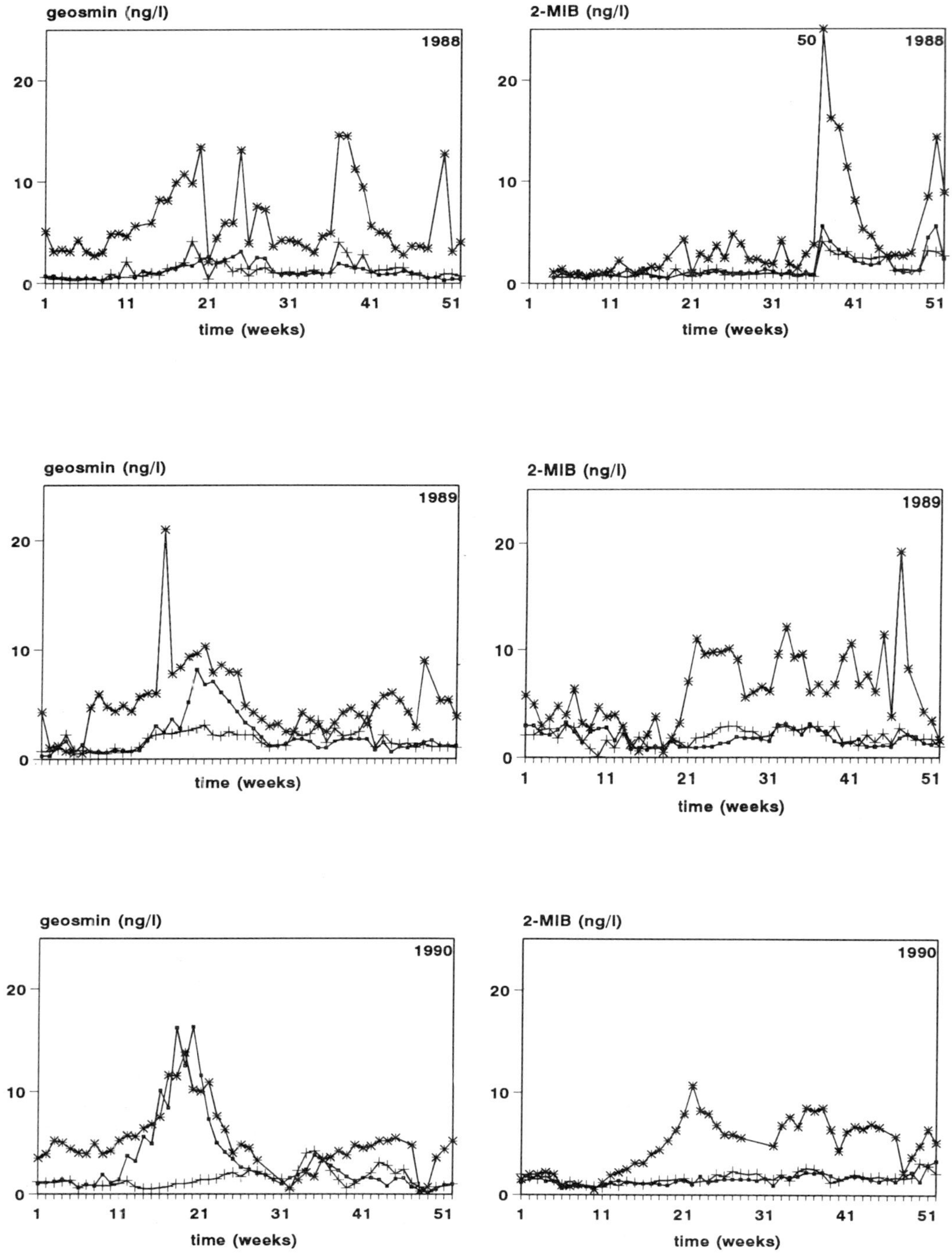

Figure 10. Concentrations (ng l^{-1}) of geosmin and 2-methylisoborneol in the River Meuse (∗), the outlet of Honderd en Dertig Reservoir (+) and the outlet of Pertusplaat Reservoir (■) in 1988, 1989 and 1990.

cyanobacteria are observed. The general conclusion of this survey is that storage of water from the River Meuse in open reservoirs, with a retention time of 4–5 months, reduces geosmin and 2-methylisoborneol concentrations by 60–70%. Autochthonous production of geosmin by benthic cyanobacteria, however, may interfere with this water quality improvement.

Quality improvement in the reservoirs

Theoretical considerations

Quality considerations already formed an integral part of the design stage of the Biesbosch project. For this reason the total volume is divided over three reservoirs. Even when the first storage reservoir is empty, the other two process reservoirs will guarantee sufficent residence time to reach the quality objectives for the abstracted water. The most important processes that contribute to quality improvement in storage reservoirs are outlined in Table 3.

Many degradation processes can be described by first-order reaction kinetics. Assuming steady-state conditions, it can be shown that the removal effect of a series of completely mixed reservoirs can be expressed as (Oskam 1982):

$$\frac{C_{out}}{C_{in}} = \left[1 + k\left(\frac{T}{n}\right)\right]^{-n}$$

where C_{out} is the concentration in outlet water (mass per volume), C_{in} is the concentration in inlet water (mass per volume), k is the rate constant (day^{-1}), T is the total residence time (days) and n is the number of reservoirs.

Table 4 gives calculated C_{out}/C_{in} ratios as a function of compartmentalisation and reaction rates (k) for a total residence time of 180 days. From Table 4 it can be concluded that slow reactions do not benefit much from compartmentalisation, but at higher reaction rates the effect becomes quite large. Apart from planological requirements this was the main reason for dividing the process volume of the Biesbosch reservoirs into two: Honderd en Dertig and

Table 3. *Quality improvement processes in reservoirs*

Type	Process
Biological	Biodegradation of organic substances Nitrification of ammonia to nitrate Die-off of faecal bacteria and viruses
Physical	Equalizing of water quality Gas exchange of oxygen and carbon-dioxide Evaporation of volatile substances Settling of suspended solids and absorbed material
Chemical	Oxidation of divalent iron and manganese Hydrolysis of organic esters and polyphosphates Photolysis of humic substances and polynuclear aromatic hydrocarbons.

Table 4. *Ratios of oulet and inlet concentrations for three different k-values and compartments (T = 180 days)*

	Number of compartments		
k	1	2	3
0.01	0.36	0.28	0.24
0.05	0.10	0.033	0.016
0.15	0.036	0.0048	0.001

Petrusplaat (Knoppert & Vreedenburgh 1982). It was also decided to include the storage volume of De Gijster as much as possible in attaining maximum benefit from quality improvement processes.

Results from practice

This section describes typical results for, respectively, a slow, intermediate and fast removal process, with rate constants for nitrification of ammonia in winter (k ~ 0.018 day^{-1}), photolysis of benzo(a)pyrene (k ~ 0.07 day^{-1}), and die-off of enteroviruses (k ~ 0.14 day^{-1}).

In order to eliminate the need for breakpoint chlorination, the delivered water from the Biesbosch reservoirs should not contain more than 0.2 mg l^{-1} ammonia in the critical winter period. A model description of the ammonia behaviour in the reservoirs – based on nitrification (k ~ 0.018 day^{-1}) and equalization – shows that the intake should be closed when ammonia in the river exceeds 1.2 mg l^{-1}. Figure 11 shows the results of this ammonia control policy for average and maximum values in river water, inlet water and successive reservoirs. The winter maximum in the delivered water usually amounts to 0.15 mg l^{-1}. If only one large reservoir with the same total residence time of about 170 days had been available, the reduction would have been only 4-fold. This would result in a maximum ammonia concentration of about 0.3 mg l^{-1}, at which level treatment problems have been experienced in the past, due to nitrite formation.

Figure 12 illustrates the effect of storage for the carcinogenic polynuclear aromatic hydrocarbon (PAH), benzo(a)pyrene. The removal process of PAHs is photolysis. The inlet concentration of 21 ng l^{-1} is reduced to 0.3 ng l^{-1} through the successive reservoirs. This 70-fold reduction would have been only 10-fold if only one large reservoir with the same total residence time of 170 days had been available. Figure 12 also shows that the concentration in the river itself was nearly double the inlet concentration. This is also caused by the application of selective water intake, when the quality does not meet the requirements. This also applies

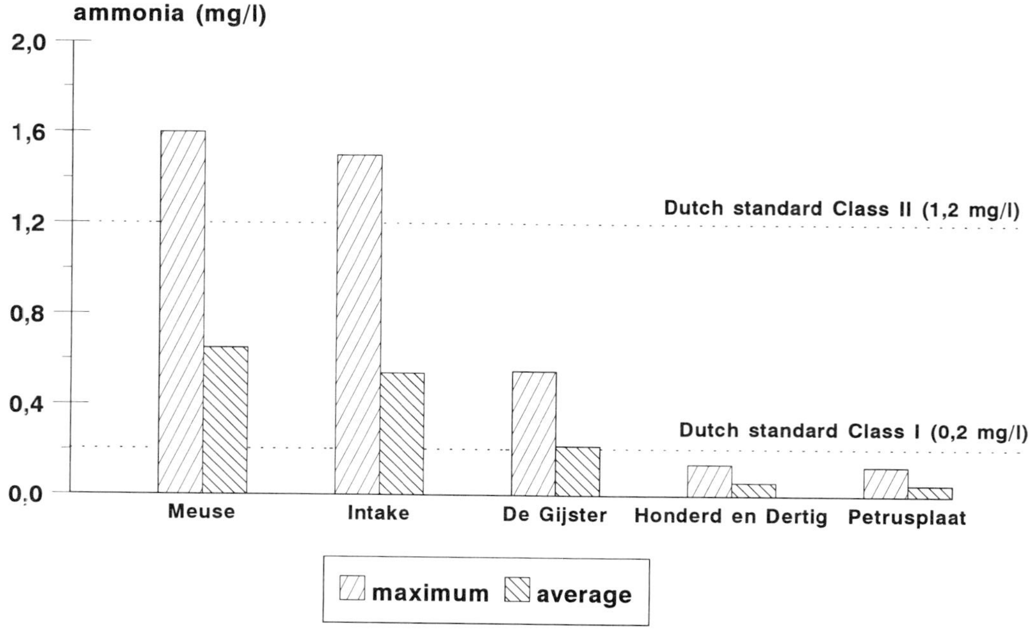

Figure 11. The effects of a selective intake policy and nitrification in the Biesbosch reservoirs.

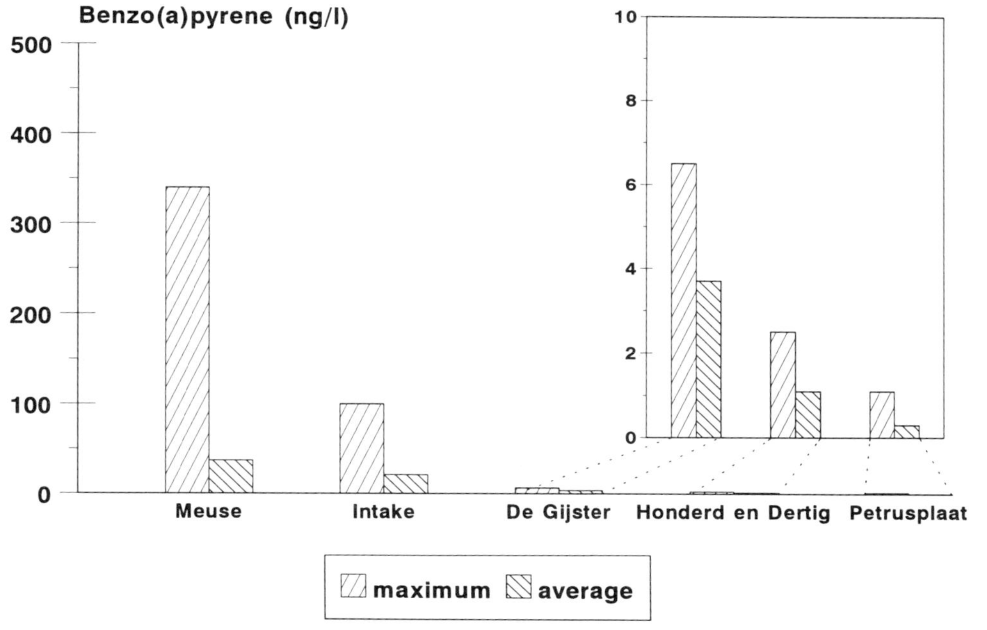

Figure 12. The effects of a selective intake policy and photolysis in the Biesbosch reservoirs.

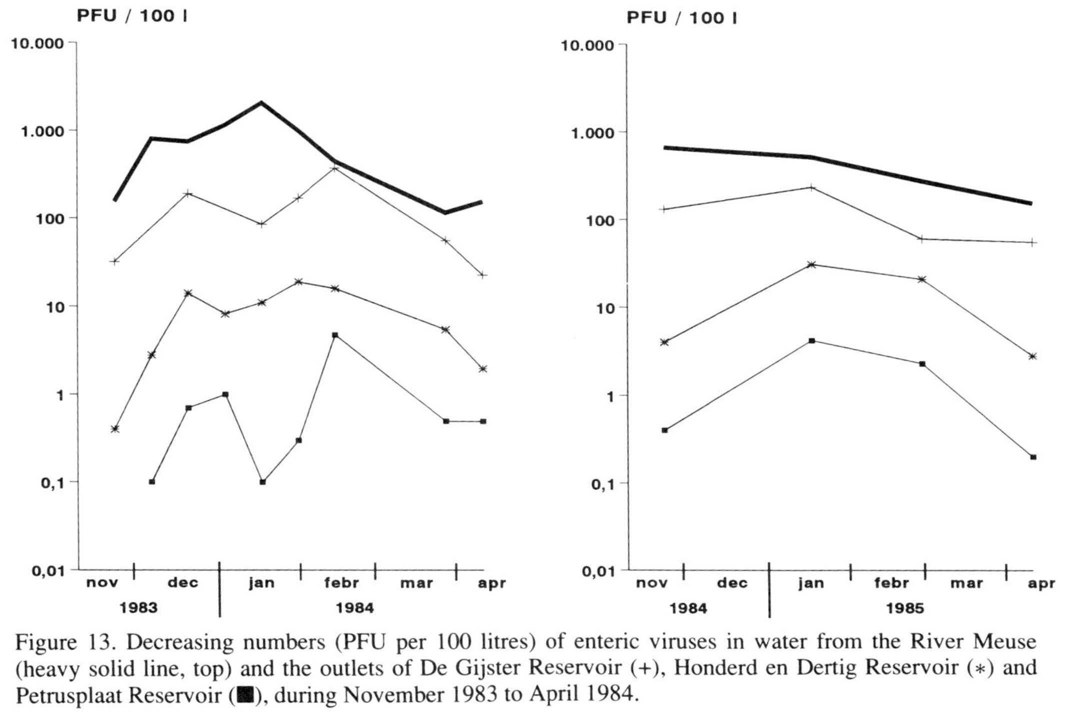

Figure 13. Decreasing numbers (PFU per 100 litres) of enteric viruses in water from the River Meuse (heavy solid line, top) and the outlets of De Gijster Reservoir (+), Honderd en Dertig Reservoir (∗) and Petrusplaat Reservoir (■), during November 1983 to April 1984.

when, under spate (flood) conditions, the river is highly loaded with suspended solids and associated impurities like hydrocarbons and heavy metals.

During the winters of 1983–84 and 1984–85, a study of virus removal was carried out. The winter period was chosen as the critical period, virus survival being better at low temperatures. The results are shown in Figure 13. The number of enteroviruses in the river water varied from 1 to 20 per litre. The overall reduction in the reservoirs was nearly 1000-fold, even while some short-circuiting between inlet and outlet water of De Gijster reservoir could be noted. An average k-value of 0.14 day^{-1} can be calculated, indicating that virus die-off is a fast process in the kinetic sense, which greatly benefits from compartmentalisation. In a single large reservoir the reduction would have been less than 30-fold and the production of virologically safe drinking water would need more elaborate treatment.

The overall quality improvement is very important. Table 5 gives an idea for some selected parameters. From these results it is apparent that the Biesbosch storage system, through its quality improvement potential, may be regarded as an integral part of the treatment of river water to provide a safe drinking water. Unfortunately the herbicides atrazine and simazine are hardly removed in the reservoirs. Due to equalizing, the river peaks of 0.5 to 1.0 μg l^{-1} in summer are reduced to maxima of 0.2 to 0.3 μg l^{-1} in the delivered water, but on average the drinking water limit of 0.1 μg l^{-1} is exceeded for both herbicides. In the meantime, all receiving waterworks are equipped with granular activated carbon filtration to remove these substances.

Table 5. *Average quality of inlet and reservoir waters, in 1991.*

Variables	Units	Inlet	De Gijester	Honderd en Dertig	Petrusplaat
Suspended solids	mg l^{-1}	0.71	0.22	0.08	0.02
Aluminium	mg l^{-1}	15	5.2	2.5	3.4
Iron	mg l^{-1}	0.60	0.20	0.05	0.01
Lead	μg l^{-1}	5.9	3.3	1.5	1.1
DOC	mg l^{-1}	4.0	3.7	3.5	1.1
COD	mg l^{-1}	17	13	11	11
Colour (Pt-Co scale)	mg l^{-1}	14	11	8	8
Chlorophyll	μg l^{-1}	27	12	6	5
Faecal coli	no. ml^{-1}	1.6	0.31	0.20	0.03
Atrazine	μg l^{-1}	0.16	–	–	0.15
Simazine	μg l^{-1}	0.13	–	–	0.11

Conclusions

(1). In "optically deep" reservoirs, artificial mixing provides an efficient and economic way of maintaining the development of planktonic algae at moderate levels.

(2). The development of geosmin-producing benthic cyanobacteria can be effectively controlled by rooting up the bottom with a harrow.

(3). Modelling approaches can be applied to management for water-quality control in storage reservoirs.

(4). With fully-mixed reservoirs, compartmentalisation of the reservoir volume leads to increased water quality improvement during storage.

The authors are grateful to Hans Waals, Henk Ketelaars and Hans Dits for preparing the text-figures and their critical comments on the manuscript. We wish to thank all co-workers of the laboratory division for their practical assistance, both in the field and laboratory, and Anneke Bos for preparing the manuscript.

References

Breeman, L. W. C. A. van, Ketelaars, H. A. M., Vissers, P. & Ebbeng, J. H. (1991). A new method to control growth of geosmin-producing benthic cyanobacteria. *Verhandlungen der Internationalen Vereinigung für Theoretische und Angewandte Limnologie*, **24**, 2168-2173.

Breeman, L. W. C. A. van, Dits, J. S. & Ketelaars, H. A. M. (1992). Production and reduction of geosmin and 2-methylisoborneol during storage of river water in deep reservoirs. *Water Science and Technology*, **25**, 233-240.

Gerber, N. N. (1983). Volatile substances from actinomycetes: their role in the odor pollution of water. *Water Science and Technology*, **15**, 115-125.

Gerber, N. N. & Lechevalier, H. A. (1965). Geosmin, an earthy-smelling substance isolated from actinomycetes. *Applied Microbiology*, **13**, 935-938.

Hoekstra, A. C., Bol, J. & Seinen, W. (1991). Cyanobacteria and their toxins in surface water used for the production of drinking water. *H_2O*, **24**, 387-393 (in Dutch).

Izaquirre, G., Hwang, C. J., Krasner, J. W. & McGuire, M. J. (1982). Geosmin and 2-methylisoborneol from cyanobacteria in three water supply systems. *Applied Environmental Microbiology*, **43**, 708-714.

Knoppert, P. L. & Vreedenburgh, E. G. H. (1982). The Biesbosch water storage scheme. *Aqua*, 0381-0385.

Lawton, L. A. & Codd, G. A. (1991). Cyanobacterial (blue-green algal) toxins and their significance in U.K. and European waters. *Journal of the Institution of Water and Environmental Management*, **5**, 460-465.

Means, E. G., Preston, A. E. & McGuire, M. J. (1984). Scuba diving: a tool for managing water quality. *Journal of the American Water Works Association*, **76**, 86-92.

Oskam, G. (1978). Light and zooplankton as algae regulating factors in eutrophic Biesbosch reservoirs. *Verhandlungen der Internationalen Vereinigung für Theoretische und Angewandte Limnologie*, **20**, 1612-1618.

Oskam, G. (1982). Quality aspects of the Biesbosch reservoirs. *Aqua*, 0447-0504.

Persson, P.-E. (1983). Off-flavours in aquatic ecosystems - an introduction. *Water Science and Technology*, **15**, 1-11.

Persson, P.-E. (1988). Odorous algal cultures in culture collection. *Water Science and Technology*, **20**, 211-213.

Ridley, J. E., Cooley, P. & Steel, J. A. P. (1966). Control of thermal stratification in Thames Valley Reservoirs. *Proceedings of the Society for Water Treatment and Examination*, **15**, 225-240.

Rosen, A. A., Mashni, E. J. & Safferman, R. J. (1970). Recent developments in the chemistry of odour in water: the cause of earthy/musty odour. *Proceedings of the Society for Water Treatment and Examination*, **19**, 106-119.

Steel, J. A. P. (1975). The management of Thames Valley Reservoirs. In *The Effects of Storage on Water Quality*. pp. 371-396. WRC Symposium.

Wood, S., Williams, S. I., White, W. R. & Jones, F. (1983). Factors influencing geosmin production by a streptomycete and their relevance to the occurrence of earthy taints in reservoirs. *Water Science and Technology*, **15**, 191-198.

Summary and concluding remarks

J. Gwynfryn Jones

Freshwater Biological Association and NERC Institute of Freshwater Ecology, Windermere Laboratory, Far Sawrey, Ambleside, Cumbria LA22 0LP, England

This meeting, entitled "Eutrophication: Research and Application to Water Supply", was the brainchild of Mr Jack Jeffery, the President of the FBA. The general feeling was that academic research on the process of eutrophication had progressed significantly over the past decade, but there had been little contact with the water managers, i.e. those whose work was at the sharp end of dealing with the associated problems. Uppermost in the minds of the organisers had been the problems related to the development of cyanobacterial (blue-green algal) blooms, but this work did not result in the exclusion of geochemical cycling, and other aspects which were relevant to reservoir management.

Mr Jeffery and I then met with Prof. Heinz Bernhardt and Mr Len Bays of IWSA, and it was clear from the outset that all agreed on the need for such a meeting. The programme was drawn up, admittedly at short notice, but we were convinced that we had achieved the correct balance, that each area of fundamental research covered would be followed by examples of application of the knowledge gained.

More than 100 delegates attended the two-day conference at the BAFTA Centre in London. One of the most gratifying aspects was the vigorous nature of the discussion which took place, not only during the formal sessions, but also during lunch breaks and the Conference Reception. This is, therefore, an entirely appropriate moment for the FBA and the IWSA to thank General Utilities for sponsoring this Reception. Although such events are seen as being, essentially, social in nature, they also provide the opportunity for the exchange of important information on the future management of our lakes and reservoirs.

Prof. Bernhardt chaired the first morning's session on the consequences of eutrophication – theory and practice. Dr Colin Reynolds (UK) gave his plenary lecture on eutrophication and the management of planktonic algae: what Vollenweider couldn't tell us. This was a detailed and critical account of the use and misuse of the "Vollenweider model" for lakes in northern temperate regions and their ability to support phytoplankton. Topics were framed and discussed in relation to the concept of phosphorus-loading and phosphorus-limitation. Dr Reynolds emphasised that we need to distinguish rate-limitation from capacity-limitation, and to understand which is more manageable and why. He also emphasised the importance of the mechanisms of internal cycling of phosphorus in lakes and reservoirs, and the need to appreciate the respective roles of physical and biotic components in the local control of algae, topics which were addressed in further detail by later speakers at the meeting.

Prof. Dieter Imboden (Switzerland) then discussed the impact of physical processes on algal growth, emphasising the effects of mixing and transport processes on the growth of algal populations in lakes. Rapid changes on time-scales of seconds, minutes and hours are particularly relevant for fast biological processes such as nutrient uptake and photosynthesis. Slower changes on time-scales of days to years are relevant for the less dynamic processes such as growth, respiration, mineralization and settling out of algal cells. Mathematical models were used to examine these points.

Prof. Gerrit van Straten (The Netherlands) dealt with the predicting power of models for eutrophication, and the problems of using mathematical models as management tools for predicting chemical and biological changes in the aquatic environment. Uncertainty, variability and "educated speculation" are necessary when developing models which are to be used for predicting the effects of measures specifically designed to bring a system into an entirely different "operating point", as is typically the case when attempting to reverse eutrophication. The even more difficult problems associated with predicting the behaviour of complex non-linear systems, some of which exhibit chaotic and therefore fundamentally unpredictable behaviour, were also discussed.

Sven Jørgensen rounded off the first session with an account of structural dynamic models of seasonal succession in algal species and populations in lakes and reservoirs. A new generation of quantitative models is being developed. These attempt to translate some important biological properties of species (survival, variation, inheritance, reproductive rates and population growth) into predictions about the survival of the fittest, where "fitness" is measured or estimated in thermodynamic terms. For this the concept of "exergy" and its calculation was introduced and explored as a means of examining changes in species composition.

The afternoon session on remedial measures and their application was chaired by Mr Jack Jeffery (UK). Prof. Brian Moss (UK) began the session with an account of the scope for biomanipulation in improving water quality, illustrated by his own elegant work on the Norfolk Broads in south-east England but also including examples from North America and Europe, in a short but valuable review. Biomanipulation involves the managed use of grazing animals (especially zooplankton and some fishes) to control large populations of edible algae, the so-called "top–down" method of control via the food web. Prof. Moss discussed the options available and their advantages, disadvantages and limitations (e.g. inedible algae may thrive and become nuisance blooms), concluding that biomanipulation is not a palliative for solving eutrophication problems at minimal expense, nor is it recommended for use in large natural lakes which have a diverse fauna and flora. However, it is a potentially useful tool in artificial reservoirs where there is a high degree of existing intervention and where water retention times are low, although the complete removal of fish is a necessary prerequisite for success.

This theme was continued by Prof. Jürgen Benndorf (Germany) in his paper on the control of indirect effects of biomanipulation. He identified three types of indirect effects which are as important as direct effects are for achieving stable population control by biomanipulation. Behavioural changes, such as a shift in the daily pattern of vertical migration by zooplankters when their predators are removed or scarce, can alter the transport of phosphorus within a lake. So too can the development of anti-predator traits, including morphological changes like the growth of helmets and neck-teeth in *Daphnia*. The most important indirect effects, however, result from resource limitation and "bottom–up" control within the lower trophic levels of the food web. These effects have to be carefully managed in conjunction with control of the basic nutrient loadings into, and cycling within, the lake or reservoir.

The last point above was pursued further by Prof. Dietrich Uhlmann and Heidemarie Horn (Germany), who examined in detail the significance of sedimentation and sediments to phytoplankton growth in drinking-water reservoirs, comparing and contrasting the oligotrophic Neunzehnhain Reservoir with the nearby mesotrophic–eutrophic Saidenbach Reservoir, in Saxony. The second session was then brought to a close by Dr Helmut Klapper (Germany), with an account of his ingenious work on calcite covering of sediments as a possible way of curtailing blue-green algae. This involves pumping natural calcite from deeper sediments to seal the phosphorus-rich surface layer and promote the co-precipitation of phosphorus from the water column, thereby reducing its internal cycling and availability to algae.

The second day of the conference began with a short session on the role of sediments in eutrophication, chaired by Mme M. Boubigot (France). Dr Ivan Heaney, Julie Corry and Jean

Lishman described the changes in water quality and sediment phosphorus of a small productive lake following decreased phosphorus-loading. Removal of some 47-67% of the external phosphorus input to Esthwaite Water (English Lake District) in treated sewage effluent, has been monitored since 1986. The effects of this large reduction in the total input have yet to be realised. Internal release from a large store of phosphorus in the sediments apparently compensates for the reduced input from external sources, and the remaining high levels of phosphorus input (including that from caged trout grown in the lake), may prevent any major reduction in algal productivity. Dr Lars Kamp-Nielsen also examined the role of sediments as sinks and stores for nutrients, with particular emphasis on phosphorus and nitrogen, using simple models for predicting events in lakes following reductions in external nutrient inputs.

Eutrophication in rivers was the topic of the session after coffee, with papers by Dr J.-P. Descy (Belgium) on the River Meuse, and by Drs T. Vandevelde and V. Da Molo (France) on the River Oise. The latter, in the Paris area, is subject to seasonal blooms of algae which generate problems for the production of drinking-water from a treatment plant on the Oise. A mathematical model has been developed to simulate variation in water quality in a pre-treatment storage reservoir, and another model is being developed to simulate events in the main river. Integration of the two models should provide a management decision-aid for optimizing control of the treatment process for potable water.

Algal blooms have caused problems in the River Meuse since the early 1980s. Plankton algae appear to start developing large populations in the French section upstream, utilizing large inputs of phosphorus from some of the tributaries. Thus co-ordinated co-operation between France and Belgium is required to tackle the problem. A mathematical model has been applied to understand the main factors that control the algal growth: underwater light, temperature, discharge and grazing by zooplankton. This last is a major loss process in summer and may trigger a seasonal succession of algal species, leading to dominance by larger-sized taxa. After periods of large population growth, decomposing algae may adversely affect the oxygen budget of the river but, conversely, algal photosynthesis is the most important source of oxygen input at periods of low discharge. Thus a reduction in algal productivity may result in a dramatic decrease in the oxygen content of heavily polluted stretches of the river, causing problems that could be as severe as those resulting from algal blooms.

Two afternoon sessions, on the control of blue-green algae, and the active management of reservoirs, were chaired by Prof. Gwynfryn Jones (UK). Prof. Tony Walsby (UK) gave an account of the control of gas-vacuolate cyanobacteria, colonial forms of which frequently cause problems in water supply. They form noxious scums on the water surface during calm, warm periods, and the gas vacuoles may prevent the colonies from sedimenting in water-treatment plants. Methods of applying pressure to collapse the gas vacuoles were briefly reviewed, including that of using explosives – apparently this works! Dr Christian Steinberg and Elke Gruhl (Germany) also considered physical measures to inhibit planktonic cyanobacteria. They described experimental work in the Fischkaltersee, a hard-water lake in upper Bavaria, which receives substantial inputs of phosphorus. An air-compressor was used to artificially destratify the lake and maintain an aerobic sediment–water interface. For a complex set of reasons (including top–down grazing effects), the overall effect was to remove or prevent development of populations of cyanobacteria, and also other members of the phytoplankton. Intermittent destratification gave fast-working results in the lake, and is less risky and cheaper than artificial control by permanent destratification.

Drs John Hilton, Tony Irish and Colin Reynolds (UK) addressed the problem of active reservoir management, based on predictions from a functional model developed from Chew Valley Lake/Reservoir in the west midlands of England. The model predicts algal growth capacity (maximum biomass) and seasonal succession of algal species. Model predictions were compared with observations in the reservoir, and model input variables were then modified to improve the predictive simulations. Good agreement was obtained by including pulsed inputs of phosphorus, derived from the sediments, at levels consistent with field observations. The uses of this particular model in several hypothetical operational scenarios were explored, and an example was given of an extension to the model which predicts chemical properties of abstracted and treated water.

Drs Gijsbert Oskam and Lambert van Breeman (The Netherlands) then described the management of Biesbosch Reservoirs for quality control with special reference to eutrophication. These are pumped-storage reservoirs fed with enriched water from the River Meuse, in which problems upstream in France and Belgium were described earlier in the day. Air injection is used to prevent stratification and further deterioration of water quality, and is an aid to controlling the algae, in conjunction with grazing zooplankton. Benthic *Oscillatoria* species are a special problem, producing toxic geosmin and growing on the inner embankments. Physical disturbance with a harrow is used to control them.

This was the first joint meeting between the FBA and the IWSA and, by all measures, it was a remarkable success. I would like to take this opportunity to thank the authors for their stimulating contributions and the IWSA for its collaboration.

The main aim of the meeting was to bring together those involved in research into aspects of the eutrophication of fresh water, and those with the responsibility of treating that water and providing a potable supply. It was clear, from the lively interchange of ideas and information, that we had succeeded in doing this.